ANATOMY AND PHYSIOLOGY: AN INTEGRATED APPROACH

First Edition

By Julian Pittman
Troy University

Bassim Hamadeh, CEO and Publisher
Michael Simpson, Vice President of Acquisitions
Jamie Giganti, Managing Editor
Jess Busch, Graphic Design Supervisor
Zina Craft, Acquisitions Editor
Monika Dziamka, Project Editor
Natalie Lakosil, Licensing Manager
Mandy Licata, Interior Designer

Copyright © 2014 by Cognella, Inc. All rights reserved. No part of this publication may be reprinted, reproduced, transmitted, or utilized in any form or by any electronic, mechanical, or other means, now known or hereafter invented, including photocopying, microfilming, and recording, or in any information retrieval system without the written permission of Cognella, Inc.

First published in the United States of America in 2014 by Cognella, Inc.

Copyright © 2013 by Depositphotos Inc./Eraxion

Trademark Notice: Product or corporate names may be trademarks or registered trademarks, and are used only for identification and explanation without intent to infringe.

Printed in the United States of America

ISBN: 978-1-62661-059-0 (pbk)/ 978-1-62661-060-6 (br)

www.cognella.com   800-200-3908

Contents

Acknowledgments ix

Foreword xi

Section One: Levels of Organization 1

Chapter One: Overview of Human Anatomy and Physiology 3

Chapter Two: Chemical Basis of Life 23

Chapter Three: Cells 33

Chapter Four: Cellular Metabolism 57

Chapter Five: Tissues, Glands, and Membranes 63

Section Two: Covering, Support, and Movement

Chapter Six: Integumentary System 75

Chapter Seven: Skeletal System 87

Chapter Eight: Joints of the Skeletal System   131

Chapter Nine: Muscular System   143

Section Three: Integration and Control Systems

Chapter Ten: Nervous System I: Structure and Function   191

Chapter Eleven: Nervous System II: Special Senses   249

Chapter Twelve: Endocrine System   279

Section Four: Circulation and Body Defense

Chapter Thirteen: Blood   297

Chapter Fourteen: Cardiovascular System   315

Chapter Fifteen: Lymphatic System and Immunity   347

Section Five: Absorption and Excretion

Chapter Sixteen: Respiratory System   363

Chapter Seventeen: Digestive System   381

Chapter Eighteen: Urinary System   409

Chapter Nineteen: Water, Electrolyte, and Acid–Base Balance   427

Section Six: Reproduction

Chapter Twenty: Reproductive Systems — 439

Credits — 463

This book is dedicated in loving memory to my grandfather, Julian A. Pittman.

Acknowledgments

I would like to thank the following individuals for the help and support that they have provided in the preparation of this text. It is not possible to adequately acknowledge the support and encouragement provided by loved ones, especially my wife, my mother, and my grandmother. They have had the patience and understanding to tolerate my absences and my frustrations. They have provided unwavering support. Finally, my thanks go to the skilled project team at Cognella for all their help, namely: Monika Dziamka, Zina Craft, Chelsey Rogers, Jess Busch, Natalie Lakosil, and Jennifer Bowen.

Foreword

Welcome to *Introduction to Anatomy and Physiology*. This text provides detailed information about the salient topics covered in a traditional first-year, two-course sequence of college-level anatomy and physiology without a substantial amount of peripheral information. This allows students to focus on the primary concepts without getting lost in ancillary information that may not be relevant to their future careers. This text provides not only the conceptual knowledge they will need, but also teaches them how to apply it. This text should also serve as a good review for anyone wanting to refresh their memory on the subject. Interested readers for this text should include allied health students in fields such as nursing, physical therapy, medical assistant, pre-medical, pre-chiropractic, and pre-vet.

Section 1
Levels of Organization

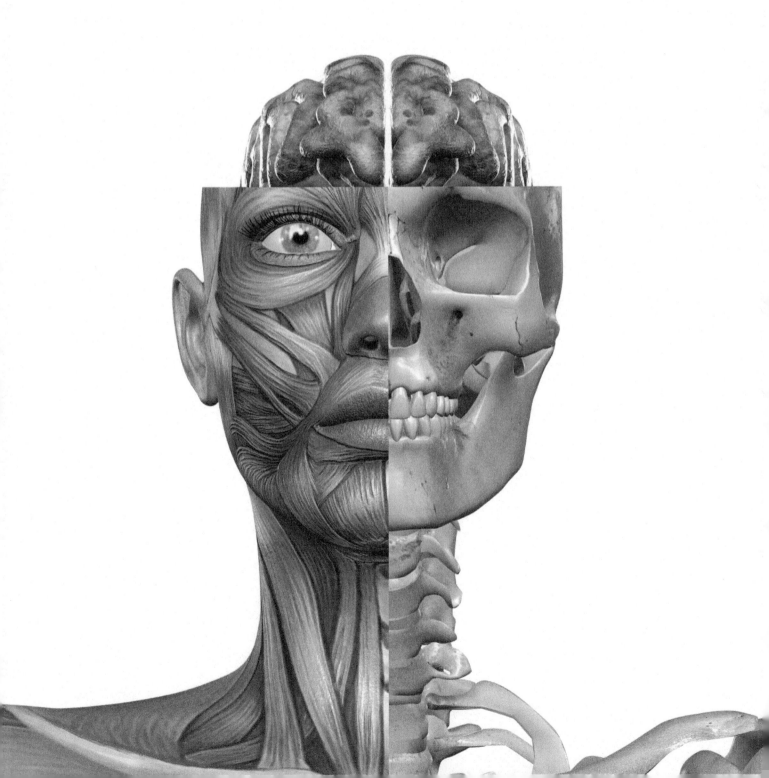

Chapter One

Overview of Human Anatomy and Physiology

Welcome to the captivating world of human anatomy and physiology! Before we start learning about the body in detail, let's go over some strategies for success in this course.

Use anatomical terminology frequently. The language of anatomy is quite foreign to many students and it takes practice to feel comfortable pronouncing and using the terms with ease. One way to practice is to simply use the terms by making up sentences. For example, you could say that your arm and wrist includes the brachial, antebrachial, and carpal regions.

Form small study groups. Study groups are a great way to practice terms and concepts. Small groups are much better than larger groups because they tend to stay more focused.

Study in smaller time periods with increased frequency. Repetition is a key to learning anatomical structures and physiological processes. It is better to study in smaller time periods than one or two long study sessions before an exam.

Look at diagrams from other sources. Sometimes students become too focused on one particular picture for a set of structures. Looking at several different pictures of the same structure helps to provide a more three-dimensional understanding of the structure. Try using Google Images as a rich source for additional diagrams and histological images. I particularly encourage this for histology lab exams.

Human anatomy is the scientific study of the body's structures. Some of these structures are very small and can be observed and analyzed only with a microscope. Other larger structures can readily be seen, manipulated, measured, and weighed. The word "anatomy" comes from a Greek root that means "to cut apart." When a body is dissected, its structures are cut apart in order to observe their physical attributes and their relationships to one another. Dissection is extensively used in medical schools, anatomy courses, and pathology labs. In order to observe structures in living people, however, a number of imaging techniques have been developed. These techniques allow clinicians to visualize structures inside the living body such as a cancerous tumor or a fractured bone. Like most scientific disciplines, anatomy has areas of specialization. Gross anatomy is the study of the larger structures of the body, those visible without the aid of magnification. Macro- means "large," thus, gross anatomy is also referred to as macroscopic anatomy. In contrast, micro- means "small," and microscopic anatomy is the study of structures that can be observed only with the use of a microscope or other magnification devices.

Microscopic anatomy includes cytology, the study of cells, and histology, the study of tissues. As the technology of microscopes has advanced, anatomists have been able to observe smaller and smaller structures of the body from slices of large structures like the heart to the three-dimensional structures of large molecules in the body.

Anatomists have two general approaches to the study of the body's structures: regional and systemic. Regional anatomy is the study of the interrelationships of all of the structures in a specific body region, such as the abdomen. Studying regional anatomy helps us appreciate the interrelationships of body structures, such as how muscles, nerves, blood vessels, and other structures work together to serve a particular body region. In contrast, systemic anatomy is the study of the structures that make up a discrete body system—that is, a group of structures that work together to perform a unique body function. For example, a systemic anatomical study of the muscular system would consider all of the skeletal muscles of the body.

While anatomy is focused on structure, physiology focuses on function. Human physiology is the scientific study of the chemistry and physics of the structures of the body and the ways in which they work together to support the functions of life. Much of the study of physiology centers on the body's propensity toward homeostasis. Homeostasis is the state of steady internal conditions maintained by living things. The study of physiology includes observation, both with the naked eye and with microscopes, as well as manipulations and measurements. However, current advances in physiology usually depend on carefully designed laboratory experiments that reveal the functions of the many structures and chemical compounds that make up the human body. Like anatomists, physiologists typically specialize in a particular branch of physiology. For example, neurophysiology is the study of the brain, spinal cord, and nerves and how these work together to perform functions as complex and diverse as vision, movement, and thought. Physiologists may work from the organ level (exploring, for example, what different parts of the brain do) to the molecular level (such as exploring how an electrochemical signal travels along nerves).

Form is closely related to function in all living things. For example, the thin flap of your eyelid can snap down to clear away dust particles and almost instantaneously slide back up to allow you to see again. At the microscopic level, the arrangement and function of the nerves and muscles that serve the eyelid allow for its quick action and retreat. At a smaller level of analysis, the function of these nerves and muscles likewise relies on the interactions of specific molecules and ions. Even the three-dimensional structure of certain molecules is essential to their function. Your study of anatomy and physiology will make more sense if you continually relate the form of the structures you are studying to their function. In fact, it can be somewhat frustrating to attempt to study anatomy without an understanding of the physiology that a body structure supports. Imagine, for example, trying to appreciate the unique arrangement of the bones of the human hand if you had no conception of the function of the hand. Fortunately, your understanding of how the human hand manipulates objects helps you appreciate the unique alignment of the thumb in opposition to the four fingers, making your hand a structure that allows you to pinch and grasp objects and type text messages, for example.

Life

How do you distinguish between something living and something non-living? In other words, how would you define a living system? All living systems have certain characteristics; living systems can move, grow, and respond to stimuli. They also need energy from food and oxygen from the air or water. These substances must be digested and absorbed so that they can be assimilated for growth, maintenance, and energy, and waste products must be excreted.

Homeostasis

Homeostasis is a particularly important concept with regard to life. Life maintains itself by virtue of homeostasis. Homeostasis refers to a system's ability to maintain a range of values. Think of how your body maintains certain levels of substances in your blood. For example, your body maintains a certain level of glucose by monitoring the glucose and then secreting certain hormones to raise or lower it. If glucose levels get too high, your body responds by secreting a hormone called insulin to lower it. If glucose levels get too low, then your body responds by secreting a hormone called glucagon to raise it. Homeostasis relies on what are referred to as feedback mechanisms. Your body has thousands of feedback systems in place that work in concert to regulate many substances. There are two types of feedback:

Negative feedback is when a system tends to reduce output or the product prevents the reaction from continuing.

Positive feedback is when a system tends to increase output.

It is helpful to think of feedback in a "stimulus/response" way. A good example of feedback is a thermostat. Let's say we set the thermostat at 70 degrees. It's summer and hot outside and the room temperature begins to rise. Once it gets above 70 degrees the thermostat senses it and turns on the air conditioner. The result is the room cools down to below 70 degrees. Now, let's review the stimulus/response.

Stimulus = room getting warmer

Response = turn on air conditioner to cool room down

Can you see that the stimulus and response are opposite? This is an example of negative feedback.

Let's say we still have our thermostat set at 70 degrees but this time it is winter and we open the window. The temperature in the room begins to lower until it gets lower than 70 degrees. The system now responds by turning on the furnace. The room then gets warmer until the temperature gets above 70 degrees. Again, let's review the stimulus response.

Stimulus = room getting colder

Response = turn on furnace to warm room up

Can you see that the stimulus and response are still opposite? So this is still an example of negative feedback. Let's say that I wired up the thermostat the wrong way. Now when the temperature in the room rises above 70 degrees, instead of turning on the air conditioner the furnace turns on and raises the room temperature. Can you see that the stimulus and response are now the same?

Stimulus = room getting warmer

Response = turn on furnace to make room even warmer

Since the stimulus and response are the same, we call this positive feedback.

Although positive feedback is needed within homeostasis, it also can be harmful at times. When you have a high fever it causes a metabolic change that can push the fever higher and higher. In rare occurrences, the body temperature reaches 113 degrees and the cellular proteins stop working and metabolism stops, resulting in death.

Sustainable systems require combinations of both kinds of feedback. Generally, with the recognition of divergence from the homeostatic condition, positive feedbacks are called into play, whereas once the homeostatic condition is approached, negative feedback is used for "fine tuning" responses. This creates a situation of "metastability," in which homeostatic conditions are maintained within fixed limits, but once these limits are exceeded, the system can shift wildly to a wholly new (and possibly less desirable) situation of homeostasis. It should be obvious from this discussion that homeostasis is not static, but very much a dynamic system. Physiology is largely a study of processes related to homeostasis. Some of the functions you will learn about in this book are not specifically about homeostasis (how muscles contract), but in order for all bodily processes to function there must be a suitable internal environment.

Pathways that Alter Homeostasis

A variety of homeostatic mechanisms maintain the internal environment within tolerable limits. Either homeostasis is maintained through a series of control mechanisms, or the body suffers various illnesses or disease. When the cells in the body begin to malfunction, the homeostatic balance becomes disrupted. Eventually this leads to disease or cell malfunction. Disease and cellular malfunction can be caused in two ways: either *deficiency* (cells not getting all they need) or *toxicity* (cells being poisoned by substances they do not need or excess).

When homeostasis is interrupted in cells, there are *pathways* to correct or worsen the problem. In addition to the internal control mechanisms, there are external influences based primarily on lifestyle choices and environmental exposures that influence our body's ability to maintain cellular health.

- **Nutrition:** If your diet is lacking in a specific vitamin or mineral, your cells will function poorly, possibly resulting in a disease condition. For example, a menstruating woman with inadequate dietary intake of iron will become anemic. Lack of hemoglobin, a molecule that requires iron, will result in reduced oxygen-carrying capacity. In mild cases symptoms may be vague (fatigue), but if the anemia is severe, the body will try to compensate by increasing cardiac output, leading to palpitations and sweatiness, and possibly heart failure.
- **Toxins:** Any substance that interferes with cellular function, causing cellular malfunction. This is accomplished through a variety of ways: chemical, plant, insecticides, and/or bites. A commonly seen example of this is drug overdoses. When a person takes too much of a drug, their vital signs begin to waver; either increasing or decreasing, these vital signs can cause problems including coma, brain damage, and even death.
- **Psychological:** Your physical health and mental health are inseparable. Our thoughts and emotions cause chemical changes to take place either for better as with meditation, or worse as with stress.
- **Physical:** Physical maintenance is essential for our cells and bodies. Adequate rest, sunlight, and exercise are examples of physical mechanisms for influencing homeostasis. Lack of sleep is related to a number of ailments such as irregular cardiac rhythms, fatigue, anxiety, and headaches.
- **Genetic/Reproductive:** Inheriting strengths and weaknesses can be part of our genetic makeup. Genes are sometimes turned off or on due to external factors that we can have some control over, but at other times little can be done to correct or improve genetic diseases. Beginning at the cellular level a variety of diseases is derived from mutated genes. For example, cancer can be genetically inherited or can be caused due to a mutation from an external source such as radiation or genes altered in a fetus when the mother uses drugs.
- **Medical:** Because of genetic differences some bodies need help in gaining or maintaining homeostasis. Through modern medicine our bodies can be given different aids, from antibodies to help fight infections, or chemotherapy to kill harmful cancer cells. Traditional and alternative medical practices have many benefits, but like any medical practice, the potential for harmful effects is present. Whether by nosocomial infections or wrong dosage of medication, homeostasis can be altered by that which is trying to fix it. Trial and error with medications can cause potential harmful reactions and possibly death if not caught quickly.

Homeostasis Throughout the Body

Each body system contributes to the homeostasis of other systems and of the entire organism. No system of the body works in isolation, and the well-being of the person depends upon the well-being of all the interacting body systems. A disruption within one system generally has consequences for several additional body systems. The following are some brief examples/explanations of how various body systems contribute to the maintenance of homeostasis:

Nervous System

Since the nervous system does not store nutrients, it must receive a continuous supply from blood. Any interruption to the flow of blood may bring brain damage or death. The nervous system maintains homeostasis by controlling and regulating the other parts of the body. A deviation from a normal set point acts as a stimulus to a receptor, which sends nerve impulses to a regulating center in the brain. The brain directs an effector to act in such a way that an adaptive response takes place. If, for example, the deviation was a lowering of body temperature, the effector acts to increase body temperature. The adaptive response returns the body to a state of normalcy and the receptor, the regulating center, and the effector temporarily cease their activities. Since the effector is regulated by the very conditions it produced, this process is called "control by negative feedback." This manner of regulating normalcy results in a fluctuation between two extreme levels. Not until body temperature drops below normal do receptors stimulate the regulating center and effectors act to raise body temperature. Regulating centers are located in the central nervous system, consisting of the brain and spinal cord. The hypothalamus is a portion of the brain particularly concerned with homeostasis; it influences the action of the medulla oblongata, a lower part of the brain, the autonomic nervous system, and the pituitary gland. The nervous system has two major portions: the central nervous system and the peripheral nervous system. The peripheral nervous system consists of the cranial and spinal nerves. The autonomic nervous system is a part of peripheral nervous system and contains motor neurons that control internal organs. It operates at the subconscious level and has two divisions, the sympathetic and parasympathetic systems. In general, the sympathetic system brings about those results we associate with emergency situations, often called "fight or flight reactions," and the parasympathetic system produces those effects necessary to our everyday existence.

Endocrine System

The endocrine system consists of glands that secrete hormones into the bloodstream. Each hormone has an effect on one or more target tissues. In this way the endocrine system regulates the metabolism and development of most body cells and body systems. To be more explicit, the endocrine system has sex hormones that can activate sebaceous glands, development of mammary glands, alter dermal blood flow, and release lipids from adipocytes. Our bone growth is regulated by several hormones, and the endocrine system helps with the mobilization of calcitonin and calcium. In the muscular system, hormones adjust muscle metabolism, energy production, and growth. In the nervous system, hormones affect neural metabolism, regulate fluid/electrolyte balance, and help with reproductive hormones that influence CNS development and behaviors. In the cardiovascular system, we need hormones that regulate the production of RBCs (red blood cells), which elevate and lower blood pressure. Hormones also have anti-inflammatory effects and stimulate the lymphatic system. In summary, the endocrine system has a regulatory effect on essentially every other body system.

Integumentary System

The integumentary system (the skin) is involved in protecting the body from invading microbes (mainly by forming a thick impenetrable layer), regulating body temperature through sweating and vasodilation/vasoconstriction, or shivering and piloerection (goose bumps), and regulating ion balances in the blood. Stimulation of mast cells also produces changes in blood flow and capillary permeability, which can affect the blood flow in the body and how it is regulated. It also helps synthesize vitamin D, which interacts with calcium and phosphorus absorption needed for bone growth, maintenance, and repair. Hair on the skin guards entrance into the nasal cavity or other orifices, preventing invaders from getting further into our bodies. Our skin also helps maintain balance by excretion of water and other solutes; the keratinized epidermis limits fluid loss through skin. It also provides mechanical protection against environmental hazards. We need to remember that our skin is integumentary; it is our first line of defense.

Skeletal System

As the structural framework for the human body, the skeletal system consists mainly of the 206 or so bones of the skeletal system but also includes cartilages, ligaments, and other connective tissues that stabilize and interconnect them. Bones work in conjunction with the muscular system to aid in posture and locomotion. Many bones of the skeleton function as levers, which change the magnitude and direction of forces generated by skeletal muscle. Protection is a pivotal role occupied by the skeletal system, as many vital organs are encased within the skeletal cavities (cranial and spinal), and bones form much of the structural basis for other body cavities (thoracic and pelvic cavities). The skeletal system also serves as an important mineral reserve. For example, if blood levels of calcium or magnesium are low and the minerals are not available in the diet, they will be taken from the bones. Also, the skeletal system provides calcium needed for all muscular contraction. Finally, red blood cells, lymphocytes, and other cells relating to the immune response are produced and stored in the bone marrow.

Muscular System

The muscular system is one of the most versatile systems in the body. The muscular system contains the heart, which constantly pumps blood through the body. The muscular system is also responsible for involuntary (e.g., goosebumps, digestion, breathing) and voluntary (e.g., walking, picking up objects) actions. Muscles also help protect organs in the body's cavities. The muscles in your body contract, which increases your body heat when you're cold. The act of shivering occurs when internal temperature drops. Muscles around vital organs contract, breaking down ATP, and thereby expending heat, which is then distributed to the rest of the body.

Cardiovascular System

The cardiovascular system, in addition to needing to maintain itself within certain levels, plays a role in maintenance of other body systems by transporting hormones (heart secretes Atrial Natriuretic Peptide and Brain Natriuretic Peptide, or ANP and BNP, respectively) and nutrients (oxygen, erythropoietin [EPO] to bones, etc.), taking away waste products, and providing all living body cells with a fresh supply of oxygen and removing carbon dioxide. Homeostasis is disturbed if the cardiovascular or lymphatic systems are not functioning correctly. Our skin, bones, muscles, lungs, digestive tract, and nervous, endocrine, lymphatic, urinary, and reproductive systems use the cardiovascular system as its "road" or "highway" as far as distribution of things such as nutrients, oxygen, waste products, hormones, drugs, etc. There are many risk factors for an unhealthy cardiovascular system. Some diseases associated are typically labeled "uncontrollable" or "controllable." The primary uncontrollable risk factors are age, gender, and a family history of heart disease, especially at an early age.

The cardiovascular system also contains sensors to monitor blood pressure, called baroreceptors, that work by detecting how stretched a blood vessel is. This information is relayed to the medulla oblongata in the brain where action is taken to raise or lower blood pressure via the autonomic nervous system.

Lymphatic System

The lymphatic system has three principal roles. First is the maintenance of blood and tissue volume. Excess fluid that leaves the capillaries when under pressure would build up and cause edema. Secondly, the lymphatic system absorbs fatty acids and triglycerides from fat digestion so that these components of digestion do not enter directly into the bloodstream. Third, the lymphatic system is involved in defending the body against invading microbes, and the immune response. This system assists in maintenance, such as bone and muscle repair after injuries. Another defense is maintaining the acidic pH of urine to fight infections in the urinary system. The tonsils are our bodies' "helpers" to defend us against infections and toxins absorbed from the digestive tract. The tonsils also protect against infections entering our lungs.

Respiratory System

The respiratory system works in conjunction with the cardiovascular system to provide oxygen to cells within every body system for cellular metabolism. The respiratory system also removes carbon dioxide. Since CO_2 is mainly transported in the plasma as bicarbonate ions, which act as a chemical buffer, the respiratory system also helps maintain proper blood pH levels, a fact that is very important for homeostasis. As a result of hyperventilation, CO_2 is decreased in blood levels. This causes the pH of body fluids to increase. If acid levels rise above 7.45, the result is respiratory alkalosis. On the other hand, too much CO_2 causes pH to fall below 7.35, which results in respiratory acidosis. The respiratory system also helps the lymphatic system by trapping pathogens and protecting deeper tissues within. Note that when you have increased thoracic space, it can provide abdominal pressure through the contraction of respiratory muscles. This can assist in defecation. Remember the lungs are the gateway for our breath of life.

Digestive System

Without a regular supply of energy and nutrients from the digestive system, all body systems would rapidly suffer. The digestive system absorbs organic substances, vitamins, ions, and water that are needed all over the body. In the skin, the digestive tract provides lipids for storage in the subcutaneous layer. Note that food undergoes three types of processes in the body: digestion, absorption, and elimination. If one of these is not working, you will have problems that will be exceedingly noticeable. Mechanics of digestion can include chemical digestion, movements, ingestion absorption, and elimination. In order to maintain a healthy and efficient digestive system, we have to remember the components involved. If these are disturbed, digestive health may be compromised.

Urinary System

Toxic nitrogenous wastes accumulate as proteins and nucleic acids are broken down and used for other purposes. The urinary system rids the body of these wastes. The urinary system is also directly involved in maintaining proper blood volume (and, indirectly, blood pressure) and ion concentration within the blood. One other contribution is that the kidneys produce a hormone (erythropoietin) that stimulates red blood cell production. The kidneys also play an important role in maintaining the correct water content of the body and the correct salt composition of extracellular fluid. External changes that lead to excess fluid loss trigger feedback mechanisms that act to inhibit fluid loss.

Reproductive System

The reproductive system is unique in that it does little to contribute to the homeostasis of the organism. Rather than being tied to the maintenance of the organism, the reproductive system relates to the maintenance of the species. Having said that, the sex hormones do have an effect on other body systems, and an imbalance can lead to various disorders (a woman whose ovaries are removed early in life is at much higher risk of osteoporosis).

Thermoregulation

Our bodies are characterized with a number of automated processes, which make them self-sustainable in the natural environment. Among these many processes are that of reproduction, adjustment with external environment, and instinct to live, which are gifted by nature to living beings. The survival of living beings greatly depends on their capability to maintain a stable body temperature irrespective of temperature of surrounding environment. This capability of maintaining body temperature is called thermoregulation. Cold-blooded animals, such as reptiles, have somewhat different means of temperature regulation than warm-blooded (or homeothermic) animals, such as humans and other mammals.

Body temperature depends on the heat produced minus the heat lost. Heat is lost by radiation, convection, and conduction, but the net loss by all three processes depends on a gradient between the body and the outside. Thus, when the external temperature is low, radiation is the most important form of heat loss. When there is a high external temperature, evaporation is the most important form of heat loss. The balance of heat produced and heat lost maintains a constant body temperature. However, temperature does vary during the day, and this set point is controlled by the hypothalamus. Body temperature is usually about 37.4°C, but does vary during the day by about 0.8°C. The lowest daily temperature is when the person is asleep. Temperature receptors are found in the skin, the great veins, the abdominal organs, and the hypothalamus. While the ones in the skin provide the sensation of coldness, the hypothalamic (central core) temperature receptors are the most important. The core body temperature is usually about 0.7–1.0°C higher than axillary or oral temperature.

When body temperature drops due to external cold, an important component of protection is vasoconstriction of skin and limb blood vessels. This drops the surface temperature, providing an insulating layer (such as the fat cell layer) between the core temperature and the external environment. Likewise, if the temperature rises, blood flow to the skin increases, maximizing the potential for loss by radiation and evaporation. Thus, if you dilated the skin blood vessels by alcohol ingestion, this might give a nice warm glow, but it would increase heat loss (if the external temperature was still low). The major adjustments in cold are to shiver to increase heat production, and constrict blood vessels in the periphery and skin. This helps to minimize heat loss through the skin, and directs blood to the vital internal organs.

Besides the daily variation in body temperature, there are other cyclic variations. In women, body temperature falls prior to ovulation and rises by about 1°C at ovulation, largely due to progesterone increasing the set point. Thyroid hormone and pyrogens also increase the set point. The basal metabolic rate (BMR) is about 30 calories/sqm/h. It is higher in children than in adults, partly as a result of different surface area to body mass ratio. Due to this relationship, young children are more likely to drop their temperature rapidly; there is greater temperature variation in children than in adults. It is increased by thyroid hormone and decreased by thyroid hormone lack. Different foods can affect BMR and the Respiratory Quotient of foods differ. Carbohydrate 1.0; Protein = 1.0; Fats = 0.7.

Levels of Organization

The body is organized according to levels of complexity. The lowest level of complexity is the atom. The highest level of complexity is the organism. The levels from lowest to highest complexity are:

1. Atom
2. Molecule
3. Cell
4. Organelle
5. Organ
6. Organ System
7. Organism (human body)

Specialized Cells of the Human Body

Although there are specialized cells (both in structure and function), within the body, all cells have similarities in their structural organization and metabolic needs (such as maintaining energy levels via conversion of carbohydrate to ATP and using genes to create and maintain proteins).

The following are some of the different types of specialized cells within the human body.

Nerve cells: Also called neurons, these cells are in the nervous system and function to process and transmit information. They are the core components of the brain, spinal cord, and peripheral nerves. They use chemical synapses that can evoke electrical signals, called action potentials, to relay signals throughout the body.

Epithelial cells: Functions include secretion, absorption, protection, transcellular transport, sensation detection, and selective permeability. Epithelium lines both the outside (skin) and the inside cavities and lumina of bodies.

Exocrine cells: These cells secrete products through ducts, such as mucus and sweat, or digestive enzymes. The products of these cells go directly to the target organ through the ducts. For example, the bile from the gallbladder is carried directly into the duodenum via the bile duct.

Endocrine cells: These cells are similar to exocrine cells, but secrete their products directly into the bloodstream instead of through a duct. The products of the endocrine cells go throughout the body in the bloodstream but act on specific organs by receptors on the cells of the target organs. For example, the hormone estrogen acts specifically on the uterus and breasts of females because there are estrogen receptors in the cells of these target organs.

Blood cells: The most common types of blood cells are:

red blood cells (erythrocytes). The primary function of red blood cells is to collect oxygen in the lungs and deliver it through the blood to the body tissues. Gas exchange is carried out by simple diffusion.

white blood cells (leukocytes). They are produced in the bone marrow and help the body to fight infectious disease and foreign objects in the immune system. White cells are found in the circulatory system, lymphatic system, spleen, and other body tissues.

General Anatomy

The human body can be divided into two basic sections. The axial section contains the head, neck, and trunk. The appendicular section contains the arms and legs, also known as the upper and lower extremities.

The body also contains hollow areas called cavities. There are 2 large cavities. One cavity is in the front portion of the body and is called the ventral cavity. The other is in the back and is called the dorsal cavity. Both cavities can be subdivided into smaller cavities. The ventral cavity can be subdivided into the thoracic and abdominopelvic cavities. The thoracic portion is in the chest area and the abdominopelvic portion is in the stomach area. The thoracic and abdominopelvic cavities are separated by the diaphragm. The dorsal cavity can also be subdivided into 2 smaller cavities. One cavity is called the cranial cavity and is located in the head. The other is called the spinal

Overview of Human Anatomy and Physiology | 13

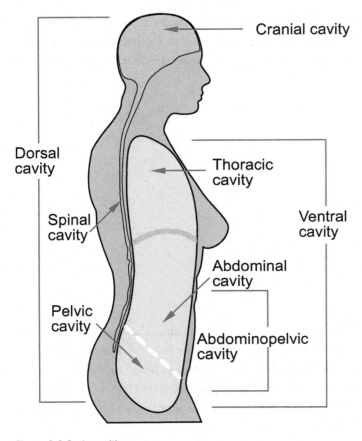

Figure 1.1 Body cavities.

canal and runs down the back. The cranial cavity contains the brain and the spinal canal contains the spinal cord (Fig. 1.1).

There are also some smaller cavities in the body.

- Oral (teeth, tongue)
- Nasal (sinuses)
- Orbital (eyes and associated muscles, nerves)
- Middle ear (middle ear bones)

Summary of Body Systems

Now let's have a cursory look at all of the body systems.

- Integumentary
- Skeletal
- Muscular
- Nervous
- Endocrine
- Lymphatic
- Digestive
- Respiratory
- Urinary
- Reproductive

The integumentary system consists of the hair, skin, nails, sweat glands, and sebaceous glands. Its function is protection of the body, secretion of waste products, production of vitamin D, and regulation of body temperature. The integumentary system also supports sensory receptors that send information to the nervous system. The skeletal system consists of the bones, ligaments, and cartilage. It provides protection and support and produces red blood cells. It also stores chemical salts. The muscular system produces movement, helps to maintain posture, and produces heat. The nervous system consists of the brain, spinal cord, and receptors. It receives sensory information, detects changes, and in response, stimulates muscles and glands. The endocrine system is a series of glands that secrete hormones. The endocrine system contains many feedback systems to help maintain homeostasis. The glands:

- Pituitary
- Thyroid
- Parathyroid
- Adrenal
- Pancreas
- Ovaries
- Testes
- Pineal
- Thymus
- Hypothalamus

The cardiovascular system includes the heart, arteries, capillaries, and veins. The function of the cardiovascular system is to transport blood. The lymphatic system includes the lymph vessels, lymph nodes, thymus, and spleen. The function of the lymphatic system is to return tissue to blood as well as transport some absorbed food molecules and defend against infection. The respiratory system consists of the nasal cavity, lungs, pharynx, larynx, trachea, and bronchi. The respiratory system supplies the body with oxygen and eliminates carbon dioxide. The digestive system:

- Mouth
- Tongue
- Teeth
- Salivary glands
- Pharynx
- Esophagus
- Liver
- Gallbladder
- Pancreas
- Intestines

The function of the digestive system is to receive, break down, and absorb food. It also eliminates wastes. The urinary system:

- Kidneys
- Ureters
- Urinary bladder
- Urethra

The function of the urinary system is to remove wastes, maintain water and electrolyte balance, and store and transport urine. The male reproductive system:

- Scrotum
- Testes
- Epididymis
- Vasa deferentia
- Seminal vesicles
- Prostate
- Bulbourethral glands
- Urethra
- Penis

The female reproductive system:

- Ovaries
- Uterine tubes
- Uterus
- Vagina
- Clitoris
- Vulva

The function of the reproductive systems is to pass genetic information down to future generations as well as produce hormones that help the body to mature.

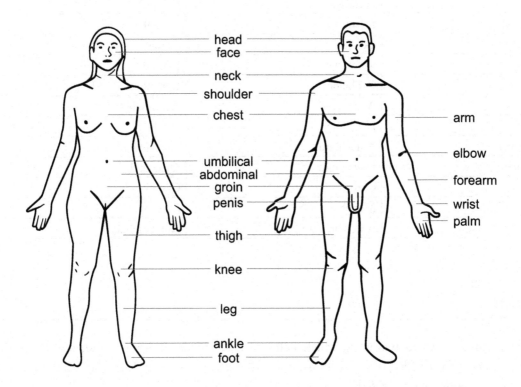

Figure 1.2 Anatomical position.

Anatomical Terminology

Now that we are familiar with the overview of the body, let's proceed with some anatomical terminology. In anatomy we always reference the body with regard to anatomical position (Fig. 1.2).

Next, we will explore some of the important regions of the body. A region is a broader area such as the upper leg (femoral region). Although a region may sound like an actual body part, it is not; remember it is an area.

Common Term	Anatomical Term	Region
Foot	Pes	Pedal
Shin	Crus	Crural
Calf	Sura	Sural
Front of knee	Patella (knee cap)	Patellar
Back of knee	Popliteus	Popliteal
Thigh	Femorus	Femoral
Groin	Inguina	Inguinal
Butt	Buttock	Gluteal
Stomach	Abdomen	Abdominal
Low back	Lumbus	Lumbar
Chest and middle back	Thorax	Thoracic
Lateral chest	Pectorus	Pectoral
Middle chest	Sternum	Sternal
Neck	Cervicis	Cervical
Chin	Mentum	Mental
Head	Cephalon	Cephalic
Shoulder	Acromion	Acromial
Arm	Brachium	Brachial
Elbow (front)	Antecubitus	Antecubital
Elbow (back)	Olecranon	Olecranal
Wrist	Carpus	Carpal
Hand	Manus	Manual
Forearm	Antebrachium	Antebrachial

Abdomen

The abdomen can be divided two ways, which helps to describe the locations of structures. In one method, the abdomen is divided into 9 sections, much like tic-tac-toe (Fig. 1.4). The other method is simpler in that the abdomen is divided into 4 sections.

Four planes are needed in order to divide the abdomen into 9 equal sections. There are 2 parasagittal planes (sometimes called lateral lines) and 2 transverse planes. The superior transverse plane is called the transpyloric plane and the inferior plane is called the transtubercular plane. The center of the 9 regions

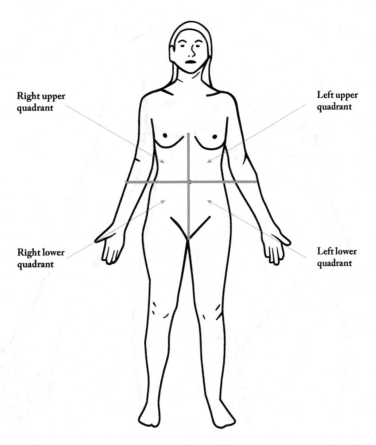

Figure 1.3 Abdominal quadrants.

is the umbilical The 3 superior regions are the epigastric and right and left hypochondriac. The middle regions are the umbilical and right and left lumbar. The lower regions are the hypogastric and right and left inguinal. The other method of dividing up the abdominal area consists of using a transverse and mid-sagittal plane intersecting at the umbilicus. This results in 4 quadrants: the right and left upper quadrants and right and left lower quadrants.

Positional Terms

Positional terminology is used in order to specify locations of anatomical structures. The positional terms usually are in pairs. For example, superior and inferior go together. Superior means above and inferior means below. So we would say that the head is superior to the chest, or to be more specific—the cephalon is superior to the thorax, and the reverse would also be true: The thorax is inferior to the cephalon.

Some other terms of importance and examples of their use:

Anterior means toward the front.
Posterior means toward the back.

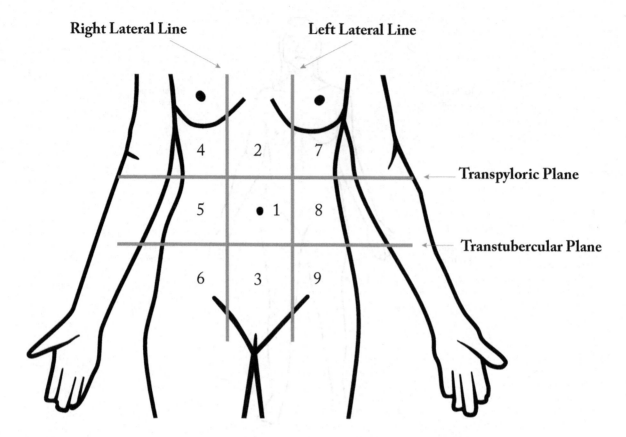

Figure 1.4 Planes dividing the abdominal region into 9 areas.
1. Umbilical
2. Epigastric
3. Hypogastric
4. Right hypochondriac
5. Right lumbar
6. Right iliac
7. Left hypochondriac
8. Left lumbar
9. Left iliac

Ex: the sternum is anterior to the heart.
OR the heart is posterior to the sternum.

Medial means toward the midline of the body.
Lateral means away from the midline.
Ex: the ears are lateral to the nose.
And…
The nose is medial to the ears.

Proximal means toward the trunk of the body.
Distal means away from the trunk.
Proximal and distal are usually used when describing structures in the extremities.
Ex: the elbow is proximal to the wrist.

The wrist is distal to the elbow.

Superficial means toward the surface.
Deep means under the surface.
Ex: the skin is superficial to the stomach.
The stomach is deep to the skin.

Ipsilateral means on the same side.
Contralateral means on the opposite side.
Ex: the right shoulder and right elbow are ipsilateral.
The right shoulder and left elbow are contralateral.

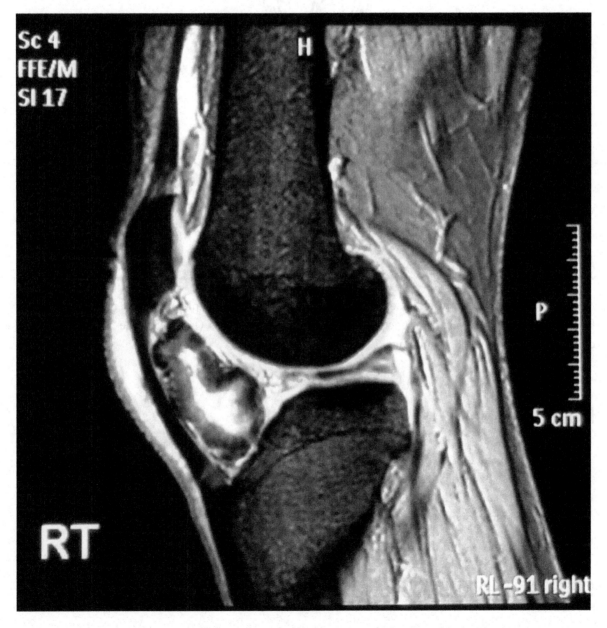

Figure 1.5 MRI image of the knee. The MRI presents a slice of the body (a sagittal section of the knee).

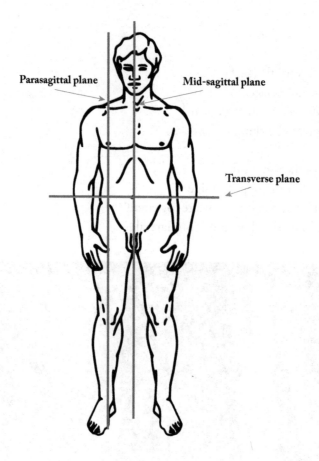

Figure 1.6 Anatomical planes.

Anatomical Planes

Anatomical planes are used for studying slices of the body. For example, magnetic resonance imaging (MRI) can "slice" the body into sections to look for abnormalities (Fig. 1.5). The anatomical planes divide the body in various ways (Figs. 1.6, 1.7).

Planes

The sagittal plane divides the body into right and left portions.
The transverse plane divides the body into superior and inferior portions.
The coronal plane divides the body into anterior and posterior portions.
The oblique plane divides the body at an angle.

CRITICAL-THINKING QUESTIONS

1. Name at least three reasons to study anatomy and physiology.

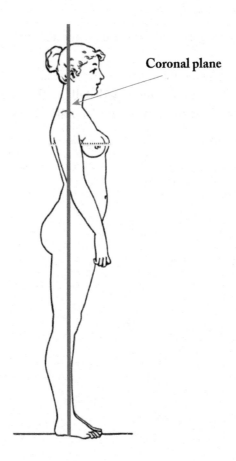

Figure 1.7 Coronal plane.

2. For whom would an appreciation of the structural characteristics of the human heart come more easily: an alien who lands on Earth, abducts a human, and dissects his heart; or an anatomy and physiology student performing a dissection of the heart on her very first day of class? Why?
3. Name the levels of organization of the human body.
4. The female ovaries and the male testes are a part of which body system? Can these organs be members of more than one organ system? Why or why not?
5. Explain why the smell of smoke when you are sitting at a campfire does not trigger alarm, but the smell of smoke in your residence hall does.
6. Identify three different ways that growth can occur in the human body.
7. On his midsummer trek through the desert, John ran out of water. Why is this particularly dangerous?
8. Identify the components of a negative feedback loop and explain what would happen if the secretion of a body chemical controlled by a negative feedback system became too great.
9. What regulatory processes would your body use if you were trapped by a blizzard in an unheated, uninsulated cabin in the woods?
10. In which direction would an MRI scanner move to produce sequential images of the body in the frontal plane, and in which direction would an MRI scanner move to produce sequential images of the body in the sagittal plane?
11. If a bullet were to penetrate a lung, which three anterior thoracic body cavities would it enter, and which layer of the serous membrane would it encounter first?

12. Can you interpret the following excerpt from a medical report?

The patient presented with a moderate injury to the cervical region on the right side with radiation of pain into the ipsilateral brachial region extending distally to the antebrachium and carpals. The lesion extends from the medial aspect of the inguinal region laterally to the lateral femoral region.

Chapter Two

Chemical Basis of Life

Basic Chemistry

The following is a review of basic chemistry concepts that are relevant to the study of anatomy and physiology. Let's begin with the building blocks of matter. All matter is made up of elements. There are 26 elements in your body. Carbon, Oxygen, Nitrogen, and Hydrogen make up about 96% of your body. The elements are made up of atoms (Fig. 2.1).

Atoms

Atoms are made from 3 basic particles. Protons are positively charged and reside in the nucleus (atom core). Electrons are negatively charged and orbit around the nucleus. Neutrons carry no charge and are located in the nucleus (Fig. 2.2).

The basic structure of the atom is a core or nucleus surrounded by electrons orbiting in "shells." These shells can hold different numbers of electrons (Fig. 2.3).

Atomic Number

The number of protons in the atom is known as the atomic number. If we add the protons and neutrons together, we have what is known as the mass number. Usually, the atom has the same number of protons and neutrons. If an atom has a different number of protons than it has neutrons, it is known as an isotope.

If an atom has equal numbers of protons and electrons, it is said to be neutral. The positive and negative charges balance each other out. If an atom has unequal numbers of protons and electrons, it will be charged and is known as an ion. In physiology we call these atoms electrolytes. The atomic mass is

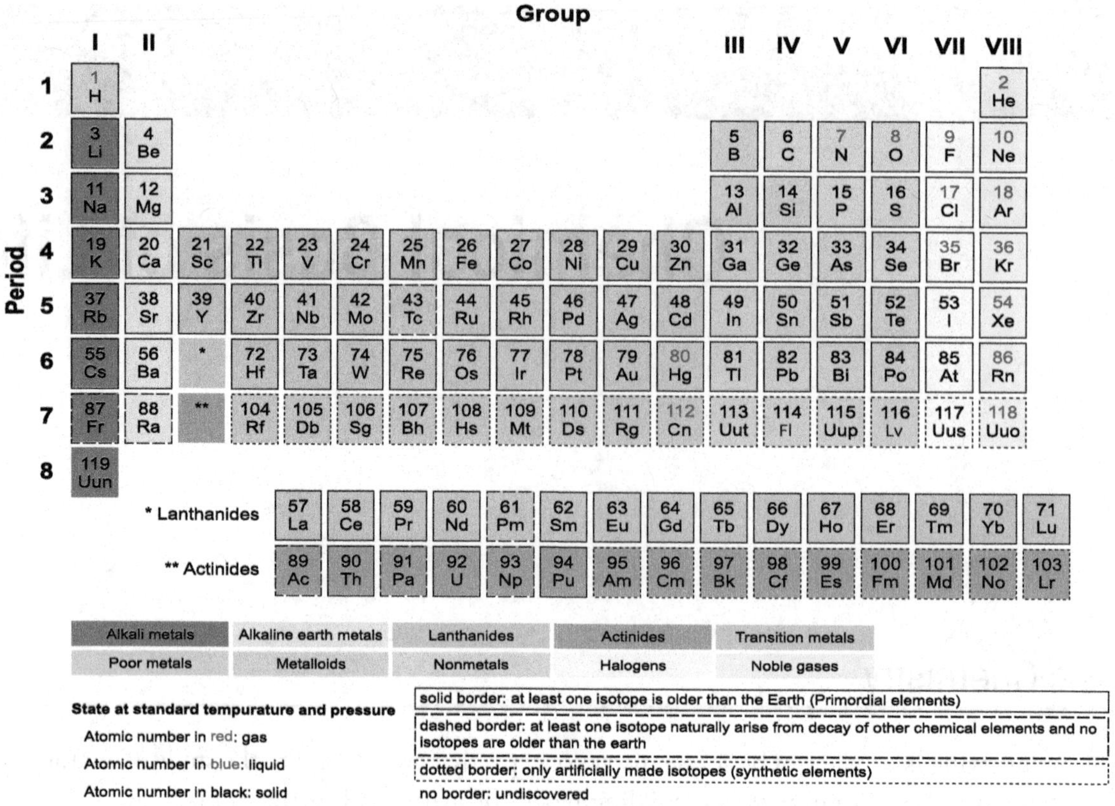

Figure 2.1 Periodic table of the elements.

the sum of the masses of protons, neutrons, and electrons and represents the average mass of all naturally occurring isotopes. Generally atoms and molecules built from atoms like to have their outer shells filled with electrons. Electrons usually fill these shells in even numbers. However, this does not always occur. If an atom or molecule has an unpaired electron in the outer shell, it is known as a free radical.

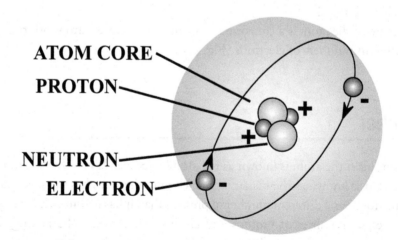

Figure 2.2 The atom. Protons and neutrons are located in the core or nucleus with electrons located in shells surrounding the nucleus.

Chemical Basis of Life | 25

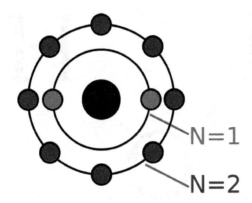

Figure 2.3 A diagram of an atom. Notice that there are 2 possible orbits or shells that carry electrons. The shells are named N1 and N2 respectively.

Chemical Bonds

Atoms are held together by forces. There are 4 fundamental forces in the universe. These are the strong and weak nuclear forces, gravity, and the electromagnetic force. The forces are carried by tiny particles. The force that accounts for chemical bonding is the electromagnetic force, which is carried by the photon.

Ionic Bonds

If an atom loses an electron and another picks up an electron, an ionic bond forms (Fig. 2.4). Some ionic bonds can "break up" or dissociate in solution. This dissociation occurs because water molecules surround the sodium and chloride atoms. The sodium chloride is then said to have dissolved in water. Ionic bonds are characterized by this exchange of electrons.

Fig. 2.4 Sodium (Na) loses an electron and Chloride (Cl) gains an electron, forming an ionic bond (NaCl).

Covalent Bonds

Some atoms can "share" electrons. In this case the bond is known as a covalent bond. The more electrons are shared, the stronger the bond. Most organic molecules are covalently bonded. Covalent bonds are stronger than ionic bonds (Fig. 2.5).

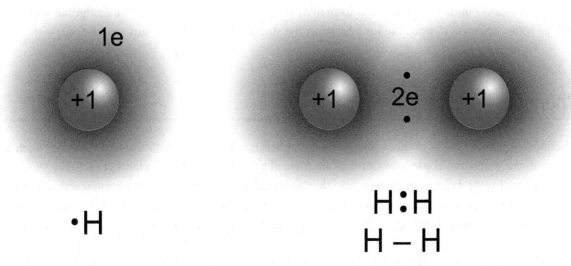

Figure 2.5 Electrons are shared in a covalent bond. Hydrogen can form a covalent bond with itself.

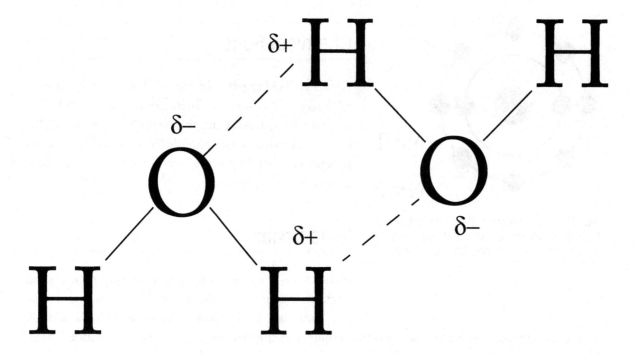

Figure 2.6 Water has a partial positive charge on the hydrogen side and a partial negative charge on the oxygen side. Notice how these 2 water molecules can bond together. The partial positive and negative sides attract, forming weak bonds called polar covalent bonds.

Polar Covalent Bonds

If the electrons are not equally shared, one side carries a greater charge. The molecule is then described as polar (Fig. 2.6).

Polar covalent bonds are weak bonds formed by the partial positive and negative charges in molecules such as water. They are responsible for a force on the surface of water called surface tension.

Chemical Reactions

Chemical bonds can break and new bonds form in chemical reactions. The substances we start with are known as the reactants. The substances that are formed are known as the products (Fig. 2.7).

Some reactions release more energy than they absorb. These are known as exergonic reactions. Others absorb more energy than they produce. These are known as endergonic reactions. Chemical reactions need a certain amount of energy to get started. The energy needed to break the bonds in a reaction is known as the activation energy.

Let's look at a theoretical reaction. Let's say that we would like to get reactants A and B together to form the product AB. We will put a little bit of A in a beaker and a little bit of B in the same beaker.

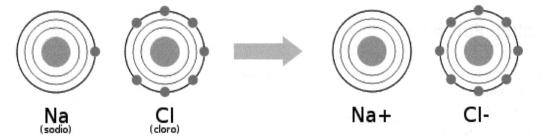

Figure 2.7 In this reaction sodium and chlorine are the reactants and sodium chloride is the product.

There is a particular probability that A and B will get together. But let's say we would like to increase our chances. We could put more A and B in the beaker (increase the concentration of A and B) or we could heat up our beaker to get A and B moving around more. Both an increase in concentration and temperature will increase our chances of getting the product AB. There is another way to get A and B together without raising the concentration or temperature. We could use a third substance known as a catalyst or enzyme. Enzymes work by virtue of their shapes. They act like templates or jigs that allow substances to either get together or break apart (Fig. 2.8). Enzymes can speed up chemical reactions by lowering the activation energy.

Cellular Metabolism

Many chemical reactions occur in the human body. These reactions are controlled by enzymes. If we add all of the reactions of the body together, we have what is called metabolism.

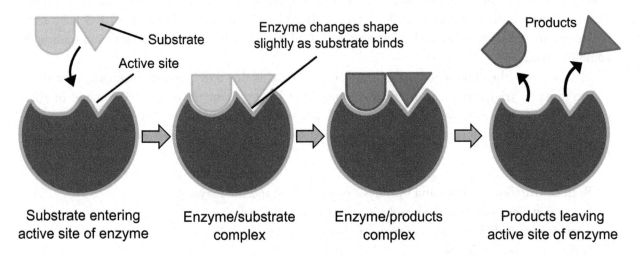

Figure 2.8 The site on the enzyme that connects to the reactants is known as the active site. The area on the reactant that connects to the enzyme is known as the substrate.

Reaction Pathways

Chemical reactions occur in specific directions or pathways where the product of one reaction may influence another:
   A + B -> AB
   AB + C -> AC + B

Metabolic Reactions

There are 2 main types of metabolic reactions. Anabolic reactions produce larger molecules from smaller ones. Catabolic reactions break larger molecules down into smaller ones. The human body uses anabolism for building substances for growth and repair. It also uses catabolism for breaking substances down and liberating energy. We can build larger organic molecules from smaller molecules through a reaction called dehydration synthesis. Dehydration synthesis is an example of an anabolic reaction.

> Dehydration means "to remove water"
> Synthesis means "to assemble"
> So we are assembling larger molecules by removing water.
> Carbohydrates, fats, and proteins are assembled via dehydration synthesis.
> Dehydration synthesis
>
> - Joins sugar molecules together to form glycogen (carbohydrate)
> - Joins fatty acid molecules and glycerol together to form fat molecules
> - Joins amino acids together to form peptides and eventually proteins

So we can build the major organic molecules by the process called dehydration synthesis. For example, if we take 2 simple sugars such as glucose and put them together, we form a more complex carbohydrate called a disaccharide (maltose). And if we take a glycerol molecule and combine it with 3 fatty acid molecules, we get a type of fat called a triglyceride. And if we take a large number of amino acids and combine them, we get a protein.

Catabolism is the opposite of anabolism. It is the breaking down of larger molecules into smaller. For example, a triglyceride can be broken down into glycerol and fatty acids. The name of this catabolic reaction is hydrolysis. Hydro means to "add water." Lysis means to "break down." So now we have 2 reactions (dehydration synthesis and hydrolysis). What causes reactions to go one way or the other?

> Both dehydration synthesis and hydrolysis require use of specific *enzymes*.
> Some enzymes cause anabolism and some enzymes cause catabolism.
> There are many different types of enzymes. Here are a few examples:
> Lipase—catabolizes lipids
> Protease—catabolizes proteins
> Amylase—catabolizes carbohydrates

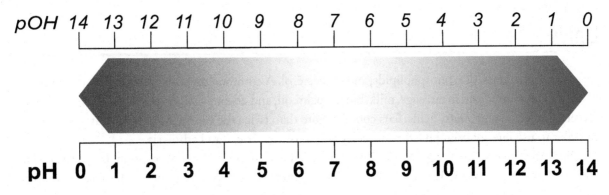

Figure 2.9. The pH scale ranges from 0 to 14.

Co-factors/Co-enzymes

Sometimes, an enzyme is not active until it combines with a non-protein molecule called a co-factor or co-enzyme. Co-enzymes are usually vitamins.

Acids and Bases

If a substance dissociates into hydrogen ions (positive) and negative ions (known as anions), it is called an acid. If a substance dissociates into hydroxide ions (OH-) and positive ions, it is called a base. Some substances dissociate into anions and cations that are not hydrogen or hydroxide. These are known as salts. A measure of the acidity or alkalinity is known as pH. If a solution has a lot of hydrogen ions, it is known as acidic. If a solution has a lot of hydroxide ions, it is known as alkaline (basic). The pH scale is a logarithmic scale and ranges from 0 to 14. Between 0 and 7 is acidic, with 7 being neutral. Between 7 and 14 is basic (Fig. 2.9).

Carbohydrates, Lipids, and Proteins

Carbohydrates are known as sugars and starches. Some examples of carbohydrates are monosaccharides, disaccharides, and polysaccharides. Monosaccharides are considered simple sugars (fructose, galactose, and glucose). Examples are honey and fruits. Disaccharides are also simple sugars. Examples are milk sugar, cane sugar, beet sugar, and molasses. Polysaccharides are considered complex sugars. Examples are starches from grains and vegetables, and glycogen from meat. Cellulose is also a carbohydrate. Cellulose cannot be digested by the human system and provides bulk to the diet. Glucose is also converted to glycogen in the liver by a process called glycogenesis and stored as a reserve. Excess glucose is stored as fat.

Lipids

Lipids consist of fats, oils, phospholipids, and cholesterol. A common dietary lipid is called a triglyceride. Triglycerides come from meat, eggs, milk, butter, palm oil, and coconut oil. A triglyceride is made up of a glycerol molecule and 3 fatty acids. Fats contain more than twice the energy of carbohydrates or proteins. In order to use energy from fats, the fat molecules must first undergo hydrolysis.

Digestion breaks down triglycerides into fatty acids and glycerol. After absorption, they travel to the bloodstream by way of the lymphatic system. Some lipids are not produced or synthesized by the liver and must be taken in by diet. These are known as essential fatty acids.

Proteins

Proteins consist of chains of amino acids. Proteins are used by body in enzymes, clotting factors, skin and hair keratins, elastin and collagen, plasma proteins, muscle components, hormones, and antibodies. Proteins also supply energy. However, your body will use the carbohydrates and lipids before using proteins for energy.

Adenosine Triphosphate

Next we will look at an energy molecule in the body known as adenosine triphosphate (ATP) (Fig. 2.10). ATP is a major source of energy for many physiological processes in the body. These include muscle contraction in skeletal muscles and the heart, production of nerve impulses, and metabolism. ATP consists of an adenosine molecule, a ribose molecule, and 3 phosphates. What is important about ATP is that it can store and release energy. Energy can be extracted from food molecules such as carbohydrates and then be used to make ATP. The energy that is available for use in the body is stored in ATP in the phosphate bond. This is called the high-energy phosphate bond. The basic reaction that releases ATP proceeds in this way:

ATP = ADP + Phosphate + Energy

Notice that ATP releases energy by giving up a phosphate. The product includes adenosine biphosphate or ADP. Energy can also be stored in ATP by adding a phosphate to ADP. This process is known as phosphorylation.

CRITICAL-THINKING QUESTIONS

1. The most abundant elements in the foods and beverages you consume are oxygen, carbon, hydrogen, and nitrogen. Why might having these elements in consumables be useful?
2. Oxygen, whose atomic number is eight, has three stable isotopes: $16O$, $17O$, and $18O$. Explain what this means in terms of the number of protons and neutrons.

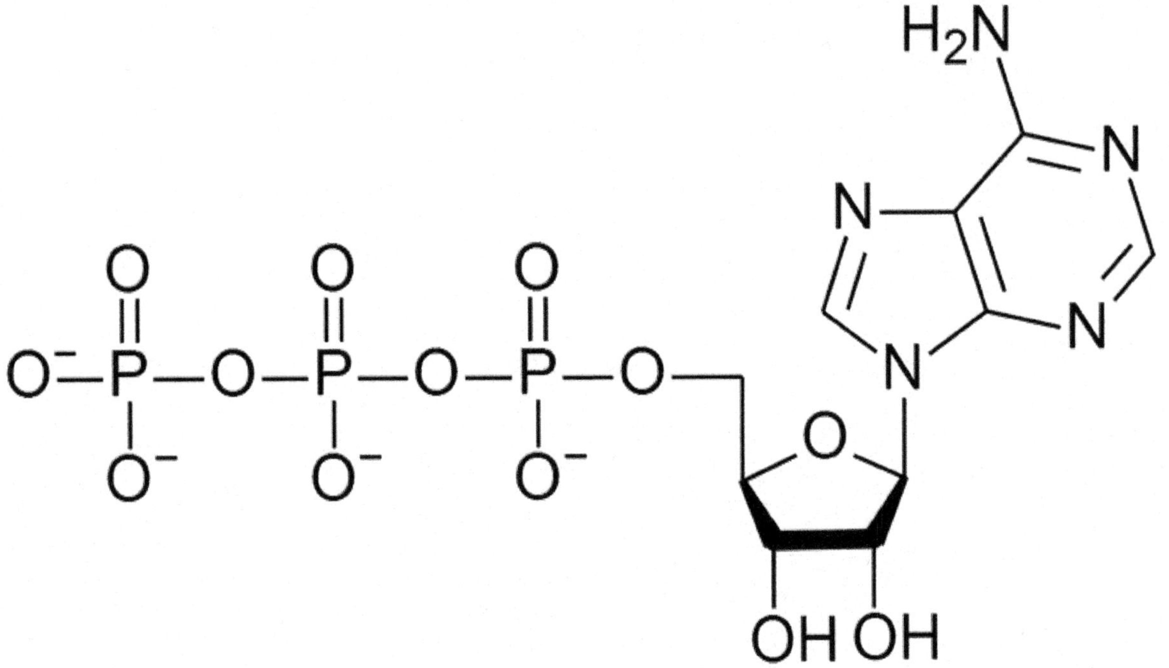

Figure 2.10 ATP contains an adenosine molecule (containing nitrogens), a ribose (ring molecule), and 3 phosphates.

3. Magnesium is an important element in the human body, especially in bones. Magnesium's atomic number is 12. Is it stable or reactive? Why? If it were to react with another atom, would it be more likely to accept or to donate one or more electrons?
4. Explain why CH_4 is one of the most common molecules found in nature. Are the bonds between the atoms ionic or covalent?
5. Could two atoms of oxygen engage in ionic bonding? Why or why not?
6. The pH of lemon juice is 2, and the pH of orange juice is 4. Which of these is more acidic, and by how much? What does this mean?
7. During a party, Eli loses a bet and is forced to drink a bottle of lemon juice. Not long thereafter, he begins complaining of having difficulty breathing, and his friends take him to the local emergency room. There, he is given an intravenous solution of bicarbonate. Why?
8. If the disaccharide maltose is formed from two glucose monosaccharides, which are hexose sugars? How many atoms of carbon, hydrogen, and oxygen does maltose contain and why?
9. Once dietary fats are digested and absorbed, why can they not be released directly into the bloodstream?

Chapter Three

Cells

Cell Structure and Function

What is a cell? A cell is a structure as well as functional unit of life. Every living thing has cells: bacteria, protozoans, fungi, plants, and animals are the primary groups (Kingdoms) of living things. Some organisms are made up of just one cell and are called unicellular (e.g., bacteria and protozoans), but animals, including human beings, are multi-cellular. An adult human body is composed of about 100,000,000,000,000 cells! Each cell has basic requirements to sustain it, and the body's organ systems are built largely around providing the many trillions of cells with those basic needs (such as oxygen, food, and waste removal). There are about 200 different kinds of specialized cells in the human body.

When many identical cells are organized together, it is called a tissue (such as muscle tissue, nervous tissue). Various tissues organized together for a common purpose are called organs (the stomach is an organ, and so is the skin, the brain, the uterus, etc.). Concepts about cell structure have changed considerably over the years. Early biologists saw cells as simple membranous sacs containing fluid and a few floating particles. Today's biologists know that cells are much more complex than this. Therefore, a strong knowledge of the various cellular organelles and their functions is important to any physiologist. If a person's cells are healthy, then that person is generally healthy. All physiological processes, disease, growth, and development can be described at the cellular level.

Specialized Cells of the Human Body

Although there are specialized cells, both in structure and function within the body, all cells have similarities in their structural organization and metabolic needs (such as maintaining energy levels via conversion of carbohydrate to ATP and using genes to create and maintain proteins). Below are some of the different types of specialized cells within the human body.

- **Nerve cells**: Also called neurons, these cells function to process and transmit information. They are the core components of the brain, spinal cord, and peripheral nerves. They use chemical synapses that can evoke electrical signals, called action potentials, to relay signals throughout the body.
- **Epithelial cells**: Functions include secretion, absorption, protection, transcellular transport, sensation detection, and selective permeability. Epithelium lines both the outside (skin) and the inside cavities and lumina of bodies.
- **Exocrine cells**: These cells secrete products through ducts, such as mucus and sweat, or digestive enzymes. The products of these cells go directly to the target organ through the ducts. For example, the bile from the gallbladder is carried directly into the duodenum via the bile duct.
- **Endocrine cells**: These cells are similar to exocrine cells, but secrete their products directly into the bloodstream instead of through a duct. Endocrine cells are found throughout the body but are concentrated in hormone-secreting glands such as the pituitary. The products of the endocrine cells go throughout the body in the bloodstream but act on specific organs by receptors on the cells of the target organs. For example, the hormone estrogen acts specifically on the uterus and breasts of females because there are estrogen receptors in the cells of these target organs.
- **Blood cells**: The most common types of blood cells are red blood cells (erythrocytes). The main function of red blood cells is to collect oxygen in the lungs and deliver it through the blood to the body tissues. Gas exchange is carried out by simple diffusion. Also there are various types of white blood cells (leukocytes). They are produced in the bone marrow and help the body to fight infectious disease and foreign objects in the immune system. White cells are found in the circulatory system, lymphatic system, spleen, and other body tissues.

Cell Size

Cells are the smallest structural and functional living units within our body, but play a big role in making our body function properly. Many cells never have a large increase in size, like eggs, after they are first formed from a parental cell. Typical stem cells reproduce, double in size, then reproduce again. Most cytosolic contents such as the endomembrane system and the cytoplasm easily scale to larger sizes in larger cells. If a cell becomes too large, the normal cellular amount of DNA may not be adequate to keep the cell supplied with RNA. Large cells often replicate their chromosomes to an abnormally high number or become multinucleated. Large cells that are primarily for nutrient storage can have a smooth surface membrane, but metabolically active large cells often have some sort of folding of the cell surface membrane in order to increase the surface area available for transport functions.

Cellular Organization

Several different molecules interact to form organelles within our body. Each type of organelle has a specific function. Organelles perform the vital functions that keep our cells alive. In this section we will be concerned primarily with studying cells that contain a nucleus; these are known as eukaryotic cells. Cells contain small structures called organelles. These are much like the organs in our bodies. The organelles have various functions that are important in maintaining the cell (Fig. 3.1).

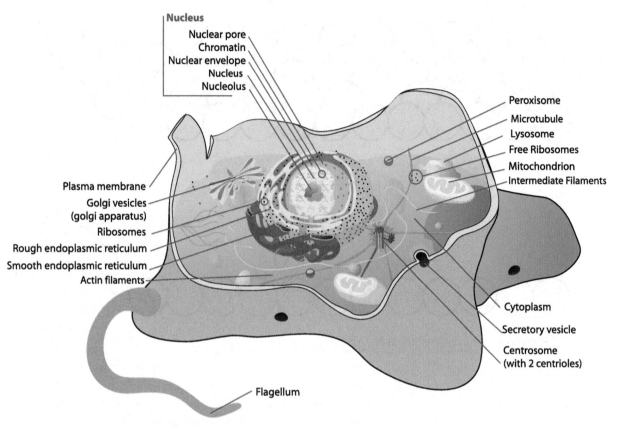

Figure 3.1 The cell contains a variety of organelles.

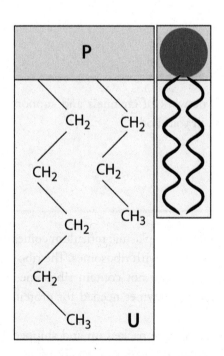

Figure 3.2 Phospholipids contain a phosphate head and a lipid tail.

Cell Membrane

We will start by looking at the cell membrane. The structure of the cell membrane has a lot to do with its function. The cell membrane is composed of molecules called phospholipids (Fig. 3.2).

The phosphate head of the phospholipid likes water, so it is called hydrophilic, while the lipid tail is called hydrophobic or "water hating." Because of the water-loving and -hating characteristics of the heads and tails, phospholipids arrange themselves in what is known as a bilayer (Fig. 3.3).

The cell membrane also allows certain substances in or out; it is "selectively permeable." For example, lipid-soluble substances can pass right through the cell membrane. Examples of lipid-soluble substances are oxygen, carbon dioxide, and steroids. Water-soluble substances cannot pass through the cell membrane and require carrier proteins in order to get into or out of the cell. The cell membrane also contains a number of proteins. Some of these proteins are imbedded on the surface of the cell and some go all the way through the cell membrane.

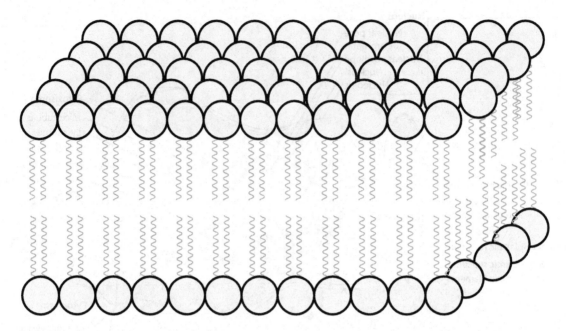

Figure 3.3 Phospholipids arrange themselves into a bilayer.

Some proteins act as channels to allow substances to pass through the membrane. Others act as receptors that receive information carried by proteins. Still others act as connection points for other cells to attach. These are known as intercellular junctions.

Cytoplasm

The cytoplasm or cytosol is the fluid inside the cell. It contains a network of channels and support structures called the cytoskeleton. This is very much like the skeleton in your body.

Endoplasmic Reticulum

Another important organelle is the endoplasmic reticulum (Fig. 3.4). The endoplasmic reticulum comes in two varieties: rough and smooth. Rough endoplasmic reticulum is studded with ribosomes. The ribosome's function is to produce proteins. Smooth endoplasmic reticulum does not contain ribosomes. It functions in making lipids. Ribosomes contain RNA, protein, and the enzymes needed for protein synthesis.

After the endoplasmic reticulum synthesizes the proteins, they need to be packed up and shipped out to other parts of the cell or to other cells. That's where the Golgi apparatus takes over. The Golgi apparatus packs up the proteins. Besides the vesicles from the Golgi apparatus, there are other vesicles containing enzymes for breaking up debris in the cell. These are called lysosomes.

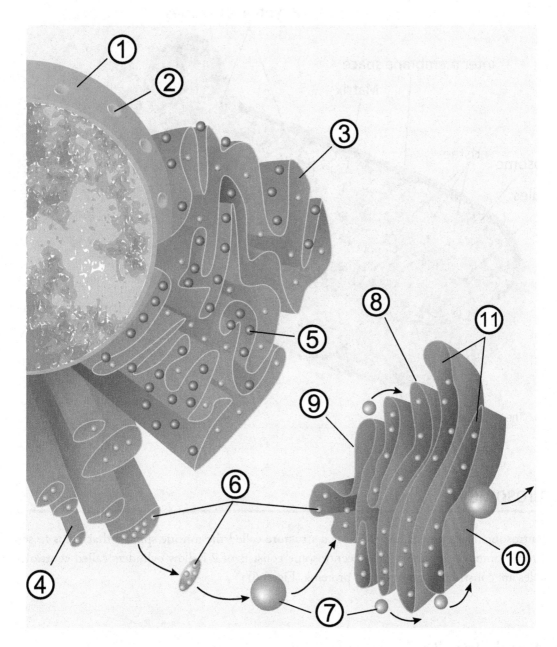

Figure 3.4 The endoplasmic reticulum (3) contains ribosomes (5) that function in making proteins. The Golgi apparatus (11) then packages the proteins in vesicles (12).

Mitochondrion

The next organelle we will explore is very important because it produces energy that is needed throughout the body. It is known as the "powerhouse" of the cell and is called the mitochondrion (Fig. 3.5). The mitochondrion takes fuel such as glucose and extracts the energy from it to make ATP. The inner portion of the mitochondrion is folded into shelves called cristae. These are studded with enzymes needed for the many chemical reactions used to make ATP.

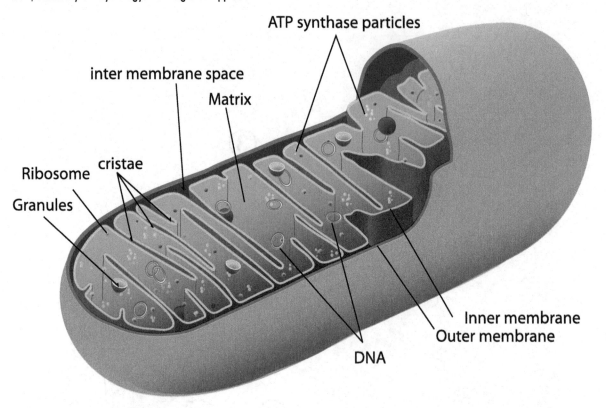

Figure 3.5 The mitochondrion.

Centrosome

The centrosome is important in producing a structure called the mitotic spindle that helps to separate the chromosomes during mitosis. The centrosome consists of 2 hollow cylinders called centrioles. The centrioles are constructed from tubular proteins (Fig. 3.6).

Cilia and Flagella

The cell contains other protein structures named cilia and flagella. Cilia and flagella are important in cellular movements (Fig. 3.7). Cilia are protein structures that move substances across cells. A flagellum is a long protein structure that moves the cell. Cells may have many cilia but will have only one flagellum.

Microfilaments and Microtubules

Microfilaments are solid protein structures that form the cytoskeleton to support the cell. Microtubules are hollow and can transport substances around the cell (Fig. 3.8).

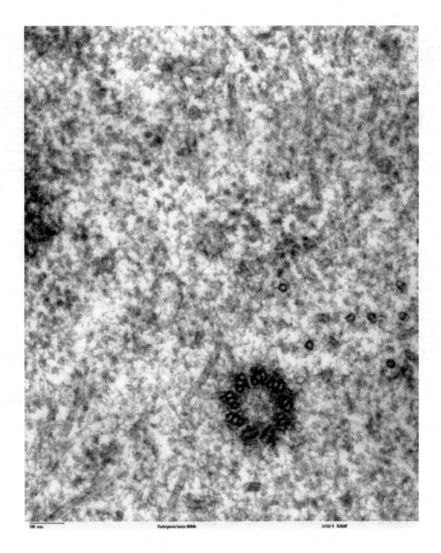

Figure 3.6. A centriole (the circular structure in the lower right-hand corner). The centriole consists of tubular proteins.

Nucleus

The nucleus contains the DNA of the cell. It is surrounded by a membrane much like the cell membrane. Inside the nucleus is the nucleolus, which contains RNA and proteins. This is where ribosomes are synthesized (Fig. 3.9).

Transport of Substances in Cells

Now that you have been introduced to some of the components of the cell, let's look at how substances move into and out of the cell. Remember that the cell membrane is made up of phospholipids. Since the membrane is composed of phospholipids, lipid-soluble substances can move across the

Figure 3.7 Notice the hair-like structures on B, E, H, and I. These are cilia. Cilia can move substances along the surface of cells.

membrane. Remember some examples of lipid-soluble substances are oxygen, carbon dioxide, and steroids. But what pushes or pulls substances across the membrane?

Diffusion

Diffusion is the movement of substances from an area of higher concentration toward an area of lower concentration until reaching equilibrium. The force that drives diffusion comes from differences in concentration called concentration gradients. The actual mechanism behind diffusion is quite complex and has to do with the second law of thermodynamics. This law states that in any given system there must be an increase in entropy. To explain this in simpler terms, substances tend to move from an organized state (concentrated state) to a more disorganized state (less-concentrated state). The process of diffusion can be illustrated by making a drink from a powdered mix. When the powder first hits the water, it is in higher concentration than its surrounding fluid. The powder will dissolve and then begin to distribute evenly throughout the glass of water. The powder is said to move from an area of higher concentration to lower concentration until it is equally distributed throughout the glass. Another example is with an aerosol spray. Let's say I stood in front of the class and sprayed a room freshener into the air. The particles would eventually distribute evenly throughout the room so that even the students in the back of the room could smell it.

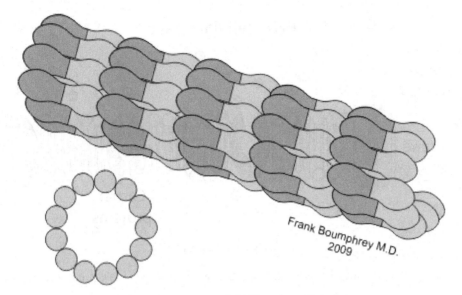

Figure 3.8 Example of a microfilament called myosin. Myosin is found in muscles.

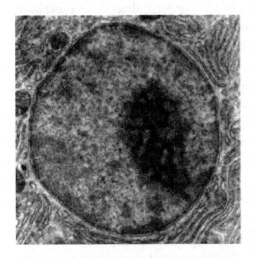

Figure 3.9 The nucleus (dark spot on the right side is the nucleolus). The nucleolus is the site of ribosome production.

Substances can diffuse at different rates. The rate of diffusion depends on a number of factors. These include:

- Molecule size—smaller molecules diffuse faster than larger molecules.
- Size of concentration gradient—the larger the difference in concentration, the faster substances will diffuse.
- Temperature—because diffusion relies on the movement of molecules, higher temperatures will cause more movement and speed up diffusion. For example, substances will diffuse faster at body temperature than room temperature.
- Distance—the shorter the distance, the faster substances will diffuse.
- Electrical forces—cells generally carry a negative charge on the inside. Negative charges will attract positive electrolytes and repel negative ones. This can speed up or slow down the rate of diffusion.

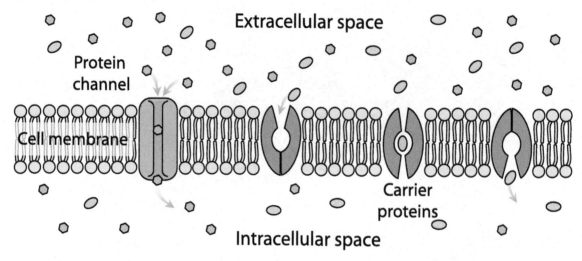

Figure 3.10 Carrier proteins allow non-lipid-soluble substances to move into and out of cells in channel-mediated diffusion.

Facilitated Diffusion

At this point, you know how lipid-soluble substances pass through cell membranes powered by diffusion, but how do non-lipid-soluble substances get into and out of the cell? The answer has to do with what is known as facilitated diffusion (Fig. 3.10). Cell membranes contain proteins. Some of these proteins go all the way through the membrane and act as channels for specific substances. Examples of substances that move via facilitated diffusion are sodium, potassium, and chloride. The force that moves substances in facilitative diffusion is the same one as in diffusion. That is, the difference in concentration or concentration gradient. Again, substances still move from areas of higher to lower concentration, but this time they move through a protein channel.

Substances moving into and out of cells by facilitated diffusion must bind to receptors on the protein channel. Once they bind, the protein changes its shape, allowing the substance into or out of the cell.

Since proteins have only a finite number of receptors, once the receptors become saturated, there cannot be movement of any additional substances. Therefore, in some cases a larger concentration gradient will not move substances at a faster rate (unlike diffusion). The rate of diffusion then partially depends on the saturation of the receptors on the protein channel. One example of a substance transported into the cell via facilitated diffusion is glucose. Glucose is used by cells to make ATP, an important energy molecule in the body. Muscle cells require glucose to make ATP for muscle contraction. In order for glucose to move into a muscle cell, it not only needs to connect to a receptor on the protein channel, but another hormone called insulin also needs to connect to a special insulin receptor on the protein. In some cases the insulin receptors become resistant to insulin, causing blood glucose levels to rise. This occurs in what is known as insulin-resistant diabetes.

Osmosis

Water moves within the human body across a variety of membranes. The membranes are called semipermeable because they allow only water to move across them, not solute. The movement of water across a semipermeable membrane has a special name: osmosis. Water moves from an area of higher

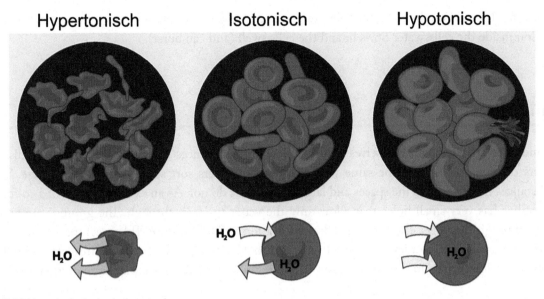

Figure 3.11 Hypertonic, isotonic, hypotonic.

concentration to lower concentration. However, we typically do not talk about concentration in terms of water. We usually talk about concentration in terms of solute. So you could think of osmosis in 2 ways:

1. Water moves across a semipermeable membrane from a higher area of concentration of water to a lower concentration of water.
2. Water moves across a semipermeable membrane from an area of lower concentration of solute to an area of higher concentration of solute.

One simple way to remember osmosis is the phrase "water follows salt" to mean that water always moves toward an area of higher concentration of solute.

Isotonic/Hypotonic/Hypertonic Solutions

The force exhibited by osmosis is called osmotic pressure. This pressure is related to the solute concentration of the solution. In chemistry we describe concentration in terms of osmolarity. However, in physiology when we are concerned with concentration with regard to cells, we use the term "tonicity." Tonicity is associated to the human cell whereby osmolarity is the number of osmoles per liter. Osmolarity depends on the number of particles of solute. For example, one mole of glucose in water would equate to 1 osmole since glucose remains as 1 molecule in water. However, one mole of sodium chloride would equate to 2 osmoles because sodium chloride dissociates in water to form 2 particles. If a solution has the same osmolarity as body fluids have, we say the solution is isotonic. The human body's osmolarity is close to .30 osmoles or 300 milliosmoles. If a solution is less concentrated than body fluids, we say the solution is hypotonic. And if the solution is more concentrated than body fluids, we say the solution is hypertonic. Tonicity is important when it comes to introducing solutions to the human body. Let's see why (Fig 3.11).

In the first image on the left, a red blood cell is placed in a hypertonic solution. Since the solution is more concentrated than the red blood cell is, the cell shrivels up or crenates. In the middle picture, the red blood cells are placed in an isotonic solution. Since the tonicity is equal, nothing happens to the cell.

In the final picture on the right, the cells are placed in a hypotonic solution. Since there is more concentration inside the cells, water flows in and the cells swell (and can burst).

Filtration

Sometimes cells arrange themselves in thin layers and substances can move between the cells. These layers or membranes work the same way as filters do. Filters sort substances based on size. Smaller substances move through the spaces and larger substances do not. As an example, think of a coffee filter. The filter has very small holes that allow only the water to move through. The grounds are too large to fit through the holes. The force that drives filtration is fluid pressure. This pressure is also known as hydrostatic pressure. In order to move substances through a filter, they must move from an area of higher pressure to lower pressure. There are many examples of filters in the body: capillaries and kidneys.

Active Transport

So far you have seen how substances move down their respective concentration gradients in diffusion and facilitative diffusion. But what if a substance needs to be moved *against* its concentration gradient? In active transport, substances are moved against their concentration gradients by carrier proteins. However, there is an energy cost to be paid for this action. So the carrier proteins use ATP as an energy source. An example of an active transport protein is the sodium-potassium pump (Fig. 3.12). Normally there is more sodium outside of the cell than in, so sodium would move from outside to in. Also, there is usually more potassium inside the cell than out, so potassium would follow its concentration gradient and move

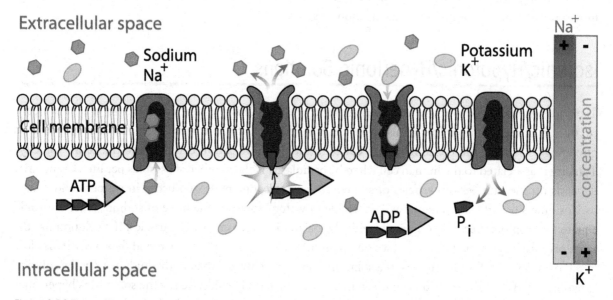

Figure 3.12 The sodium-potassium pump. Sodium moves out of the cell against its gradient while potassium moves into the cell (also against its gradient). The carrier protein must use energy in the form of ATP. One ATP moves 3 sodiums out of the cell and 2 potassiums in.

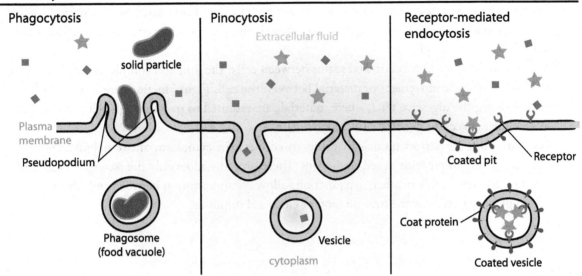

Figure 3.13 In phagocytosis the cell reaches out and engulfs a particle, bringing it into the cell for destruction. The cell brings in fluid via pinocytosis. Substances attach to receptors on the cell membrane in receptor-mediated endocytosis.

out of the cell. However, we want to move these molecules against their concentration gradients. This can be done, but energy must be used to do so. Energy is used by the pump in the form of ATP. The sodium potassium pump is vital and works to maintain and establish the concentration gradients that keep us alive. Cystic fibrosis is a genetic disorder that results in a mutated chloride ion channel. By not regulating chloride secretion properly, water flow across the airway surface is reduced and the mucus becomes dehydrated and thick.

Other Transport Mechanisms

Other ways that substances can move into and out of cells include cotransport, exocytosis, and endocytosis (Fig. 3.13). In endocytosis, substances enter cells via vesicles. There are 3 types of endocytosis: phagocytosis, pinocytosis, and receptor-mediated endocytosis. All involve the cell membrane wrapping around and engulfing a vesicle. In pinocytosis a cell can take in a small droplet of fluid. In phagocytosis the cell takes in a solid then uses a lysosome to break down the solid. In receptor-mediated endocytosis, substances bind to receptors on the cell membrane. The membrane responds by forming a vesicle and taking the substance into the cell. Substances can exit cells by same method (exocytosis). Note: certain hormones are able to target specific cells by receptor-mediated endocytosis (we will discuss this later in the text).

Cell Junctions

The plasma membranes of adjacent cells are usually separated by extracellular fluids that allow transport of nutrients to and wastes from the bloodstream. In certain tissues, however, the membranes of adjacent cells may join and form a junction. Three kinds of cell junctions are recognized:

- **Desmosomes** are protein attachments between adjacent cells. Inside the plasma membrane, a desmosome bears a disc-shaped structure from which protein fibers extend into the cytoplasm. Desmosomes act like spot welds to hold together tissues that undergo considerable stress, such as our skin or heart muscle.
- **Tight junctions** are tightly stitched seams between cells. The junction completely encircles each cell, preventing the movement of material between the cell. Tight junctions are characteristic of cells lining the digestive tract, where materials are required to pass through cells, rather than intercellular spaces, to penetrate the bloodstream.
- **Gap junctions** are narrow tunnels that directly connect the cytoplasm of two neighboring cells, consisting of proteins called connexons. These proteins allow only the passage of ions and small molecules. In this manner, gap junctions allow communication between cells through the exchange of materials or the transmission of electrical impulses.

Mitosis

We all begin as one cell. This one cell becomes many (trillions) cells by dividing in a process known as mitosis. In mitosis the genetic material (DNA) is carried on to daughter cells. Some cells in the body do not divide. Once they are lost they must be replenished through differentiation of stem cells. Other cells divide all the time. Examples of cells that do not divide are nerve cells known as neurons (although there are some exceptions to this rule) and red blood cells. An example of a cell that divides is the epithelial cell, located in the skin and digestive system. Cells that undergo mitosis follow a cycle of rest followed by cell division. During the rest phase the cell gets ready to divide. The rest phase is called interphase. During interphase the cell carries out processes of growth, metabolism, and DNA replication. There is another process of cell division known as meiosis in which the new cells have only one half of the DNA of the original cell. We will investigate meiosis in detail when we cover the reproductive system.

Interphase

Interphase is the preparatory phase for the cell undergoing mitosis. Interphase consists of 3 sub-phases known as G1, S, and G2. During the G1 phase the cell produces copies of its organelles such as the mitochondria, endoplasmic reticulum, Golgi, and ribosomes. The centrioles begin to replicate. Some cells complete G1 in 8–12 hours. Next the cell enters the S phase. During the S phase the cell replicates its DNA and ends up with 2 sets of identical chromosomes. DNA replication occurs during the S phase of interphase. During replication the 2 strands of DNA unwind and separate at the hydrogen bonds (between the bases). A new sequence of nucleotides then attaches to each individual strand, with new hydrogen bonds forming. Some cells complete the S phase in 6–8 hours. Finally the cell enters the G2 phase. The centrioles complete their replication and complete protein synthesis. Some cells complete the G2 phase in 2–5 hours. When the cell is finished, it enters the M or mitosis phase.

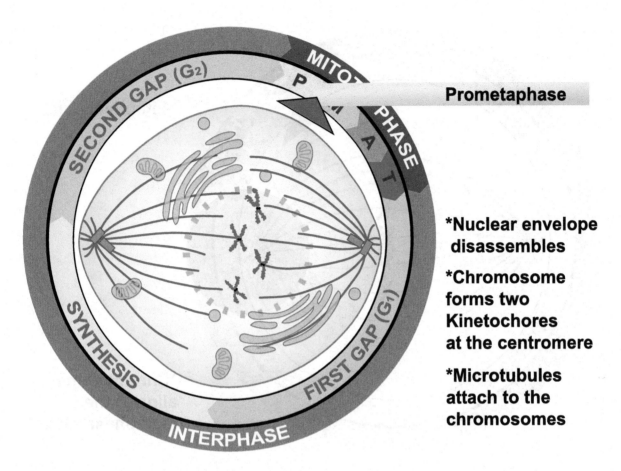

Figure 3.14. In prophase (prometaphase) the cell begins to divide.

Prophase

Generally the DNA of the cell is in the form of chromatin, which is loosely organized. During prophase the genetic material forms tightly coiled chromosomes. Remember that there are 2 sets of identical chromosomes. Each set is bound together by a central structure called a centromere (Fig. 3.14). During prophase the nuclear membrane begins to break up and disappear. The nucleolus also disappears as the chromosomes form. A system of microtubules forms the mitotic spindle at opposite ends of the cell. The centromere is surrounded by a protein structure called a kinetochore. The spindle fibers attach themselves to the kinetochores.

Metaphase

During metaphase the chromosomes all line up in the center of the cell. This area is called the metaphase plate (Fig. 3.15).

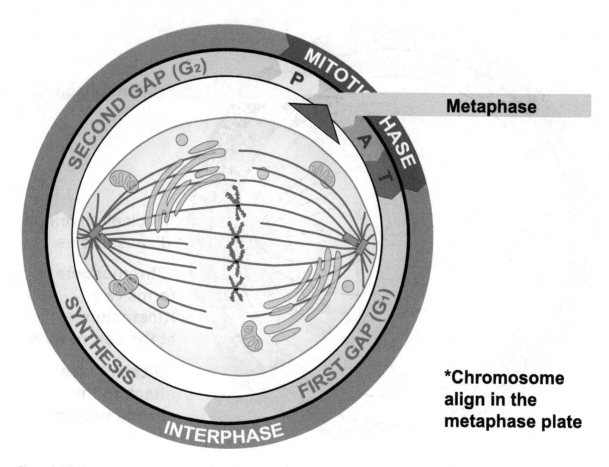

Figure 3.15. Metaphase.

Anaphase

During anaphase the spindle fibers shorten and the centromeres divide, separating the pair of chromosomes. The chromosomes move to opposite sides of the cell (Fig. 3.16).

Telophase

The final stage of mitosis is telophase. The nuclear membrane and nucleolus begin to reappear. The mitotic spindle breaks up and the chromosomes uncoil. Two daughter cells are now present (Fig. 3.17).

Protein Synthesis: Transcription and Translation

Proteins are vital information carriers in the body. Proteins consist of long chains of building blocks called amino acids. Proteins carry information by virtue of the sequence of amino acids. The information flows from DNA to RNA to protein (Fig. 3.18). This process is known as the central dogma of biology.

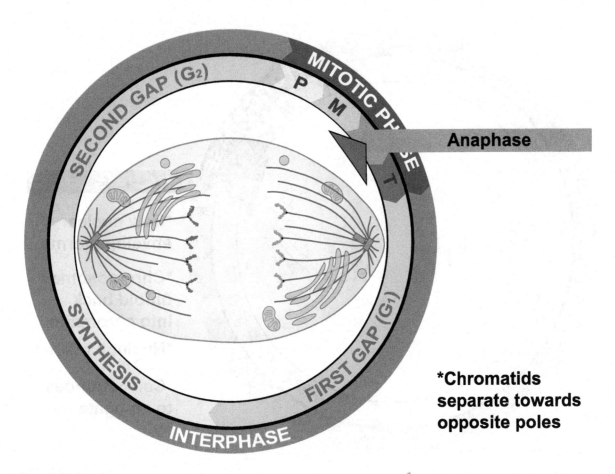

Figure 3.16. Anaphase.

The method by which information flows is in the two processes called transcription and translation. The information is transcribed from DNA to RNA, then translated to proteins.

DNA Structure

DNA consists of 3 main parts:

1. 5-carbon sugar (called a deoxyribose sugar)
2. Phosphate
3. Nitrogen-containing base

Each 3-part structure is called a nucleotide. The nucleotides connect via the phosphates to form 2 strands, much like a ladder. The 2 strands of nucleotides wrap around each other to form a double helix.

There are 4 bases in DNA:

1. Adenine (A)
2. Cytosine (C)
3. Thymine (T)
4. Guanine (G)

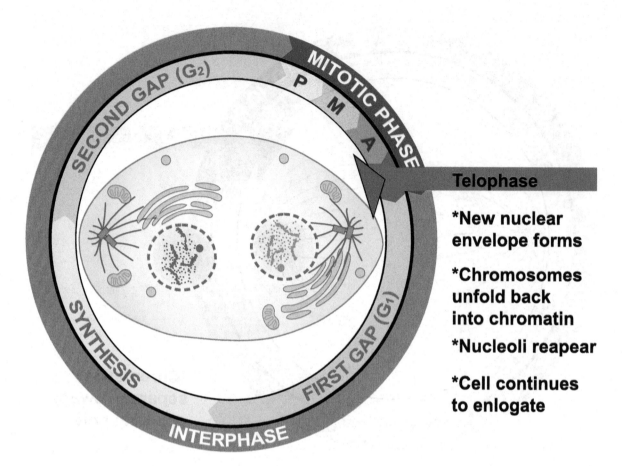

Figure 3.17 Telophase.

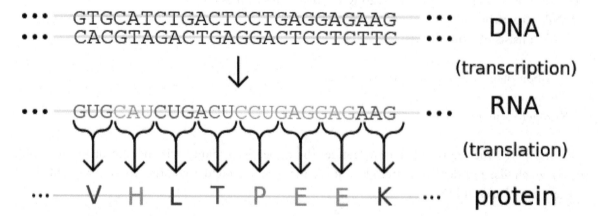

Figure 3.18 Information flows from DNA (transcription) to RNA (translation) to Protein. Each 3-base sequence codes for an amino acid.

The bases can form pairs. One base of the pair "fits" into the other. We say the bases are complementary. Cytosine pairs with guanine and adenine pairs with thymine (Fig. 3.19). The information in DNA is encoded in the sequence of bases. Each 3-base sequence along a strand of DNA is called a triplet code and can code for a specific amino acid. There are also start (promoter) and stop (terminator) codes (Fig. 3.20).

An area on DNA that contains the information for producing a specific trait is called a gene. All of the genes in DNA are known as the genome. The human genome contains 35,000 to 45,000 genes.

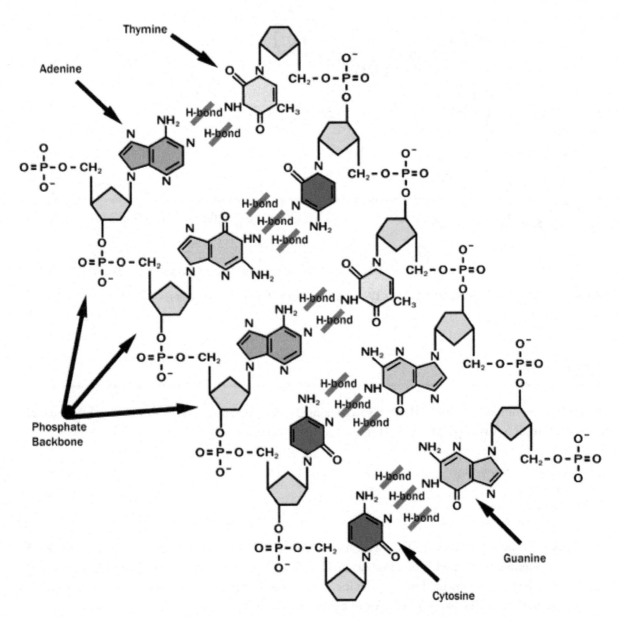

Figure 3.19 DNA consists of a 5-carbon deoxyribose sugar (turquoise), a phosphate, and 4 possible nitrogen bases. The bases connect to each other via hydrogen bonds.

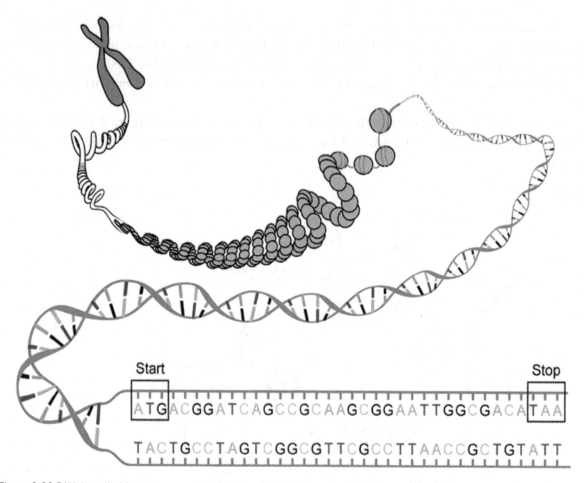

Figure 3.20 DNA is coiled into chromosomes. DNA contains information in its base pairs.

RNA

RNA is very similar to DNA. RNA also contains 3 parts, but unlike DNA's double-stranded-helix arrangement, RNA is single stranded (Fig. 3.30). RNA contains:

- 5-carbon ribose sugar
- Phosphate
- Nitrogen bases

RNA has 3 of the 4 bases in DNA with one exception. RNA contains uracil instead of thymine. So instead of adenine pairing with thymine (as in DNA), adenine pairs with uracil in RNA. In RNA the bases form the following pairs: Uracil–Adenine, Cytosine–Guanine. Here is an example of base-pairing during transcription. The bases on DNA can be read by mRNA.

```
DNA mRNA
A-----------U
T-----------A
```

```
C----------G
G----------C
C----------G
T----------A
```

RNA also comes in different types. There is the RNA that reads the information from DNA called messenger RNA. There is also a type of RNA that assembles the amino acid sequence in translation called transfer RNA. And, there is ribosomal RNA.

Transcription

During transcription, the information encoded by the sequence of bases on DNA needs to get out to the protein-making machinery at the ribosome. The first step in translation is to expose the information on DNA. A special enzyme called RNA polymerase helps to unwind the DNA to expose the bases. Next a single-stranded messenger RNA (mRNA) molecule "reads" the sequence of bases. The first 3-base code is a start code followed by codes for various amino acids. Some amino acids have more than one code so there is some redundancy in the coding. The 3-base code on mRNA is known as the codon (Figs. 3.21, 3.22). The mRNA then takes its message out of the nucleus by moving through a nuclear pore and delivers it to the protein-making machinery known as the ribosome.

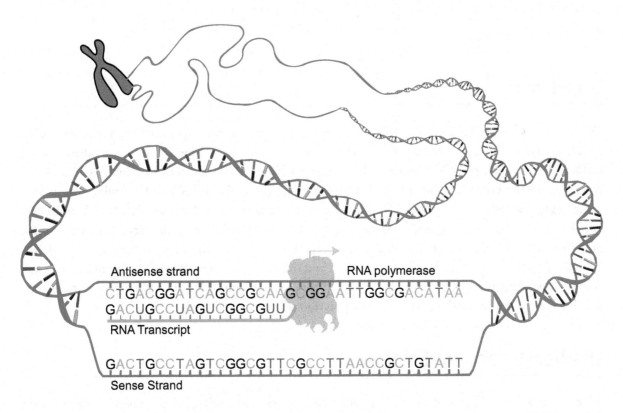

Figure 3.21 Transcription. RNA polymerase helps to unwind DNA, exposing the bases. Messenger RNA (RNA Transcript) reads the exposed bases with its complementary bases.

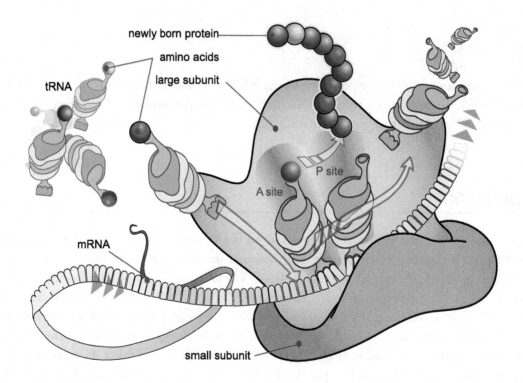

Figure 3.22. The ribosome is the site of polypeptide and protein synthesis. It consists of large and small subunits. Here mRNA meets up with transfer RNA (tRNA). The tRNA has 2 sites. One portion reads the code on the mRNA. The other has an amino acid–binding site. The amino acids are bound together to form a large string.

Translation

Translation occurs at the ribosome. Ribosomes contain large and small subunits. The messenger RNA (mRNA) carries the information for making the protein to the ribosome. There is a binding site for mRNA on the small ribosomal subunit. The transfer RNA (tRNA) attaches to a binding site on the large ribosomal subunit. Transfer RNA reads the information on the mRNA and assembles the protein accordingly. The 3-base code on the tRNA is known as the anticodon. There are 20 mRNA molecules, one for each of the 20 amino acids. On the other end of the tRNA is an amino acid–binding site. The protein is assembled one amino acid at a time. The process stops when a stop code is reached. The completed protein then detaches from the tRNA and the ribosome splits into its 2 subunits. Sometimes another ribosome attaches to the same strand of mRNA and transcribes it. When several ribosomes attach to one strand of mRNA, we call this a polyribosome.

The New Science of Epigenetics

There is an exciting new field in biology called epigenetics that explores the connection between the environment and the expression of certain genetic traits. In other words, it was long thought that information flowed down a one-way street from DNA to RNA to protein. In fact, as we mentioned earlier in this chapter, this concept of information flow was deemed the central dogma of biology. However, in the past few years, it has been discovered that there is feedback loop beyond the random mutations of natural selection that influences whether or not certain genes are expressed. This provides new hope for those carrying disease-causing genes. Proponents of epigenetics, such as Bruce Lipton (author of *Biology of Belief*), go as far as to say that since beliefs affect behavior and behavior contributes to environmental influences, then changing one's beliefs and behaviors can affect how genes are expressed. Positive behaviors can then be passed down through generations, much like genetic traits. One recent study investigated the effects of environment on the expression of rheumatoid arthritis (RA). This study, conducted at the Karolinska Institutet, looked at the work environment. Researchers looked at what is called psychosocial workload. In particular they found a correlation between low decision latitude and RA. In other words, jobs in which workers have little input in decision making had an increased incidence of RA. Lack of control in the workplace has also been associated with high blood pressure and heart attacks. Other behavioral factors that have been associated with an increase in RA include smoking and drinking alcohol. (Karolinska Institutet (2008, September 28). "Working Environment Is One Cause of Rheumatoid Arthritis.")

CRITICAL-THINKING QUESTIONS

1. What materials can easily diffuse through the lipid bilayer, and why?
2. Why is receptor-mediated endocytosis said to be more selective than phagocytosis or pinocytosis?
3. What do osmosis, diffusion, filtration, and the movement of ions away from like charge all have in common? In what way do they differ?
4. Explain why the structures of the ER, mitochondria, and Golgi apparatus assist their respective functions.
5. Compare and contrast lysosomes with peroxisomes: name at least two similarities and one difference.
6. Explain in your own words why DNA replication is said to be "semiconservative."
7. Why is it important that DNA replication take place before cell division? What would happen if cell division of a body cell took place without DNA replication, or when DNA replication was incomplete?
8. Briefly explain the similarities between transcription and DNA replication.
9. Contrast transcription and translation. Name at least three differences between the two processes.
10. What would happen if anaphase proceeded even though the sister chromatids were not properly attached to their respective microtubules and lined up at the metaphase plate?
11. What materials can easily diffuse through the lipid bilayer, and why?
12. Why is it important that DNA replication take place before cell division? What would happen if cell division of a body cell took place without DNA replication, or when DNA replication was incomplete?
13. Explain what would happen to a cell if it lost its nucleus.
14. Explain what would happen to a cell if all the lysosomes ruptured.
15. Explain what would happen to a cell if the phospholipids in the cell membrane were dissolved.
16. Explain what would happen to a cell if the cell began losing its mitochondria.
17. Explain what would happen to a cell if the transfer RNA molecules are selectively destroyed by viruses.
18. Adriamycin is a chemotherapeutic drug that binds to DNA and blocks messenger RNA synthesis. Explain why this drug is fatal to a cell.

19. Lysosomes remove nonfunctional cell parts. Explain how this function is important to the overall health of the cell.
20. You work for the JP Pharmaceutical Company. Your latest assignment is to design a drug that interferes with translation in cells. You decide the easiest way to do this is to have your drug target those organelles and molecules involved in translation. List the potential targets of your new drug

Chapter Four

Cellular Metabolism

Cells need energy to keep us alive. Energy is needed to allow us to breathe, keep our hearts pumping, muscles contracting, and conduct thousands of other vital processes in our bodies. Cells manufacture and store energy in the form of ATP (adenosine triphosphate). There are about one billion molecules of ATP in a typical cell. As mentioned in Chapter 2, ATP consists of an adenine molecule, a ribose sugar, and 3 phosphates.

Energy is stored in ATP in the high-energy phosphate bond. When energy is needed, the phosphate breaks off the ATP, liberating the energy, and leaving adenosine diphosphate (ADP). When energy is stored, a phosphate is added to ADP to make ATP. The storage and release of energy in ATP is controlled by enzymes.

When energy is liberated from ATP: ATP -> ADP + P + Energy
When energy is stored in ATP: ADP + P -> ATP

We call this phosphorylation (adding a phosphate) of ADP to make ATP.

Metabolism is the sum of all biochemical reactions in the body. There are 2 basic reactions: anabolic and catabolic. In anabolic reactions larger molecules are made from smaller molecules. Examples of anabolic reactions are the synthesis of proteins from amino acids and the construction of phospholipids from fatty acids. Catabolic reactions are characterized by the breaking down of larger molecules into smaller molecules. Every time you consume a food, your body uses catabolic reactions to break the food down into smaller molecules. For example, proteins are broken down into amino acids and complex carbohydrates are broken down into simple carbohydrates.

A general model for an anabolic reaction is:
A + B -> AB
For a catabolic reaction:
AB -> A + B

Many of the metabolic reactions in the body have to do with the production and storage of ATP. Energy is extracted from the foods we eat and then used to store energy in ATP.

One way to look at how energy is produced in the body is to examine where the majority of energy comes from in different activities. Let's use an example to illustrate this concept. Let's say Jay is going on a long bicycle ride. Jay wants to make good time so he pedals vigorously. During the first 30 seconds of Jay's ride, the majority of energy comes from a process known as the ATP-phosphocreatine system. There are molecules of phosphocreatine (PCr) stored near Jay's muscles. The PCr contains a phosphate that easily lends itself to phosphorylate ADP to make ATP to power Jay's muscles. There is only a small supply of PCr so the energy lasts for only about 30 seconds. The amount of PCr is what limits the system. Jay continues to ride beyond 30 seconds. During the next 180 seconds of intense activity, the majority of Jay's energy comes from the anaerobic portion of a reaction known as glycolysis. Glycolysis is a reaction that breaks down glucose and extracts the energy for making ATP. Glycolysis is a bit more complicated than the ATP-PCr system.

Glycolysis occurs in the cytoplasm of the cell. Glucose enters the cell and is split into two 3-carbon molecules called glyceraldehyde-3-phosphate. ATP is used to prime the reactions and the 2 molecules of glyceraldehyde-3-phosphate are converted to 2 molecules of pyruvic acid (pyruvate). The following is a summary of the steps of glycolysis. First, glucose needs a bit of help to get things going so a couple of ATPs are needed to get it ready.

Glucose + 2ATPs -> Fructose, 1, 6 biphosphate

This molecule is then split into 2 molecules. You might remember that glucose is a 6-carbon molecule. When it splits, it forms two 3-carbon molecules.

Fructose 1, 6 biphosphate-> 2 glyceraldehyde-3-phosphate

The 2 molecules of glyceraldehyde-3-phosphate lose hydrogen atoms (oxidize) and gain phosphates to form 2 molecules of 1,3-diphosphoglycerate. A byproduct of this reaction is the formation of 2NADH$_2$ molecules that will be used later. The 2 molecules of 1,3-diphosphoglycerate each give up one phosphate to phosphorylate ADP, making 2 ATPs and converting to 2 molecules of 3-phosphoglycerate. The phosphates move to another carbon, forming 2-phosphoglycerate. Water is removed from these molecules (dehydration), forming 2 molecules of phosphoenolpyruvate. The phosphates are removed and added to ADP to make 2 more ATPs, forming 2 molecules of pyruvate. If the pyruvate is converted to lactic acid, the 2 molecules of NADH$_2$ are used. If not, pyruvate is converted to acetyl-coenzyme A and enters the Krebs cycle (coming up next). The conversion of pyruvate to acetyl coenzyme A produces 2 more NADH$_2$ molecules. Notice that 2 ATP molecules are used to get the reactions started and 4 ATPs are produced. This results in a net gain of 2 ATPs. Glycolysis also produces 2 molecules of NADH which can be used by the electron transport chain to phosphorylate ADP to make ATP. If oxygen is not present the 2 molecules of pyruvate are converted to lactic acid. The 2 molecules of NADH are used in this process. This is what happens to Jay during the next 30–180 seconds of cycling. As Jay continues to cycle beyond 180 seconds, his body switches to the aerobic energy systems. The pyruvate is now converted to acetyl coenzyme A, which enters the Krebs cycle. The Krebs cycle occurs in the mitochondrion and consists of a number of reactions using enzymes located in the cristae of the mitochondrion. Each turn of the Krebs cycle produces the following products: 1 molecule of ATP, 3 molecules of NADH, 1 molecule of FADH2.

Since 2 molecules of pyruvate enter the Krebs cycle, the reactions occur twice, producing twice the amount of product. Since we began with one molecule of glucose, at this point we end up with 2 molecules of ATP, 6 molecules of NADH, and 2 molecules of FADH2.

The NADH and FADH carry energy that can be used to phosphorylate ADP. This occurs in the electron transport chain, which is also located in the mitochondrion. The electron transport chain is a series of enzymes that pass electrons from one to another. The enzymes pass the electrons along to lower energy levels. The energy is extracted to power an enzyme complex known as ATP synthase, which phosphorylates ADP. Notice that NADH2 and FADH2 enter the electron transport chain at different levels. The electron transport chain is going to extract some energy from these molecules and use it to make ATPs by adding a phosphate to ADP. NADH2 has enough energy to make 3 ATPs while FADH2 only has enough energy for 2 ATPs. We know that energy is stored in the phosphate bond in ATP but how do these molecules of NADH2 and FADH2 store energy? The energy is stored in electrons. Remember that electrons are tiny objects that "orbit" around the nucleus of an atom. If an atom is *excited* (takes on energy) the electrons move to the outer orbital shells. If the atom *calms down* (by releasing energy) the electrons move to the inner shells. It turns out that NADH2, for example, *donates* electrons to the system. It is called oxidation when a molecule loses electrons. When it does it loses 2 hydrogen atoms (which are also called protons): NADH2-> NAD+ + 2H+

As we stated earlier the electron transport chain is located in the mitochondrion. On the inside of the mighty mitochondrion is a folded membrane called the cristae. The inside of the membrane is called the matrix, while the area outside of the membrane is called the intramembranous space. Located in the membrane is a set of 5 special proteins. The big picture is that the first 4 proteins extract the energy from the electrons from NADH2 and FADH2 and use this energy to pump protons across the membrane (from the matrix to the intramembranous space). The protons build up on one side of the membrane, forming a proton gradient. The proton gradient is used by the fifth membrane protein to add phosphates to ADP. This is called phosphorylation of ADP, which makes ATP.

Protein I

NADH2 encounters the first protein and loses 2 hydrogens and 2 electrons (oxidized). The 2 hydrogens are picked up by a molecule in the protein (FMN->FMNH2), which then passes the electrons from the hydrogens to iron (Fe). The hydrogens pick up their lost electrons and are transferred to a third molecule (ubiquinone, aka coenzyme Q10). The hydrogens again separate from their electrons and the electrons are again passed to iron. Iron again passes the electrons to another ubiquinone located outside of the protein and in the membrane. In order to do this the hydrogens must recombine with their lost electrons. In summary, the first membrane protein acts as an active transport pump that pumps hydrogens from the matrix, across the membrane, and into the intramembranous space of the mitochondrion.

Protein II

This protein works with FADH2 generated from the Krebs cycle. The hydrogens are then stripped from FADH2, forming FAD+ and are combined with iron. The iron transfers the electrons to ubiquinone where they recombine with their hydrogens.

Protein III

The membrane-bound ubiquinone releases electrons that are picked up by the third protein. This protein passes the electrons via an electron carrier called cytochrome C. The cytochrome C transports electrons to the fourth protein.

Protein IV

At the fourth protein the electrons combine with hydrogen and oxygen (from breathing) to form water. We say that oxygen is the final electron acceptor. The energy allowing the electrons to move from carrier to carrier is used to pump protons (H+) across the membrane. NADH2 can pump more protons than FADH2 can. The protons build up and form a proton gradient. This gradient is used by the final protein to make ATP.

Protein V

The final protein is actually an enzyme called ATP synthase. This enzyme uses the proton gradient to add a phosphate to ADP (this is called phosphorylation). Since NADH2 releases its electrons on one side of the inner membrane, the hydrogens (protons) build up. This creates a proton gradient whereby the protons move from one side of the membrane to the other.

Total ATPs from 1 Molecule of Glucose

Perhaps you may be pondering just how many molecules of ATP can be produced from one molecule of glucose. If we add up all of the energy producing molecules we get the following:

Glycolysis = net gain of 2 ATP
Krebs cycle (2 turns) = 2 ATP

Remember that there is enough energy in one molecule of NADH2 to make 3 ATPs. Likewise there is enough energy in one molecule of FADH2 to make 2 ATPs. So let's total them up.

NADH2s from conversion of pyruvate to acetyl coenzyme A = 2 3 3 = 6 ATPs
NADH2s from glycolysis = 2 3 3 = 6 ATPs
NADH2s from Krebs (2 turns) = 6 3 3 = 18 ATPs
FADH2s from Krebs (2 turns) = 2 3 2 = 4 ATPs

So the grand total is: 2+2+6+6+18+4 = 38 ATPs from one molecule of glucose!

Summary

The Krebs cycle (Fig. 4.1) and glycolysis do not require oxygen directly but are still part of the aerobic metabolism of glucose. This is due to the use of oxygen by the electron transport chain. The last enzyme in the chain gives up a pair of electrons that combine with hydrogen ions and oxygen to form water. Oxygen is the last electron acceptor in the chain. So when Jay cycles longer than 3 minutes, most of the energy comes from the aerobic metabolism systems. These same systems provide energy when Jay is at rest (Fig. 4.2).

Cellular Metabolism | 61

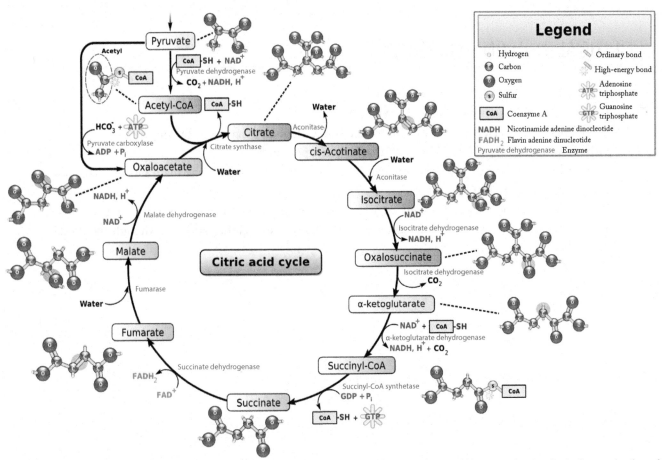

Figure 4.1 Krebs (Citric Acid cycle). The important events comprise the entrance of acetyl-CoA and its conversion to citrate; the production of NADH + H when isocitrate is converted to oxalosuccinate, alpha-ketoglutarate is converted to succinyl-CoA, and malate to oxaloacetate; the production of FADH2 when succinate is converted to fumarate; and the production of GTP (or ATP).

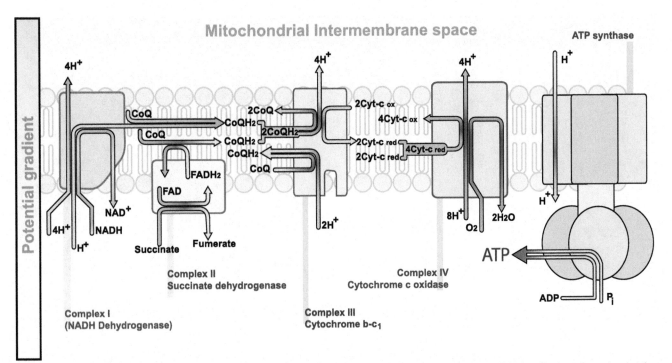

Figure 4.2 Electron Transport Chain. Electrons are passed from NADH and FADH2 to electron carrier molecules. H+ ions are released from NADH and FADH2 and "pumped" into the outer mitochondrial membrane, creating a proton gradient. The proton pumps use electrons to move H+ ions and then pass electrons to the next carrier. The final carrier forms water by combining H+ with oxygen. The proton gradient powers the ATP synthase molecule to phosphorylate ADP.

CRITICAL-THINKING QUESTIONS

1. Describe how metabolism can be altered.
2. Explain how glucose is metabolized to yield ATP.
3. If the breath of a person who has diabetes smells like alcohol, what could this mean?
4. How can the ingestion of food increase the body temperature?
5. Weight loss and weight gain are complex processes. What are some of the main factors that influence weight gain in people?
6. Some low-fat or non-fat foods contain a large amount of sugar to replace the fat content of the food. Discuss how this leads to increased fat in the body (and weight gain) even though the item is non-fat.

Chapter Five

Tissues, Glands, and Membranes

The body is filled with many diverse types of tissues. The study of tissues is known as histology. When studying tissues it helps to think about the relationship between the structure of a tissue and its function. Individuals who study histology spend a significant amount of time looking in microscopes at the various body tissues. There are 4 main categories of tissues in the human body:

- A. Epithelium
- B. Connective
- C. Muscle
- D. Nervous

Epithelium is a tissue that covers other structures (Fig. 5.1). Therefore, one side is always exposed to the outside (which could still be inside the body). You will see epithelial tissue covering the inside of body cavities and organs. The outer or superficial portion of your skin is an epithelial tissue. Epithelial tissue does not have a blood supply. Therefore nutrients must enter the tissue by diffusion. Epithelial tissue is also anchored to other structures via a basement membrane. Epithelial tissue is categorized according to the shape of the cells and number of layers (Fig. 5.2). There are 3 basic shapes of epithelial cells:

- A. Squamous cells are flattened.
- B. Cuboidal cells look like cubes.
- C. Columnar cells are rectangular.

Simple squamous epithelium is a very thin tissue. It is often the site of substance transport. Simple squamous epithelium is found in capillaries, the kidneys, and the alveoli of the lungs. Simple cuboidal epithelium is slightly thicker than simple squamous. It is found lining tubular ducts. Examples of where simple cuboidal epithelium is found are the ovaries, kidney tubules, ducts (Fig. 5.3). Simple columnar epithelium is even thicker than simple cuboidal and is found in the urethra, pharynx, and vas deferens.

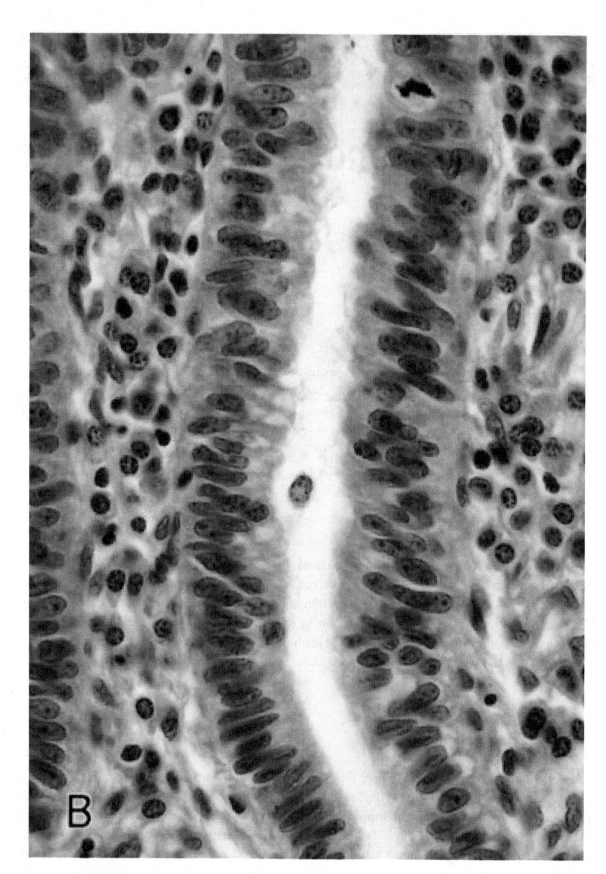

Figure 5.1 Example of epithelium.

Types of Epithelium

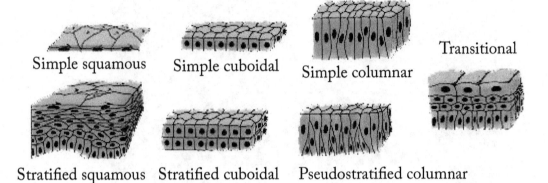

Figure 5.2 Different types of epithelia.
A. Simple columnar
B. Simple columnar with cilia
C. Stratified squamous
D. Simple squamous
E. Transitional
F. Pseudostratified columnar
G. Cuboidal
H. Choanocytes
I. Pseudostratified columnar with cilia

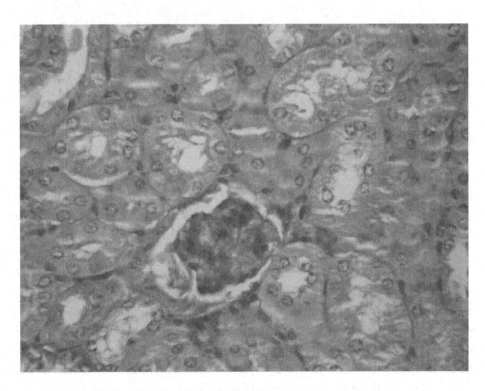

Figure 5.3 Simple cuboidal epithelium lines the tubules in the kidneys.

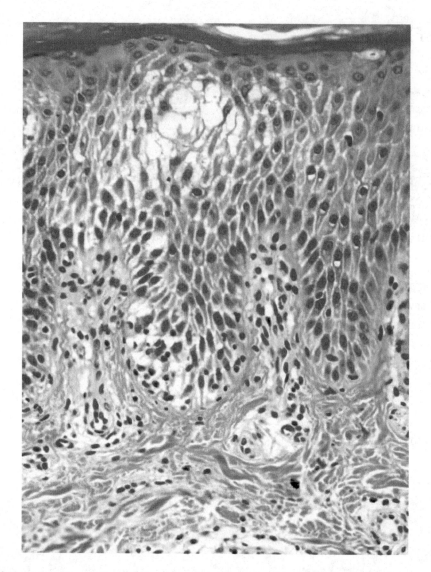

Figure 5.4 Stratified squamous epithelium. Notice the cells in the deeper layers are rounded and become flattened at the surface.

Stratified squamous epithelium also has a defined structure; the cells in the deeper layers are more cuboidal in shape. The more superficial layers contain flattened cells. You can find stratified squamous in the superficial layer of the skin known as the epidermis, and also in the oral cavity, anal canal, and vagina (Fig. 5.4).

Stratified cuboidal epithelium also lines ducts. It can be found in the ducts of mammary glands, sweat glands, salivary glands, and the pancreas. Columnar epithelium can also be stratified. This tissue is found in the vas deferens and pharynx. It provides a thicker lining for some tubular structures in the body. There are some special cases of epithelium, however, and we will investigate these next.

The first special case is called pseudostratified columnar epithelium. It is called pseudostratified because it looks stratified (more than one layer). It looks stratified because the nuclei of the cells are at various levels, but in reality there is only one layer (Fig. 5.5).

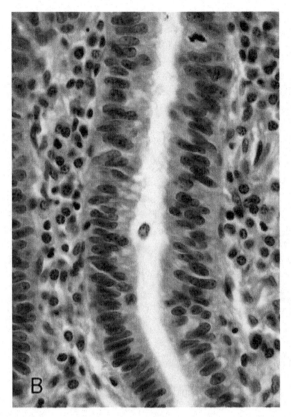

Figure 5.5 Pseudostratified columnar epithelium. Notice the nuclei appear to exist at different levels.

Pseudostratified columnar epithelium is found in the linings of the respiratory system. Another special case of epithelium is called transitional. Transitional epithelium is found in the urinary bladder. Transitional epithelium looks somewhat like stratified squamous, but there is a difference. In stratified squamous the cells are rounder in the deeper layers and then flatten out near the top. In transitional epithelium the cells are also rounder in the deeper layers; however, the cells remain rounded in the superficial layers.

Glandular Epithelium

To conclude our discussion regarding epithelium, we will next look at glandular epithelium. Glandular epithelium can secrete substances into the bloodstream (endocrine glands) or into ducts (exocrine glands). Exocrine glands can be classified by method of secretion. Merocrine glands release substances via exocytosis. Apocrine glands secrete substances by losing a small portion of the cell body. Holocrine glands secrete substances by releasing the entire cell.

Connective Tissue

Connective tissue is the most abundant tissue in the body. It consists of special cells called fibroblasts surrounded by a matrix of intercellular material. Fibroblasts secrete protein fibers into the matrix. The matrix can contain fibers such as collagen, elastic, and reticular fibers. Other cells can exist in connective tissue such as macrophages and mast cells (both are types of white blood cells). Macrophages destroy bacteria and debris by phagocytosis. Mast cells release heparin (an anticoagulant) and histamine (promotes inflammatory reactions). There are 3 types of fibers produced by fibroblasts. Collagenous fibers are thick protein fibers that are strong but not flexible. They have a high tensile strength, which means they can take a lot of force along their long axis. Ligaments and tendons have a high number of collagenous fibers. Elastic fibers are also made of protein. Elastic fibers do not have a very high tensile strength but are very flexible. They are found in areas like the vocal cords and air passages of the respiratory system. Reticular fibers are thin protein fibers that form branching networks.

There are 5 basic types of connective tissue:

1. Loose
2. Dense

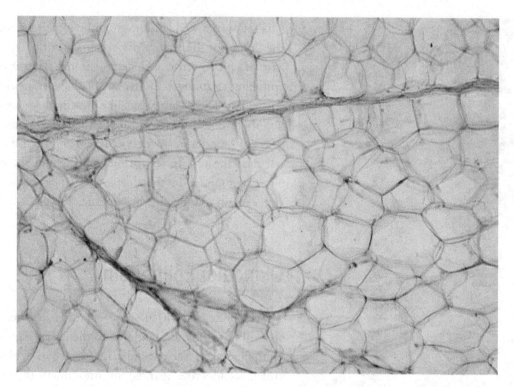

Figure 5.6 Adipose connective tissue consists of adipocytes. These large cells contain lipid.

3. Adipose
4. Reticular
5. Elastic

There is also a category called "special" connective tissue that includes blood, bone, and cartilage. Loose connective tissue is not very well-organized tissue. It contains fibroblasts, matrix, and some fibers scattered about. It is found in the dermis and subcutaneous layers of the skin as well as surrounding muscles. Sometimes it is referred to as fascia. Adipose connective tissue consists of cells containing lipid (fat) called adipocytes. The lipid is used to store energy to be used by the body. Adipose tissue is also found around some organs and joints. It forms a cushion for shock absorption. Adipose tissue also insulates the body (Fig. 5.6).

Reticular connective tissue consists of a thin supportive network of collagen fibers. It is found supporting the walls of the liver, spleen, and lymphatic system. Dense connective tissue contains thick collagenous fibers. It is found in ligaments and tendons, which have a high tensile strength. Dense connective tissue has a poor blood supply, which is why tendons and ligaments do not heal well. There are also some elastic fibers and fibroblasts. Elastic connective tissue contains more elastic fibers than collagen fibers. Elastic connective tissue is found in attachments between vertebrae and in walls of some hollow internal organs.

Blood is considered a liquid connective tissue. Blood contains a fluid matrix called plasma along with cells called formed elements. Blood contains red blood cells (erythrocytes), white blood cells (leukocytes), and platelets. It transports gases such as oxygen and carbon dioxide and functions in clotting and immunity (Fig. 5.7).

Bone is the most rigid of connective tissues. Its hardness comes from mineral salts such as calcium phosphate and calcium carbonate. It is highly organized into units called Haversian systems. The principal cell of bone is the osteocyte (Fig. 5.8).

There are 3 types of cartilage: hyaline, elastic, and fibrocartilage. Cartilage contains cells called chondrocytes imbedded in a matrix. There are also elastic and collagen fibers. Cartilage is rigid and strong so it can provide support and protection. It also forms a structural model for developing bones. The matrix in cartilage consists of a chondromucoprotein substance. Cartilage has no direct blood supply, so nutrients must enter by diffusion. Since the nutrients for cartilage diffuse into the tissue, the tissue needs water to help move these substances in. As humans age, cartilage tends to "dry up" or become dehydrated, which lends to degeneration of the tissue. The cartilage cells or chondrocytes also do not divide very frequently, which also contributes to poor healing. Chondrocytes are surrounded by a space called a lacuna. Hyaline cartilage has the characteristic chondrocytes in lacunae arrangement along with a "ground glass" appearance to the matrix. It is found at the ends of bones, soft part of the nose, larynx, and trachea. Hyaline cartilage serves as a model for bone growth (Fig. 5.9).

Elastic cartilage also has the characteristic chondrocyte in lacunae along with elastic fibers. This cartilage is found in the larynx and the ear. Fibrocartilage is characterized by rows of chondrocytes (in lacunae). It is a very strong cartilage and is found in the intervertebral discs.

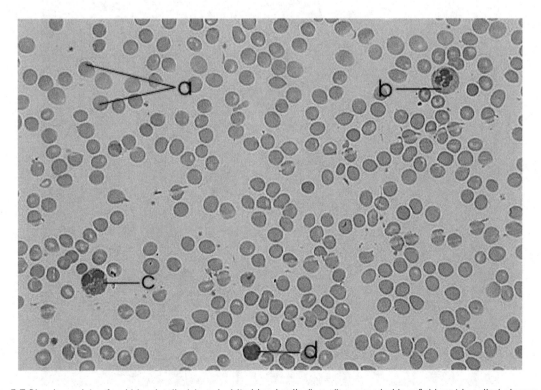

Figure 5.7 Blood consists of red blood cells (a) and white blood cells (b, c, d) suspended in a fluid matrix called plasma.

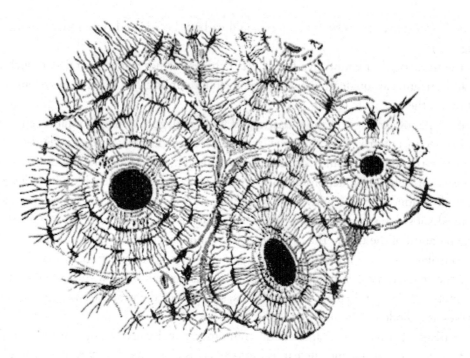

Figure 5.8 Bone is highly organized and contains structural units called Haversian systems. These are oriented along the long axis of bones to give them strength.

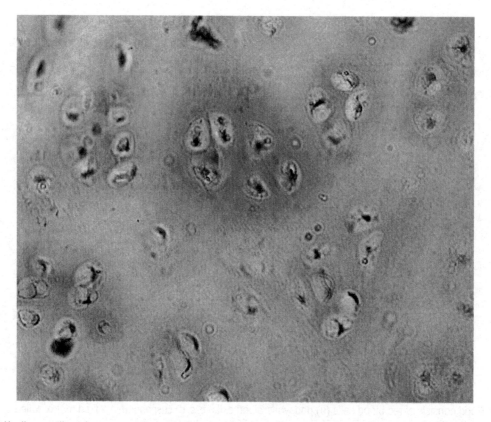

Figure 5.9 Hyaline cartilage has a smooth matrix. The chondrocytes are surrounded by a space called a lacuna.

 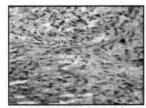 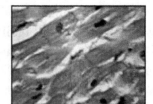

Skeletal muscle   Smooth muscle   Cardiac muscle

Figure 5.10 Three types of muscle tissue.

Muscle Tissue

There are 3 types of muscle tissue (Fig. 5.10): skeletal, smooth, and cardiac. Skeletal muscle is striated. The striations are caused by the density of overlapping protein filaments called actin and myosin. The high concentration of protein filaments creates an optical illusion when light is shown on the muscle tissue. The filaments break up the light into light and dark areas causing the striated appearance. Smooth muscle is more loosely organized with less concentrated protein filaments, therefore smooth muscle is not striated. Smooth muscle is found in organs such as in the gastrointestinal system and the arteries. Cardiac muscle is also striated but has a unique structure called an intercalated disc. The discs are special intercellular junctions that allow electrochemical impulses to be conveyed across the tissue.

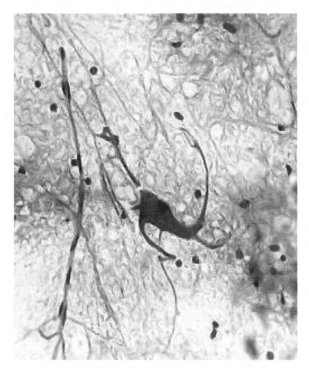

Figure 5.11 Nervous tissue contains cells called neurons (stained dark blue) and supportive glial cells.

Nervous Tissue

The last tissue we will investigate is nervous tissue. Nervous tissue consists of nervous system cells called neurons and supportive cells called glia (Fig. 5.11). We will discuss this tissue type in more detail in Chapter 10.

CRITICAL-THINKING QUESTIONS

1. Identify the four types of tissue in the body and describe the major functions of each tissue.
2. The structure of a tissue is usually optimized for its function. Describe how the structure of individual cells and tissue arrangement of the intestine lining matches its primary function of absorbing nutrients.

3. One of the main functions of connective tissue is to integrate organs and organ systems in the body. Discuss how blood fulfills this role.
4. Why does an injury to cartilage, especially hyaline cartilage, heal much more slowly than a bone fracture?
5. Why does skeletal muscle appear striated?
6. Indicate the type of cell membrane modification necessary to reinforce epithelial layers subjected to stress.
7. Indicate the type of cell membrane modification necessary to allow the epithelial layer to stretch.
8. Why is it advantageous for large artery walls to contain dense irregular elastic connective tissue?
9. Heavy smoking destroys the pseudostratified ciliated epithelium lining the trachea. Speculate what might happen if these cells are replaced by nonciliated epithelium during the course of tissue repair.

Section Two

Covering, Support, and Movement

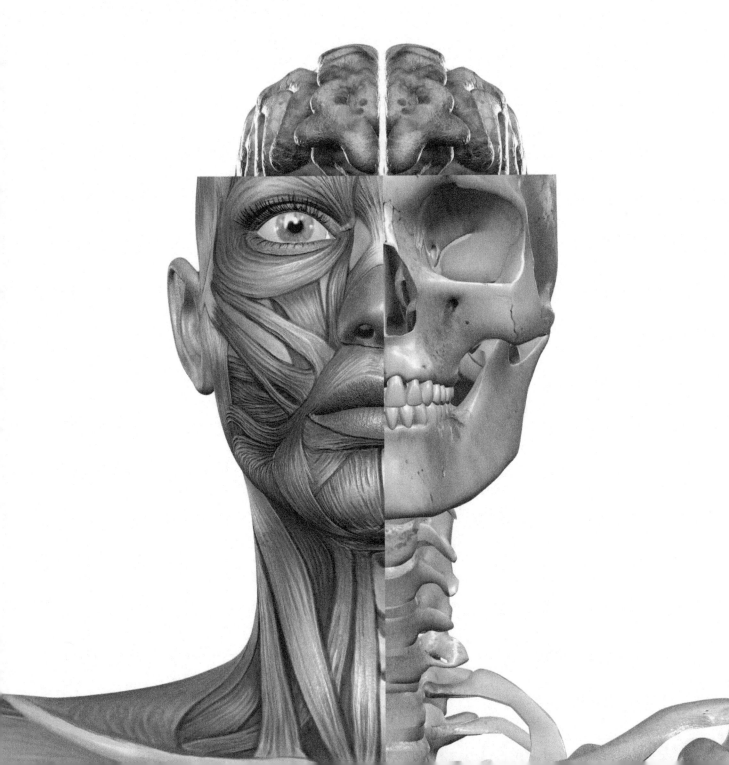

Chapter Six

Integumentary System

The integumentary system consists of the skin, hair, nails, subcutaneous tissue (below the skin), and assorted glands. Skin is our largest organ; adults carry some 8 pounds (3.6 kilograms) and 22 square feet (2 square meters) of it. The most evident function of the integumentary system is the protection that the skin gives to underlying tissues. The skin not only keeps most harmful substances out, but also prevents the loss of fluids. A major function of the subcutaneous tissue is to connect the skin to underlying tissues such as muscles. Hair on the scalp provides the head with insulation from cold. The hair of eyelashes and eyebrows helps keep dust and perspiration out of the eyes, and the hair in our nostrils helps keep dust out of the nasal cavities. Any other hair on our bodies no longer serves a function, but is an evolutionary remnant. Nails protect the tips of fingers and toes from mechanical injury. Fingernails give the fingers greater ability to pick up small objects.

There are four types of glands in the integumentary system: sudoriferous glands, sebaceous glands, ceruminous glands, and mammary glands. Sudoriferous glands are sweat-producing glands. These are important to help maintain body temperature. Sebaceous glands are oil-producing glands that help inhibit bacteria, keep us waterproof, and prevent our hair and skin from drying out. Ceruminous glands produce earwax, which keeps the outer surface of the eardrum pliable and prevents drying. Mammary glands produce milk.

As the interface with the surroundings, the skin plays the most important role in protecting against pathogens. Its other main functions are insulation and temperature regulation, sensation, and vitamin D and B synthesis. Skin has pigmentation—melanin, provided by melanocytes—which absorbs some of the potentially dangerous radiation in sunlight. It also contains DNA-repair enzymes that reverse UV damage, and people who lack the genes for these enzymes suffer high rates of skin cancer. One form predominantly produced by UV light, malignant melanoma, is particularly invasive, causing it to spread quickly, and can often be deadly. Human skin pigmentation varies among populations in a striking manner. This has sometimes led to the classification of people(s) on the basis of skin color. Damaged skin will try to heal by forming scar tissue, often giving rise to discoloration and depigmentation of the skin.

Layers of the Skin

The skin contains 2 layers and a subcutaneous layer. The superficial layer is called the epidermis. The epidermis consists of stratified epithelium tissue arranged in layers called strata. Deep to the epidermis is the dermis. The dermis consists of loose connective tissue and a number of other structures we will investigate later. The deepest layer is the subcutaneous layer that consists of loose connective tissue and adipose tissue along with blood vessels and nerves (Fig. 6.1). The epidermis consists of stratified squamous epithelium arranged in layers or strata. The layers are:

- Stratum corneum
- Stratum lucidum
- Stratum granulosum
- Stratum spinosum
- Stratum basale

The epidermis is anchored to the dermis by a basement membrane. The epidermis does not contain any blood vessels. The cells of the stratum basale are nourished by the blood vessels in the dermis. These cells can divide and move toward the surface, pushing the old cells off the superficial layers.

The stratum corneum is the most superficial layer of the epidermis. It consists of cells that have been hardened with keratin. Keratin is secreted by cells—located in the deep layers of the epidermis—called keratinocytes. The stratum lucidum is an additional layer that is found only in the palms of the hands and soles of the feet. It provides an added thickness to these layers. The stratum granulosum contains cells that have lost their nuclei. These cells remain active and secrete keratin. The cells contain granules in their cytoplasm that harbor keratin. The stratum spinosum contains cells called prickle cells. These cells have small radiating processes that connect with other cells. Keratin is synthesized in this layer. The stratum basale or basal cell layer contains epidermal stem cells. This is the deepest layer of the epidermis. It consists of one layer of cells that divide and begin their migration to the superficial layers. This is the layer where basal cell cancer develops. As we have seen, there many keratinocytes located in the epidermis. An example of an abnormality of keratinocytes is known as psoriasis—keratinocytes abnormally divide rapidly and migrate from stratum basale to stratum corneum. Many immature cells reach the stratum corneum, producing flaky, silvery scales (mostly on knees, elbows, and scalp). The epidermis also responds to the environment. Friction causes the formation

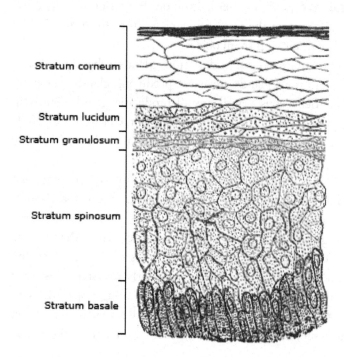

Figure 6.1 The layers of the epidermis.

of corns and calluses. Another kind of cell found in the epidermis is the melanocyte. This cell produces the pigment melanin that gives skin its color. Melanocytes are located in the deepest portion of the epidermis and superficial dermis. The color of the skin results from the activity of the melanocytes, not the number. Melanocytes respond to UVB radiation. Melanocytes are also found in the hair and middle layer of the eye. A condition known as malignant melanoma can develop in melanocytes. The skin also helps to synthesize vitamin D. Vitamin D (cholecalciferol) is synthesized when a precursor molecule known as 7-dehydrocholesterol absorbs ultraviolet radiation. This molecule then travels to the liver and kidney where it is converted to the active form of vitamin D (1,25-hydroxycholecalciferol) (Fig. 6.2). Vitamin D is an important substance in the body. It functions to help the body absorb calcium. It also works to help in calcium transport in the intestines.

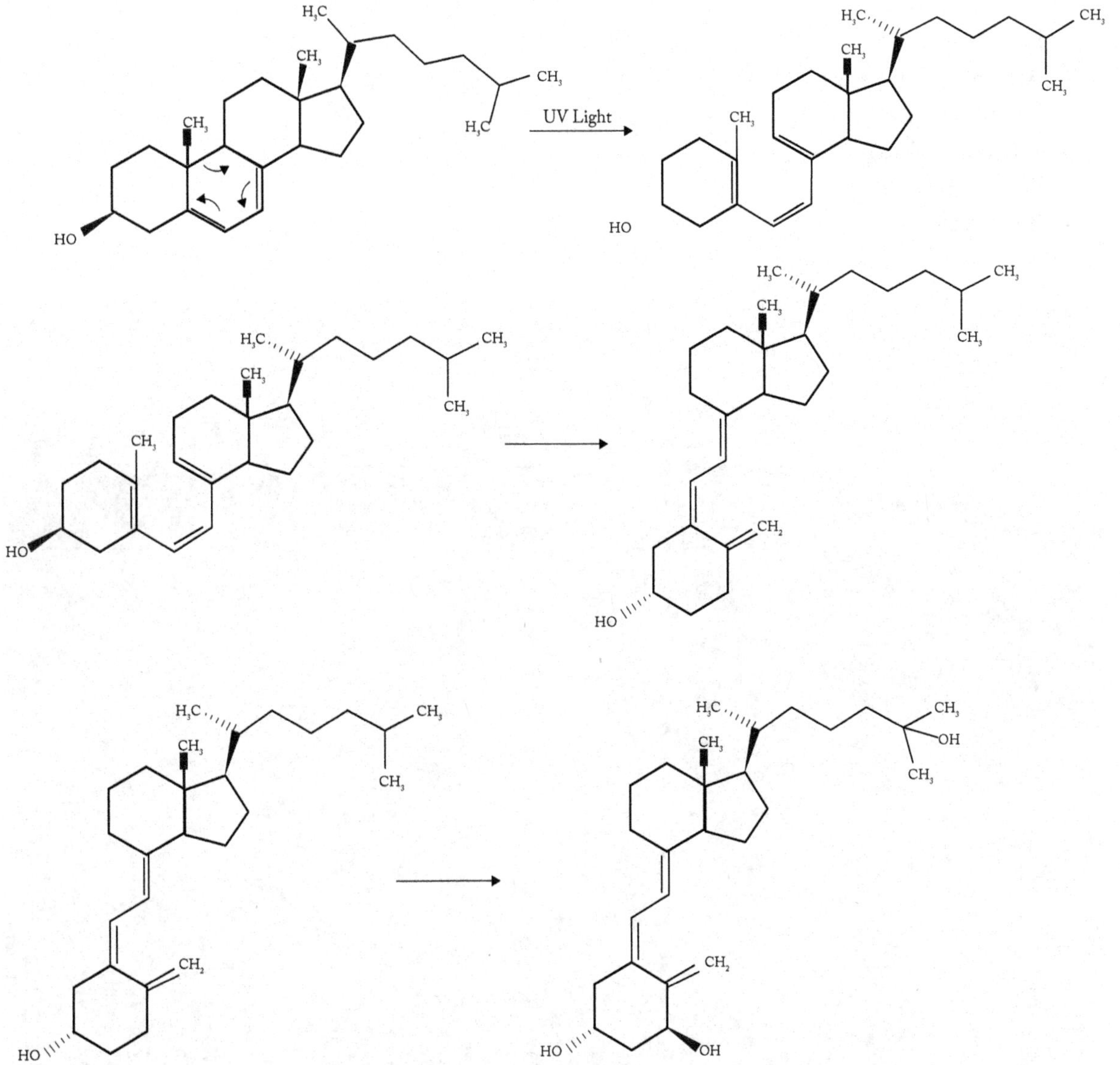

Figure 6.2 Vitamin D synthesis begins in the skin with the conversion of 7-dehydrocholesterol to cholecalciferol (row 1). Cholecalciferol travels to the liver and is converted to 25-hydrocholecalciferol (row 2); 25-hydrocholecalciferol in turn travels to the kidneys and is converted to the active form of vitamin D (1,25-hydroxycholecalciferol).

The Dermis

The middle layer of the integument is known as the dermis. The dermis consists of loose connective tissue and houses a number of accessory structures of the skin. The dermis connects to the epidermis by means of wavy structures called dermal papillae (Fig. 6.3).

Structures of the Dermis

The dermis contains a variety of accessory structures of the integument (Fig. 6.4). These are:

- Hair follicles
- Arrector pili muscles
- Sweat glands
- Sebaceous glands
- Sensory receptors
- Blood vessels

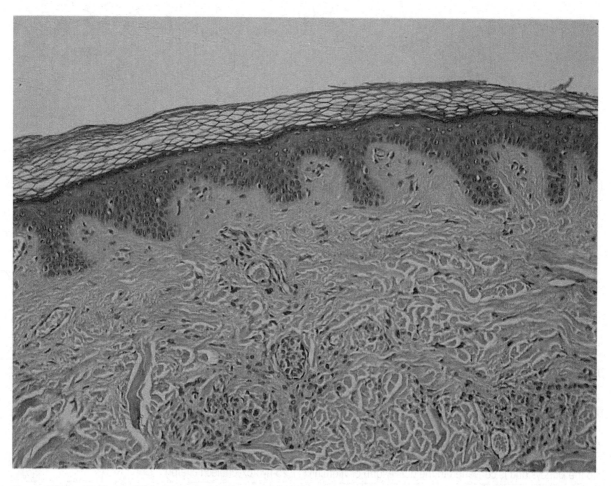

Figure 6.3 The integument. The epidermis (dark) connects to the dermis via the wavy structures called dermal papillae.

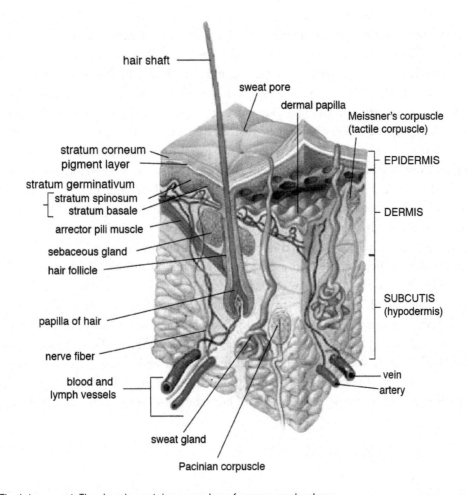

Figure 6.4 The integument. The dermis contains a number of accessory structures.

Clinical Application

The patch drug-delivery system:. The transdermal patch is an increasingly popular drug-delivery system. These patches are designed so that the drug molecules diffuse through the epidermis to the blood vessels in the dermis layer. A typical patch works well for small lipid-soluble molecules (for example, estrogen, nitroglycerin, and nicotine) that can make their way between epidermal cells.

Hair Follicles/Sebaceous Glands

The human body has approximately 2.5 million hairs on its surface. Hair is not found on the palms of the hands and soles of the feet, or on the lips, parts of the external genitalia, and sides of the feet and fingers. Hair is produced by hair follicles. Hair is not alive and develops from old dead cells that are pushed outward by new cells. The cells contain keratin for hardness and melanin for color. Hairs can be very sensitive. This is due to a tiny plexus of nerves that surround each hair follicle. Hair is so sensitive that you can feel the movement of even a single hair. A band of smooth muscle is connected to each hair follicle.

This structure is called an arrector pili muscle and is capable of moving each follicle, causing it to stand up in times of sympathetic nervous system activity such as emotional stress. Hair begins to grow at the base of the hair follicle in a structure called the hair bulb. The hair bulb is surrounded by a hair papilla that contains blood vessels and nerves. The cells of the hair bulb divide and push the cells toward the surface along the hair root and shaft. Hair grows at a rate of about .33 mm per day. Adults normally lose about 50 hairs per day. A loss of over 100 hairs per day will cause a net loss of hair. This can happen especially in males due to changing levels of sex hormones (male pattern baldness). There are 2 types of hair. Vellus hairs are the fine hairs located on much of your body's surface. Terminal hairs are thicker, more pigmented, and are found on your head as well as genital and axillary regions. A small gland surrounds each hair follicle. This gland is called a sebaceous gland. The sebaceous glands secrete an oily substance known as sebum. The substance is secreted in response to contraction of the arrector pili muscle. Sebum contains triglyceride, protein, cholesterol, and some electrolytes. Sebum makes the hair more flexible and hydrated.

Pathological impacts on hair: Drugs used in cancer chemotherapy frequently cause a temporary loss of hair, noticeable on the head and eyebrows, because they kill all rapidly dividing cells, not just the cancerous ones. Other diseases and traumas can cause temporary or permanent loss of hair, either generally or in patches. The hair shafts may also store certain poisons for years, even decades, after death. In the case of Col. Lafayette Baker, who died July 3, 1868, use of an atomic absorption spectrophotometer showed the man was killed by white arsenic. The prime suspect was Wallace Pollock, Baker's brother-in-law. According to Dr. Ray A. Neff, Pollack had laced Baker's beer with arsenic over a period of months, and a century or so later, minute traces of it showed up in the dead man's hair. Mrs. Baker's diary seems to confirm that it was indeed arsenic, as she writes of how she found some vials of it inside her brother's suit coat one day.

Sweat Glands

Sweat glands are also located in the dermis. There are 2 types of sweat glands. Apocrine sweat glands secrete their substances into the hair follicles. The secretions of apocrine glands can develop odor. The odor can increase because the secretion acts as a nutrient for bacteria that enhance the odor. Apocrine glands begin to secrete substances at puberty and are located in the axillar and genital regions. Exocrine sweat glands secrete their substances directly onto the surface of the skin. They are coiled tubular glands that secrete a substance that consists mostly of water with a trace of some electrolytes and a peptide with antibiotic properties. The exocrine sweat glands' primary function is to help to regulate body temperature. The sweat can evaporate and carry away heat. The sweat also excretes water and electrolytes.

Nails

Nails occur at the distal portions of the fingers and toes. The visible portion of the nail is called the nail body and it sits over the nail bed. The nail begins deep in the skin proximal to where it is seen. It extends distally to beyond an area of thickened epidermis called the hyponychium. The nail begins to grow at the nail root, which is close to the bone. A portion of the superficial epidermis (stratum corneum) extends over the proximal portion of the nail, forming the eponychium or cuticle. Blood vessels deep to the nail give it a pink color. These vessels may be obscured, leaving a white area known as the lunula. Nails contain keratinized cells that are pushed from the root to the distal portions. The nails can reflect health problems, which include:

- Bluish nails = circulatory problems
- White nail = anemia
- Pigmented spot under nail = possible melanoma
- Horizontal grooves = malnutrition
- Clubbing = heart, lungs, liver problems
- Red streaks = rheumatoid arthritis, ulcer, high blood pressure
- Spoon nails = iron deficiency anemia

Nail Diseases

Nail diseases are in a separate category from diseases of the skin. Although nails are a skin appendage, they have their own signs and symptoms that may relate to other medical conditions. Nail conditions that show signs of infection or inflammation require medical assistance and cannot be treated at a beauty parlor. Deformity or disease of the nails may be referred to as onychosis. There are many diseases that can occur with the fingernails and toenails. The most common of these diseases are ingrown nails and fungal infections.

Ingrown Nails

Onychocryptosis, commonly known as "ingrown nails" (unguis incarnatus), can affect either the fingers or the toes. In this condition, the nail cuts into one or both sides of the nail bed, resulting in inflammation and possibly infection. The relative rarity of this condition in the fingers suggests that pressure from the ground or shoe against the toe is a prime factor. The movements involved in walking or other physical disturbances can contribute to the problem. Mild onychocryptosis, particularly in the absence of infection, can be treated by trimming and rounding the nail. More advanced cases, which usually include infection, are treated by surgically excising the ingrowing portion of the nail down to its bony origin and cauterizing the matrix, or "root," to prevent recurrence. This surgery is called matricectomy. The best results are achieved by cauterizing the matrix with phenol. Another method, which is much less effective, is excision of the matrix, sometimes called a "cold steel procedure."

Nail Fungus

An infection of nail fungus (onychomycosis) occurs when fungi infect one or more of your nails. Onychomycosis generally begins as a white or yellow spot under the tip of the fingernail or toenail. As the nail fungus spreads deeper into the nail, it may cause the nail to discolor, thicken, and develop crumbling edges—an unsightly and potentially painful problem. Infections of nail fungus account for about half of all nail disorders. These infections usually develop on nails continually exposed to warm, moist environments, such as sweaty shoes or shower floors. Nail fungus isn't the same as athlete's foot, which affects primarily the skin of the feet, but at times the two may coexist and can be caused by the same type of fungus. An infection with nail fungus may be difficult to treat, and infections may recur. But medications are available to help clear up nail fungus permanently.

Pliability

Brittleness is associated with iron deficiency, thyroid problems, impaired kidney function, circulation problems, and biotin deficiency. Splitting and fraying are associated with psoriasis, folic acid, and protein and/or vitamin C deficiency. Unusual thickness is associated with circulation problems. Thinning nails and itchy skin are associated with lichen planus.

Shape and Texture

Clubbing, or nails that curve down around the fingertips with nail beds that bulge, is associated with oxygen deprivation and lung, heart, or liver disease. Spooning, or nails that grow upwards, is associated with iron or B12 deficiency. Flatness can indicate a B12 vitamin deficiency or Raynaud's disease. Pitting of the nails is associated with psoriasis. Horizontal ridges indicate stress, and Beau's lines are associated with many serious conditions. Vertical ridges are associated with arthritis. Vertical grooves are associated with kidney disorders, aging, and iron deficiency. Beading is associated with rheumatoid arthritis. Nails that resemble hammered brass are associated with (or portend) hair loss. Short small beds are associated with heart disease.

Coloration of the nail bed: Mee's lines are associated with arsenic or thallium poisoning, and renal failure. White lines across the nail are associated with heart disease, liver disease, or a history of a recent high fever. Opaque white nails with a dark band at the fingertip are associated with cancer, cirrhosis, congestive heart failure, diabetes, and aging. Paleness or whitening is associated with liver or kidney disease and anemia. Yellowing of the nail bed is associated with chronic bronchitis, lymphatic problems, diabetes, and liver disorders. Brown or copper nail beds are associated with arsenic or copper poisoning, and local fungal infection. Gray nail beds are associated with arthritis, edema, malnutrition, post-operative effects, glaucoma, and cardiopulmonary disease. Redness is associated with heart conditions. Dark nails are associated with B12 deficiency. Stains of the nail plate (not the nail bed) are associated with nail polish, smoking, and henna use.

Markings

Pink and white nails are associated with kidney disease. Parallel white lines in the nails are associated with hypoalbuminemia. Red skin at the base of the nail is associated with connective tissue disorders. Blue lunulae are associated with silver poisoning or lung disorder. Blue nail beds (much like blue skin) are associated with poor oxygenation of the blood (asthma, emphysema, etc.). Small white patches are associated with zinc or calcium deficiency or malabsorption, parasites, or local injury. Receded lunulae (fewer than 8) are associated with poor circulation, shallow breathing habits, or thyroid dysfunction. Large lunulae (more than 25% of the thumbnail) are associated with high blood pressure.

Temperature Regulation

The skin is very important in regulating body temperature. The skin helps keep in heat produced by skeletal muscles and liver cells. When the body gets too hot, the skin opens up the sweat pores so that the sweat can carry the excess heat away by evaporation. Heat can be lost by the body in a number of ways. Heat always moves along a gradient from warmer to cooler temperatures. Heat can radiate from the body to the surrounding areas at lower temperatures. In conduction, heat moves via molecules from the warmer body to cooler objects. An example of conduction would be to lean against a cooler concrete wall. The heat flows from your body into the wall. In convection, heat moves via air molecules circulating around body. In evaporation, fluid on the surface of the body carries heat away. Body temperature is regulated primarily by an area in the brain known as the hypothalamus. The hypothalamus sets the body's temperature and controls it by opening and closing sweat glands and contracting muscles. Let's say that it is a hot summer day and you are working hard mowing the lawn. As your body's temperature rises, the hypothalamus senses this and sends a message to your sweat glands to open. The sweat evaporates off your skin and you begin to cool down. Now let's say that you've finished mowing the lawn and you go inside your air conditioned home. Your body's temperature will begin to drop. The hypothalamus senses this and sends a message to your sweat glands to close. If your body's temperature continues to drop, the hypothalamus may send a message to your muscles to contract or shiver. The muscles will generate heat to help maintain your core temperature. In more severe cases of cold, your blood vessels will constrict in your extremities in an attempt to conserve heat at the core of your body for survival. If your core body temperature continues to drop, you may develop a condition called hypothermia. You will progress from feeling cold to shivering, experiencing mental confusion, lethargy, loss of reflexes and eventually loss of consciousness and shutting down of organs. Conversely if your core body temperature increases too much, you can develop hyperthermia. This can develop in humid conditions because of lack of evaporation. The signs of hyperthermia include lightheadedness, dizziness, headaches, muscle cramps, fatigue, and nausea.

Skin Repair

The skin has remarkable healing properties. It can heal cuts, bruises, and burns. A cut is known as a laceration. If the cut extends only into the epidermis, the epidermal cells will divide rapidly to repair the skin. If the cut extends into the dermis, broken blood vessels form a clot. The clot forms from fibrin, which is a product of blood cells. Fibroblasts collect in the injured area and grow new collagen fibers.

Burns are described in 3 categories. First-degree burns are known as superficial partial-thickness burns. Only the epidermis is affected. First-degree burns usually heal quickly because growth occurs from the deeper layers of the dermis. Second-degree burns are known as deep partial-thickness burns. In second-degree burns the epidermis and some of the dermis is damaged. Fluid accumulates between the dermis and outer layer of epidermis, forming blisters. The skin becomes discolored, from dark red to waxy white. Healing depends on the accessory organs of skin because new cell growth emerges from these layers. Third-degree burns involve the epidermis, dermis, and accessory organs. Third-degree burns are called full-thickness burns. In third-degree burns there is no new cell growth from the damaged area. Growth can occur only from the margins of the burn. Skin substitutes can be used to cover the skin while healing. These include amniotic and artificial membranes and cultured epithelial cells. Skin grafts can also be used.

Skin Diseases

The integumentary system is also susceptible to a variety of diseases, disorders, and injuries. These range from annoying but relatively benign bacterial or fungal infections, which are categorized as disorders, to skin cancer and severe burns, which can be fatal. In this section, you will learn about several of the most common skin conditions.

One of the most talked-about diseases is skin cancer. Cancer is a broad term that describes diseases caused by abnormal cells in the body dividing uncontrollably. Most cancers are identified by the organ or tissue in which the cancer originates. One common form of cancer is skin cancer. The Skin Cancer Foundation reports that one in five Americans will experience some type of skin cancer in their lifetime. The degradation of the ozone layer in the atmosphere and the resulting increase in exposure to UV radiation has contributed to its rise. Overexposure to UV radiation damages DNA, which can lead to the formation of cancerous lesions. Although melanin offers some protection against DNA damage from the sun, often it is not enough. The fact that cancers can also occur on areas of the body that are normally not exposed to UV radiation suggests that there are additional factors that can lead to cancerous lesions. In general, cancers result from an accumulation of DNA mutations. These mutations can result in cell populations that do not die when they should and uncontrolled cell proliferation that leads to tumors. Although many tumors are benign (harmless), some produce cells that can mobilize and establish tumors in other organs of the body; this process is referred to as metastasis. Cancers are characterized by their ability to metastasize.

Basal Cell Carcinoma

Basal cell carcinoma is a form of cancer that affects the mitotically active stem cells in the stratum basale of the epidermis. It is the most common of all cancers that occur in the United States and is frequently found on the head, neck, arms, and back, which are areas that are most susceptible to long-term sun exposure. Although UV rays are the main culprit, exposure to other agents, such as radiation and arsenic, can also lead to this type of cancer. Wounds on the skin due to open sores, tattoos, burns, etc. may be predisposing factors as well. Basal cell carcinomas start in the stratum basale and usually spread along this boundary. At some point, they begin to grow toward the surface and become an uneven patch, bump, growth, or scar on the skin surface. Like most cancers, basal cell carcinomas respond best to treatment when caught early. Treatment options include surgery, freezing (cryosurgery), and topical ointments.

Squamous Cell Carcinoma

Squamous cell carcinoma is a cancer that affects the keratinocytes of the stratum spinosum and presents as lesions commonly found on the scalp, ears, and hands. It is the second most common skin cancer. The American Cancer Society reports that two of 10 skin cancers are squamous cell carcinomas, and it is more aggressive than basal cell carcinoma. If not removed, these carcinomas can metastasize. Surgery and radiation are used to cure squamous cell carcinoma.

Melanoma

A melanoma is a cancer characterized by the uncontrolled growth of melanocytes, the pigment-producing cells in the epidermis. Typically, a melanoma develops from a mole. It is the most fatal of all skin cancers, as it is highly metastatic and can be difficult to detect before it has spread to other organs. Melanomas usually appear as asymmetrical brown and black patches with uneven borders and a raised surface. Treatment typically involves surgical excision and immunotherapy.

Doctors often give their patients the following ABCDE mnemonic to help with the diagnosis of early-stage melanoma. If you observe a mole on your body displaying these signs, consult a doctor.

- **A**symmetry—the two sides are not symmetrical
- **B**orders—the edges are irregular in shape

- **C**olor—the color is varied shades of brown or black
- **D**iameter—it is larger than 6 mm (0.24 in)
- **E**volving—its shape has changed

Some specialists cite the following additional signs for the most serious form, nodular melanoma:

- **E**levated—it is raised on the skin surface
- **F**irm—it feels hard to the touch
- **G**rowing—it is getting larger

Skin Disorders

Two common skin disorders are eczema and acne. Eczema is an inflammatory condition and occurs in individuals of all ages. Acne involves the clogging of pores, which can lead to infection and inflammation, and is often seen in adolescents. Other disorders, not discussed here, include seborrheic dermatitis (on the scalp), psoriasis, cold sores, impetigo, scabies, hives, and warts.

Eczema

Eczema is an allergic reaction that manifests as dry, itchy patches of skin that resemble rashes. It may be accompanied by swelling of the skin, flaking, and in severe cases, bleeding. Many who suffer from eczema have antibodies against dust mites in their blood, but the link between eczema and allergy to dust mites has not been proven. Symptoms are usually managed with moisturizers, corticosteroid creams, and immunosuppressants.

Acne

Acne is a skin disturbance that typically occurs on areas of the skin that are rich in sebaceous glands (face and back). It is most common along with the onset of puberty due to associated hormonal changes, but can also occur in infants and continue into adulthood. Hormones, such as androgens, stimulate the release of sebum. An overproduction and accumulation of sebum along with keratin can block hair follicles. This plug is initially white. The sebum, when oxidized by exposure to air, turns black. Acne results from infection by acne-causing bacteria (*Propionibacterium* and *Staphylococcus*), which can lead to redness and potential scarring due to the natural wound-healing process.

Injuries

Scars and Keloids

Most cuts or wounds, with the exception of those that only scratch the surface (the epidermis), lead to scar formation. A scar is collagen-rich skin formed after the process of wound healing that differs from normal skin. Scarring occurs in cases in which there is repair of skin damage, but the skin fails to regenerate the original skin structure. Fibroblasts generate scar tissue in the form of collagen, and the bulk of repair is due to the basket-weave pattern generated by collagen fibers and does not result in regeneration of the typical cellular structure of skin. Instead, the tissue is fibrous in nature and does not allow for the regeneration of accessory structures, such as hair follicles, sweat glands, or sebaceous glands.

Sometimes there is an overproduction of scar tissue, because the process of collagen formation does not stop when the wound is healed; this results in the formation of a raised or hypertrophic scar called a keloid. In contrast, scars that result from acne and chickenpox have a sunken appearance and are called atrophic scars. Scarring of skin after wound healing is a natural process and does not need to be treated further. Application of mineral oil and lotions may reduce the formation of scar tissue. However, modern cosmetic procedures, such as dermabrasion, laser treatments, and filler injections have been invented as remedies for severe scarring. All of these procedures try to reorganize the structure of the epidermis and underlying collagen tissue to make it look more natural.

Bedsores and Stretch Marks

Skin and its underlying tissue can be affected by excessive pressure. One example of this is called a bedsore. Bedsores, also called decubitus ulcers, are caused by constant, long-term, unrelieved pressure on certain body parts that are bony, reducing blood flow to the area and leading to necrosis (tissue death). Bedsores are most common in elderly patients who have debilitating conditions that cause them to be immobile. Most hospitals and long-term care facilities have the practice of turning the patients every few hours to prevent the incidence of bedsores. If left untreated by removal of necrotized tissue, bedsores can be fatal if they become infected. The skin can also be affected by pressure associated with rapid growth.

A stretch mark results when the dermis is stretched beyond its limits of elasticity, as the skin stretches to accommodate the excess pressure. Stretch marks usually accompany rapid weight gain during puberty and pregnancy. They initially have a reddish hue, but lighten over time. Other than for cosmetic reasons, treatment of stretch marks is not required. They occur most commonly over the hips and abdomen.

CRITICAL-THINKING QUESTIONS

1. What determines the color of skin, and what is the process that darkens skin when it is exposed to UV light?
2. Cells of the epidermis derive from stem cells of the stratum basale. Describe how the cells change as they become integrated into the different layers of the epidermis.
3. Explain the differences between eccrine and apocrine sweat glands.
4. Describe the structure and composition of nails.
5. Why do people sweat excessively when exercising outside on a hot day?
6. Explain your skin's response to a drop in body core temperature.
7. Why do scars look different from surrounding skin?
8. What might happen to the skin color of an individual whose diet was mainly carrots, tomatoes, and other yellow vegetables?
9. Both patient A and patient B suffered burns to their legs when their space heater exploded. Patient A suffered a partial-thickness burn while patient B suffered a full-thickness burn. Predict the effect their burns will have on skin regeneration and hair growth.

Chapter Seven

Skeletal System

The skeletal system not only helps to provide movement and support but also serves as a storage area for calcium and inorganic salts and a source of blood cells. The adult human body has 206 bones in a variety of shapes and sizes. Basically there are 4 types of bones categorized according to shape:

- Long bones have a long longitudinal axis (Fig. 7.1).
- Short bones have a short longitudinal axis and are more cube-like.
- Flat bones are thin and curved, such as some of the bones of the skull.
- Irregular bones are often found in groups and have a variety of shapes and sizes.

There are also 2 types of bone tissue in different amounts in bones. Compact bone (sometimes called cortical bone) is very dense. Cancellous bone (sometimes called spongy bone) looks more like a trabeculated matrix. It is found in the central regions of some of the skull bones or at ends (epiphyses) of long bones. The bone-forming cells (osteocytes) get their nutrients by diffusion.

Bone Structure

Compact bone is organized according to structural units called Haversian systems or osteons (Fig. 7.2). These are located along the lines of force and line up along the long axis of the bone. The Haversian systems are connected together and form an interconnected structure that provides support and strength to bones. Haversian systems contain a central canal (Haversian canal) that serves as a pathway for blood vessels and nerves. The bone is deposited along concentric rings called lamellae. Along the lamellae are small openings called lacunae. The lacunae contain fluid and bone cells called osteocytes. Radiating out in all directions from lacunae are small canals called canaliculi. Haversian systems are interconnected by a series of larger canals called Volkmann's canals (perforating canals).

Long Bone

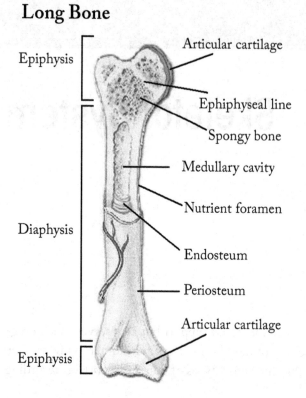

Figure 7.1 Parts of a long bone. Notice the long shaft or diaphysis in the middle of the bone. The diaphysis contains compact bone surrounding a medullary cavity containing bone marrow. On each end is an epiphysis containing cancellous or spongy bone. The epiphyseal line is a remnant of the growth plate. The epiphyses also contain hyaline cartilage for forming joints with other bones. Surrounding the bone is a membrane called the periosteum. The periosteum contains blood vessels and cells that help to repair and restore bone.

- Proximal end of humerus
- Ribs
- Bodies of vertebrae
- Pelvis
- Femur

Bone Cells

There are 3 basic types of cells in bone. Osteoblasts undergo mitosis and secrete a substance that acts as the framework for bone. Once this substance (osteoid) is secreted, minerals can deposit and form hardened bone. Osteoblasts respond to certain bone-forming hormones as well as from physical stress. Osteocytes are mature osteoblasts that cannot divide by mitosis (Fig. 7.3). Osteocytes reside in lacunae. Osteoclasts are capable of demineralizing bone. They free up calcium from bone to make it available to the body depending on the body's needs.

Bone Marrow

Bone marrow is located in the medullary (marrow) cavity of long bones and in some spongy bones. There are 2 kinds of marrow. Red marrow exists in the bones of infants and children. It is called red because it contains a large number of red blood cells. In adults the red marrow is replaced by yellow marrow. It is called yellow because it contains a large proportion of fat cells. Yellow marrow decreases its ability to form new red blood cells. However, not all adult bones contain yellow marrow. The following bones continue to contain red marrow and produce red blood cells:

Bone Growth

Bones begin to grow during fetal development and complete the growth process during young adulthood. There are 2 bone-forming processes. Flat bones called intramembranous bones develop in sheet-like layers. Tubular bones called endochondral bones develop from cartilage templates.

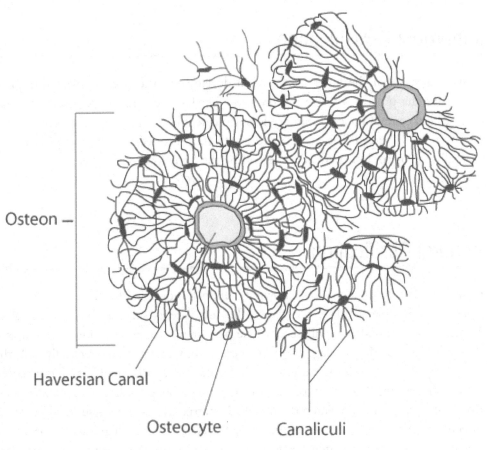

Figure 7.2 Haversian system.

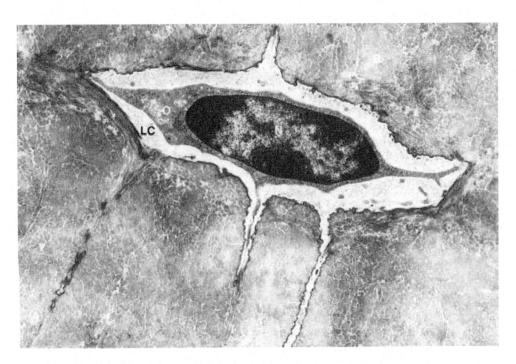

Figure 7.3 Osteocytes are mature osteoblasts that reside in a lacuna. They are surrounded by bony matrix.

Intramembranous Ossification

Flat bones such as some of the bones of the skull develop from a process called intramembranous ossification. During this process bones form from sheet-like layers of connective tissue. These layers have a vascular supply and contain bone-forming cells called osteoblasts. The osteoblasts secrete bony matrix in all directions around the cell. The matrix unites with that secreted by other osteoblasts as the bone forms. Eventually the osteoblasts may be walled off by the bony matrix. At this point the osteoblast is called an osteocyte.

Endochondral Ossification

Tubular bones develop from a process known as endochondral ossification. During this process bones develop from hyaline cartilage templates. The template is surrounded by an area called the perichondrium. The perichondrium will become the periosteum (outer covering of bone) as the bone develops. Chondrocytes in the cartilage begin to secrete bony matrix and eventually wall themselves off in lacunae. Next, blood vessels extend into the bone and transport osteoblasts and osteoclasts from the perichondrium, forming a primary ossification center. The bone continues to grow in a cylindrical fashion. Eventually blood vessels enter the calcified matrix of the epiphyses and form secondary ossification centers. Osteoclasts remove matrix from the center of the diaphysis to form a medullary cavity. The secondary ossification centers form about 1 month before birth. Bone continues to form from the cartilage until all of the cartilage is replaced except for the epiphyseal plates. These will complete their calcification in young adulthood (Fig. 7.4).

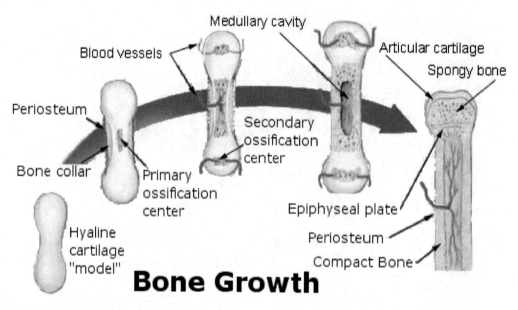

Figure 7.4 Endochondral ossification.

Epiphyseal Plates

Bone grows longitudinally as the epiphyseal plates secrete bony matrix. There are 4 zones in epiphyseal plates:

1. Zone of resting cartilage. This zone contains chondrocytes that do not divide rapidly.
2. Zone of proliferation. This zone contains active chondrocytes that produce new cartilage.
3. Zone of hypertrophy. In this zone chondrocytes from the zone of proliferation mature and enlarge.
4. Zone of calcification. In this zone the enlarged chondrocytes are replaced by osteoblasts from the endosteum. The osteoblasts secrete bone that calcifies the area.

As the chondrocytes produce cartilage and hypertrophy, the bone grows on the diaphyseal side of the plate. The plate remains the same thickness because ossification on both sides of the plate occurs at the same rate. The epiphyseal plates complete their growth and ossify between the ages of 12 and 25 years depending on the bone.

Bone Growth Factors

The length of bones and subsequent height of an individual are determined genetically. However, there are other factors that affect the expression of genes that in turn can affect bone growth. These include certain hormones, nutrition, and exercise. Growth hormone is a hormone secreted by the anterior portion of the pituitary gland. Growth hormone stimulates protein synthesis and growth of cells in the entire body, including bones. Thyroxine is secreted by the thyroid gland and increases osteoblastic activity in bones. Calcitriol is secreted by the kidneys and helps the digestive tract absorb calcium. The synthesis of calcitriol depends on vitamin D (see the integumentary system chapter). Sex hormones from the ovaries and testes also stimulate osteoblastic activity. Vitamins such as D, C, A, K, and B12 are also important in bone growth. Vitamin C is required for collagen synthesis and stimulates osteoblastic activity. A lack of vitamin D can lead to a condition called rickets in children or osteomalacia in adults. Rickets is characterized by malformed bones. Calcium and phosphorus must be adequately supplied by the diet for use in bony matrix. Vitamin A stimulates osteoblastic activity and vitamins K and B12 are needed for protein synthesis in bone cells. Bone grows according to the imposed demands of the body. This is known as Wolf's law. In other words the body produces bone along lines of force. For example, weight-bearing exercises will increase the strength of bones. Likewise bones that are cast during the healing process for fractures will be weaker. This is one reason that weight-bearing exercise is recommended for those predisposed to osteoporosis.

Fractures

There are many ways bones can break or fracture. Closed fractures are contained within the body. Closed fractures are also called "simple" and are contained within the surrounding tissues that help them to heal. Open fractures protrude through the skin and are more dangerous because of the risk of infection and bleeding. Complete fractures go all the way through a bone and incomplete fractures go only partially through a bone (Figs. 7.5–7.7).

Healing Fractures

Bone has remarkable healing properties. Bones heal from fractures in about 6 weeks. Shortly after a fracture occurs, a hematoma forms. This fracture hematoma produces a fibrous network for repair. Next cells of the periosteum and endosteum undergo rapid mitosis, with new cells moving into the damaged area.

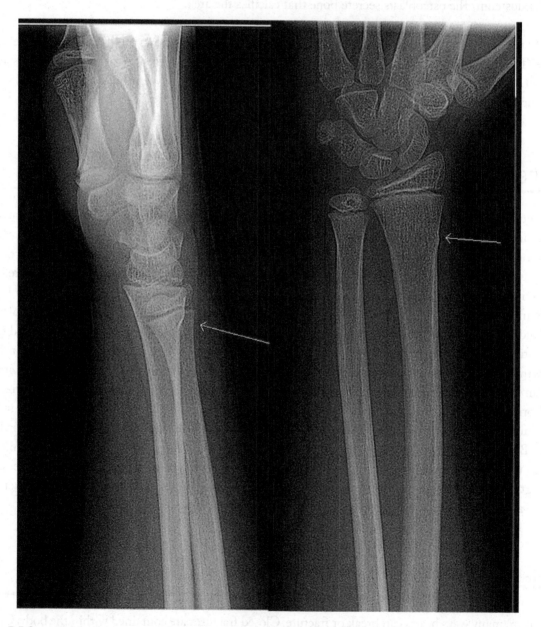

Figure 7.5 Greenstick fractures occur on the convex side of the bone and are incomplete.

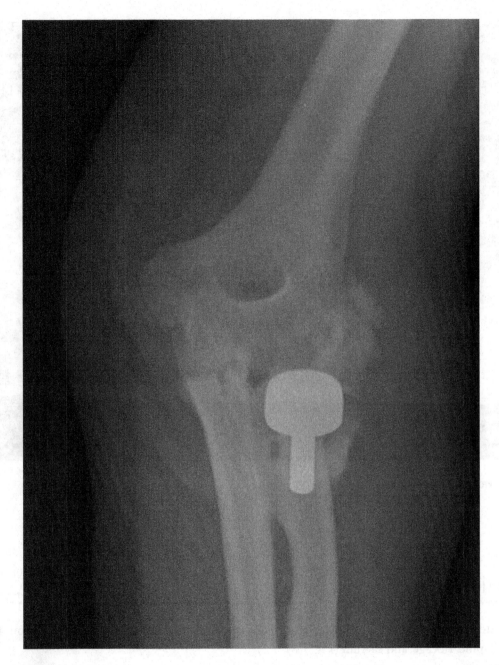

Figure 7.6 Comminuted fractures can be described as a shattering of bone. This picture illustrates a comminuted fracture of the elbow with some hardware.

The new cells form a callus. The external portion of the callus consisting of cartilage and bone extends around the damaged area. The internal portion is located in the marrow cavity. Cells in the callus differentiate into osteoblasts and begin to secrete bony matrix. Spongy bone forms and replaces the cartilage of the external callus. This provides strength to the damaged area. Finally, osteoblasts and osteoclasts work to remodel the damaged area for up to a year. The callus disappears, leaving only a remnant line. The bone is as strong as it was previous to the fracture.

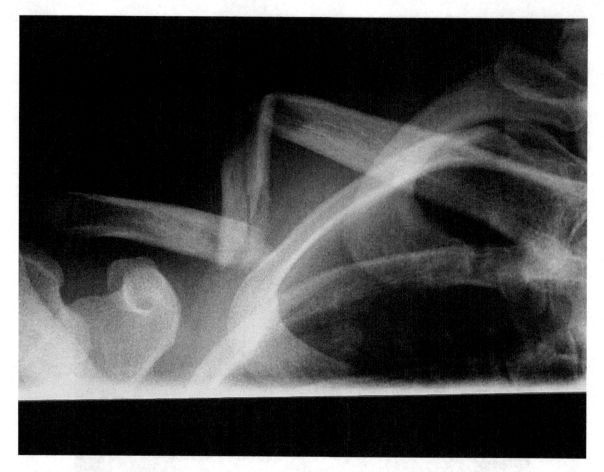
Figure 7.7 Fracture of the clavicle.

The Axial Skeleton

The skeleton is divided into 2 sections: the axial and appendicular sections. The axial skeleton includes the skull, spine, ribcage, and sacrum (Fig. 7.8).

The Bones of the Skeleton

The Skull

The skull contains the brain and sensory structures such as the eyes, ears, nasal passages, and mouth. There are 22 bones in the skull with 8 forming the cranium (Figs. 7.9, 7.10). The 8 bones of the cranium:

1. Frontal
2. Occipital

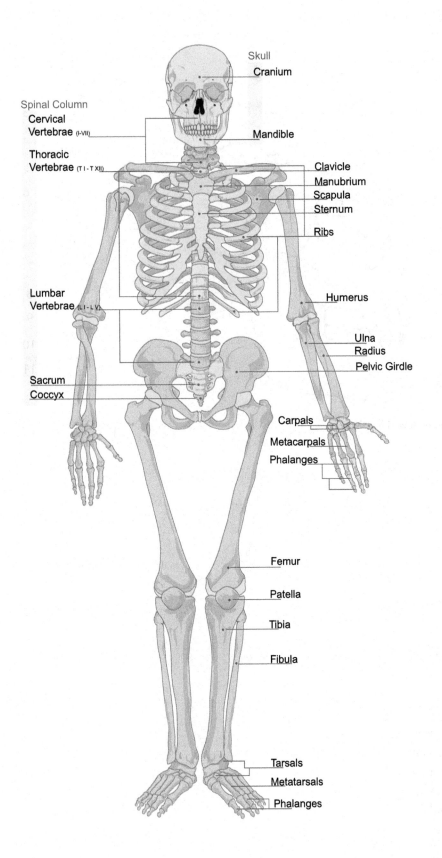

Figure 7.8. The skeleton.

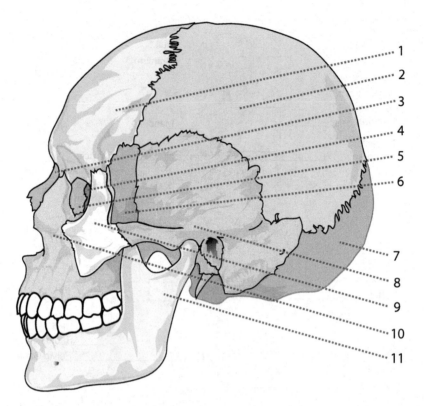

Figure 7.9 Lateral view of the skull.
1. Frontal
2. Parietal
3. Nasal
4. Lacrimal
5. Ethmoid
6. Sphenoid
7. Occipital
8. Temporal
9. Zygomatic
10. Maxilla
11. Mandible

3. Right Parietal
4. Left Parietal
5. Right Temporal
6. Left Temporal
7. Sphenoid
8. Ethmoid

The bones are held together by special joints called sutures. These joints are considered immovable and are composed of dense fibrous connective tissue (Figs. 7.11, 7.12).

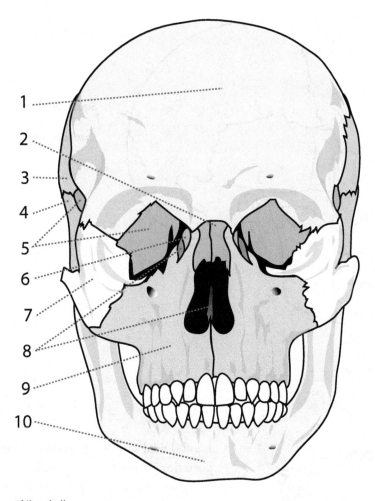

Figure 7.10 Anterior view of the skull.
1. Frontal
2. Nasal
3. Parietal
4. Temporal
5. Sphenoid
6. Ethmoid
7. Zygomatic
8. Ethmoid
9. Maxilla
10. Mandible

The types of sutures:

- Sagittal suture—connects the parietal bones at the top of the skull. It lies in the sagittal plane.
- Coronal suture—connects both parietal bones to the frontal bone on the top of the skull. It lies in a coronal plane.

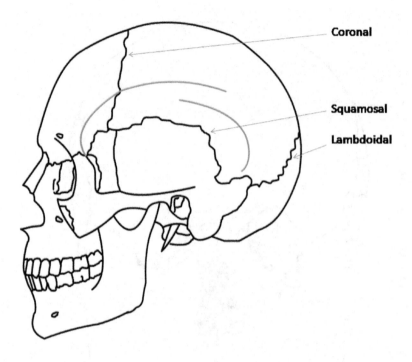

Figure 7.11 Sutures of the skull.

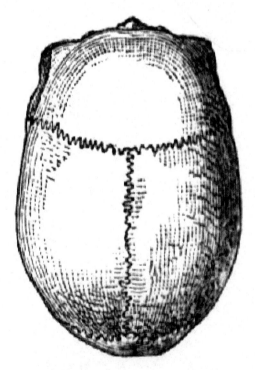

Figure 7.12 The coronal suture unites the frontal and parietal bones. The sagittal suture unites both parietal bones. Both sutures run in their respective planes.

- Lambdoidal suture—connects the occipital bone to the posterior portions of the parietal bones.
- Squamosal suture—connects the parietal bones to the temporal bones.

Bony Landmarks

There are a myriad of landmarks located on the skull. We will examine just a sampling of these landmarks. These structures include bumps, ridges, grooves, and holes. A tubercle is a rounded bump or process. Most of these bumps are sites for muscle and ligament attachments. A tuberosity is a rounded bump that has a more gradual slope. A styloid process is a pointy process. A trochanter is a very large bump. These are found on the femur bones. A condyle is a large rounded process. A foramen is a hole for arteries, veins, and nerves. A nutrient foramen does not go all the way through a bone. This is where blood vessels enter the bone to provide substances for maintenance, growth, and repair. A suture is a joint uniting at least 2 bones. A sinus is a hollow cavity within a bone.

Bones of the skull

Frontal Bone

The frontal bone is located on the anterosuperior aspect of the skull. If forms the anterior portion of the cranium and the superior portion of the orbits. It also contains sinuses (frontal sinuses) that secrete mucus to help flush the nasal cavity (Fig. 7.13). The supraorbital margin is a thickened process above the orbits that helps to protect the eye. The lacrimal fossa, located on the superior and lateral aspect of the orbit, is a small landmark for the lacrimal (tear) gland. The supraorbital foramen is a passageway for blood vessels supplying the frontal sinus, eyebrow, and eyelid.

Parietal Bones

The parietal bones are paired bones that form the lateral margins of the cranium. They articulate with the frontal bone via the coronal suture. The right and left parietal bones also connect via the sagittal suture. Both parietal bones connect with the occipital bone via the lambdoidal suture and with the temporal bones via the squamosal sutures (Fig. 7.14).

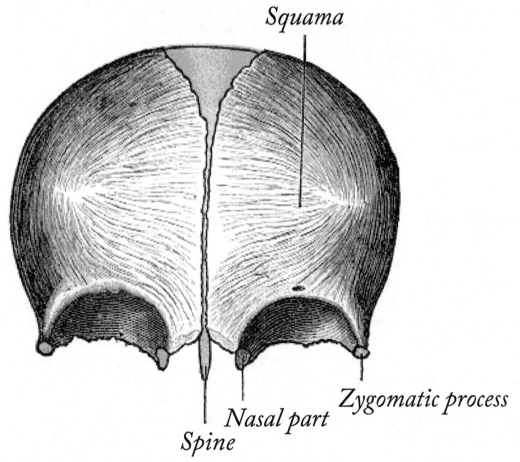

Figure 7.13 Frontal bone.

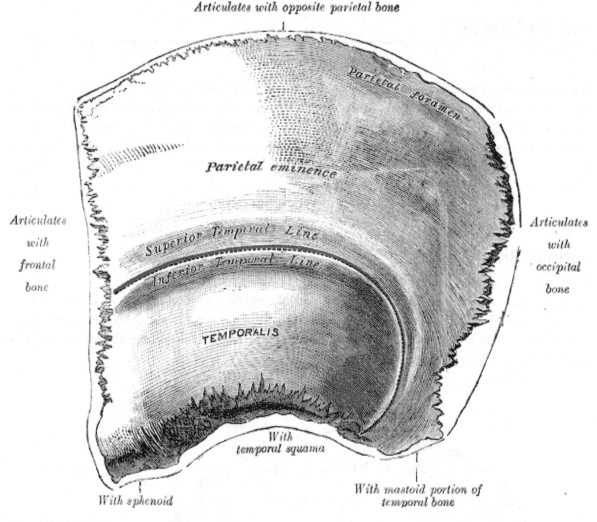

Figure 7.14 Parietal bone.

Occipital Bone

The occipital bone forms the posterior and posteroinferior margins of the cranium. The occipital bone articulates with the parietal, temporal, and sphenoid, and first cervical vertebra (Fig. 7.15). The occipital condyles are rounded processes that articulate with the first cervical vertebra (atlas) of the neck. The foramen magnum is a passageway for the spinal cord. The jugular foramen lies between the occipital and temporal bones and provides a passageway for the internal jugular vein.

The temporal bones form the inferior-lateral margins of the cranium. They house the inner ear structures and articulate with the mandible (Fig. 7.16). The zygomatic process forms the posterior portion of the zygomatic arch. It articulates with the temporal process of the zygomatic bone. The mastoid process is a site of muscle attachments for some of the neck muscles. It also contains small air cavities called air cells that connect with the middle ear. These can be a site of infection called mastoiditis. The styloid process is a pointed process that attaches to ligaments that support the hyoid bone. The external auditory meatus (canal) is a tube-like structure that houses structures for the external and middle ear. The carotid

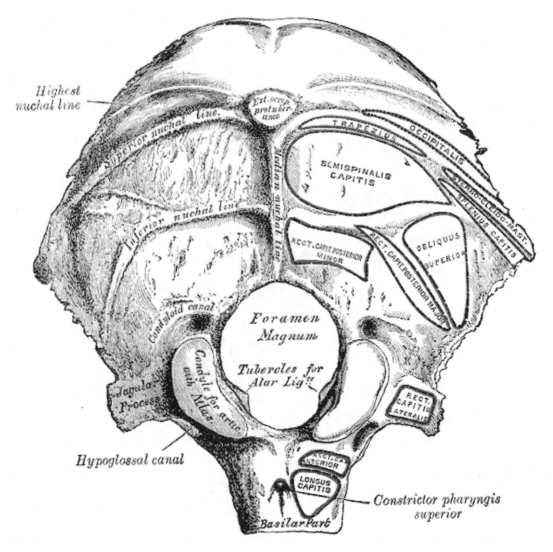

Figure 7.15 Occipital bone.

canal is a passageway for the internal carotid artery that supplies the brain. The foramen lacerum is a narrow slit-like structure located between the temporal and sphenoid bones. It carries small blood vessels that supply the inner portion of the cranium.

Sphenoid

The sphenoid bone forms part of the inferior portion of the cranium. It is visible on the lateral aspect of the skull although most of the bone resides inside of the skull (Fig. 7.17). The sella turcica (Turkish saddle) is a groove in the central region of the sphenoid. The pituitary gland resides in the sella turcica. The lesser wings extend laterally and are anterior to the sella turcica. The greater wings are lateral to the sella turcica and form part of the floor of the cranium. The optic canals are passageways for the optic nerves.

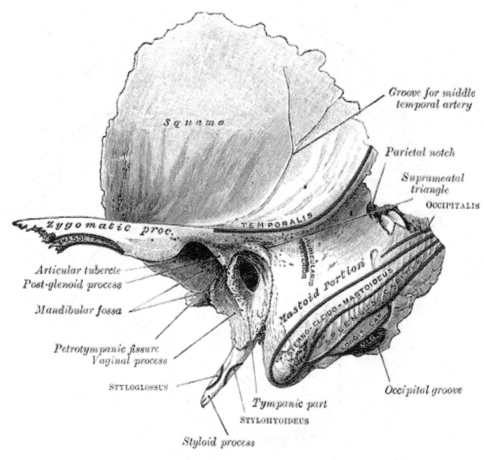

Figure 7.16 Temporal bone.

Ethmoid

The ethmoid bone is located in the anterior and medial cranium. The ethmoid bone also forms the roof of the nasal cavity and the superior portion of the nasal septum. It contains sinuses that secrete mucus to help flush the nasal cavity (Fig. 7.18). The crista galli is a ridge of bone that extends superiorly. A portion of the membrane that surrounds the brain called the dura mater attaches to this ridge. The cribriform plate is a perforated section of bone. Fibers from the olfactory nerve pass through these holes on their way to the frontal lobe of the brain. The perpendicular plate is a ridge of bone extending inferiorly and forming the superior portion of the nasal septum.

Maxilla

The maxilla is located and the anterior aspect of the skull. It is superior to the mandible and inferior to the frontal bone. It forms the upper jaw. The maxilla is actually 2 bones that have fused (Fig. 7.19). The alveolar process holds the upper teeth. The infraorbital foramen provides passage for the infraorbital

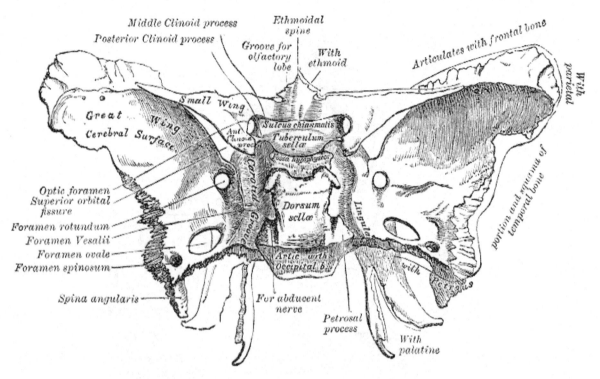

Figure 7.17 The sphenoid bone. The sella turcica is labeled "dorsum sella" in this image.

artery and nerve. The palatine process forms the anterior portion of the hard palate. The maxillary sinus is a hollow area lined with a mucous membrane. This cavity opens to the nasal passages.

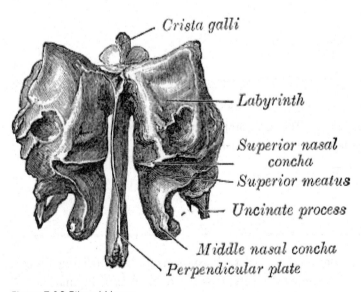

Figure 7.18 Ethmoid bone.

Mandible

The mandible forms the lower jaw. It is actually 2 bones that have fused. The alveolar process holds the lower teeth. The mandibular foramen provides passage for the inferior alveolar nerve (a division of the trigeminal nerve). It is located on the medial aspect (inside) of the mandible. The mental foramen contains fibers of the inferior alveolar nerve. The condyles form the lateral part of the temporomandibular joint (TMJ). They articulate with the temporal bone. The mental protuberance is a ridge of bone that extends anteriorly and is located in the central region of the mandible; it forms the chin.

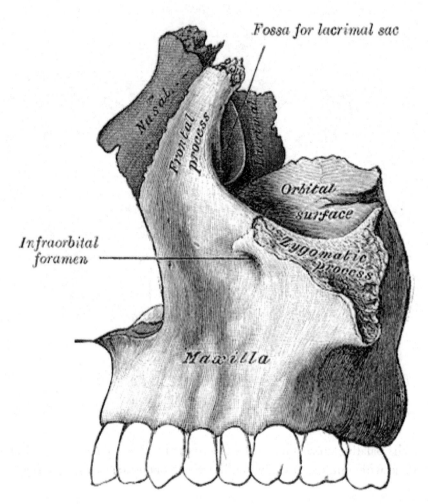

Figure 7.19 Maxilla.

Zygomatics

The zygomatic bones are located in the anterior portion of the skull. They connect with the maxilla, frontal, and temporal bones and form the cheeks. The temporal process is an extension of bone that connects with the zygomatic process of the temporal bone to form the zygomatic arch.

Palatine

The palatine bone is one of the bones that form the hard palate. It connects with the palatine process of the maxilla to form the posterior portion of the hard palate. It is located between the maxilla and sphenoid bones.

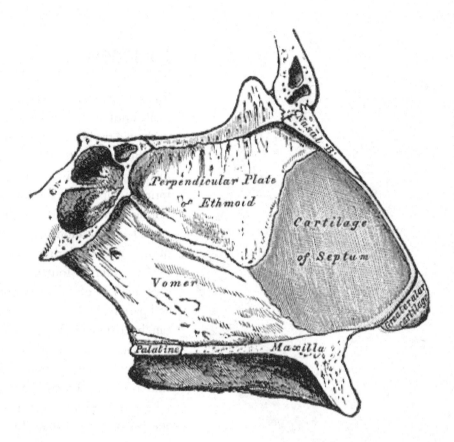

Figure 7.20 Vomer bone.

Vomer

The vomer bone forms the inferior aspect of the nasal septum. It articulates with the ethmoid, sphenoid, palatines, and maxillary bones (Fig. 7.20).

Bones of the Orbit

The orbit is formed by the following bones

- Frontal
- Lacrimal
- Maxilla
- Zygomatic
- Palatine
- Sphenoid
- Ethmoid

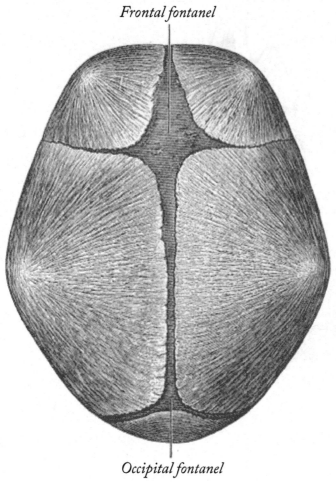

Figure 7.21 Superior view of fontanels.

Fontanels

The skeletal system does not completely ossify until the mid-twenties. This is most evident in the bones that form from intramembranous ossification in the skull. The membrane from which the skull bones form is palpable in the infant skull and is called a fontanel. The fontanels serve a critical role in allowing for compression of the fetal skull during birth (Fig. 7.21).

The anterior fontanel is located at the junction of the developing frontal and parietal bones. The anterior fontanel can be palpated for up to the age of 2 years. The posterior or occipital fontanel is located at the junction of the parietal and occipital bones. There are also sphenoidal and mastoid fontanels on the lateral sides of the skull. The sphenoidal fontanel is located at the junction of the frontal, parietal, temporal, and sphenoid bones. The mastoid fontanel is located at the junction of the parietal, temporal, and occipital bones. The remaining fontanels usually ossify by the end of the first year.

The Spine

The spine consists of 25 vertebra "stacked" one on the other, forming a column. The spine provides support for the head and trunk and houses the spinal cord. It articulates superiorly with the head and inferiorly with the sacrum. There are 3 basic sections of the spine. The cervical spine consists of 7 vertebrae and has 2 unique vertebrae called the atlas and axis. The thoracic spine consists of 12 vertebrae that articulate with ribs. The lumbar spine consists of 5 large vertebrae. The vertebrae are numbered according to their location from top to bottom. For example, C2 is the second cervical vertebra, T5 is the fifth thoracic vertebra, and L5 is the fifth lumbar vertebra (Figs. 7.22, 7.23).

Morphology

Most vertebrae have a similar construction with some slight differences. Vertebrae generally consist of a body with 2 strut-like structures called pedicles extending laterally, connecting to transverse

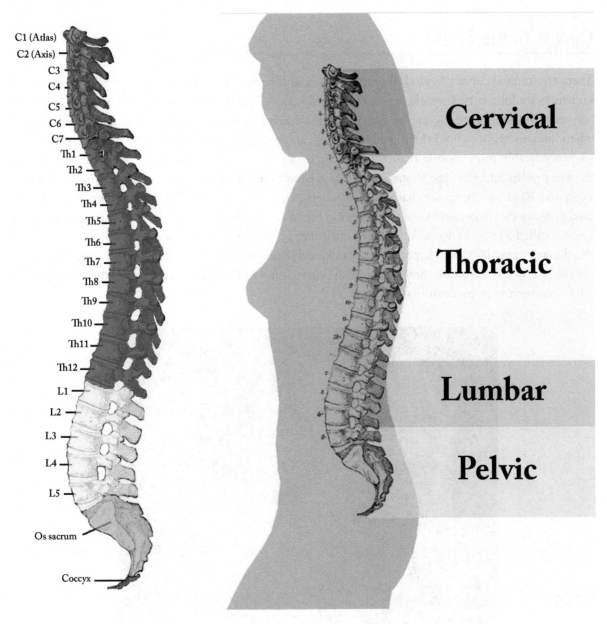

Figure 7.22 The spine.

Figure 7.23 The spine is divided into cervical, thoracic, lumbar, and pelvic sections.

processes. Structures called laminae complete the ring and fuse at the spinous processes. Bones of the cervical spine have small bodies and large-appearing spinal canals. They have foramen in their transverse processes that contain the vertebral artery and vein. They also have a forked or bifid spinous process. The atlas appears as a ring of bone. The axis has a large process extending superiorly, called the dens or odontoid process. The thoracic vertebrae are larger than the cervical vertebrae. Their bodies are larger and contain flat spots known as articulating facets, which serve as connection points for ribs. The lumbar vertebrae are the largest because they bear the most weight. Their spinal canals appear smaller.

Curves of the Spine

There are 4 spinal curves: cervical, thoracic, lumbar, and pelvic. The cervical and lumbar curves are both known as lordotic curves (example = cervical lordosis). A lordotic curve is characterized by having its convexity anterior. Lordotic curves are considered secondary curves because they develop after birth when humans begin to hold their heads up, sit up, and walk. The cervical and lumbar areas of the spine are considerably more mobile than the thoracic or a pelvic area because of the latter's connection to the bony pelvis and ribs. The thoracic and pelvic curves are called kyphotic curves (example = thoracic kyphosis). Kyphotic curves are characterized as being concave anteriorly. The kyphotic curves are considered primary curves because they are present at birth. An increased curvature of the cervical or lumbar spine is called a hyperlordosis. A decreased curvature is called a hypolordosis. An increased curvature of the thoracic spine is called a hyperkyphosis and a decreased curvature is called a hypokyphosis. A lateral curvature is called a scoliosis. Sometimes, if the curve is not severe, the curves are described as increased lateral convexities or concavities (Fig. 7.24).

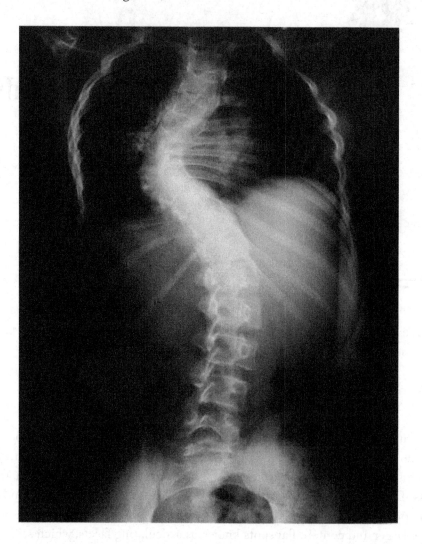

Figure 7.24 Scoliosis, which is characterized by the presence of lateral curves in the spine.

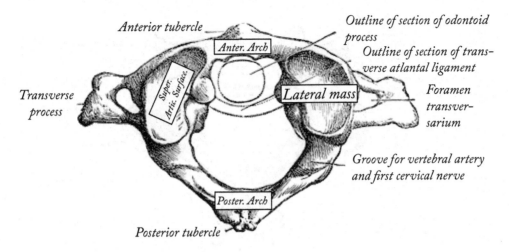

Figure 7.25 The atlas.

Individual Bones of the Spine

The atlas or C1 is the most superior vertebra. It appears as a ring of bone and articulates with the occipital condyles of the occipital bone superiorly and C2 inferiorly. The atlas contains foramen in the transverse processes that extend laterally (Fig. 7.25).

Axis (C2): The axis is a uniquely shaped vertebra. It has a small body, transverse foramen, and a large superior extending process known as the dens or odontoid process (Fig. 7.26).

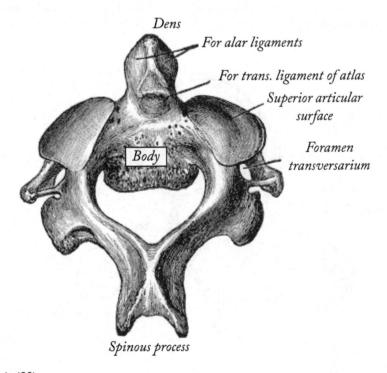

Figure 7.26 The axis (C2).

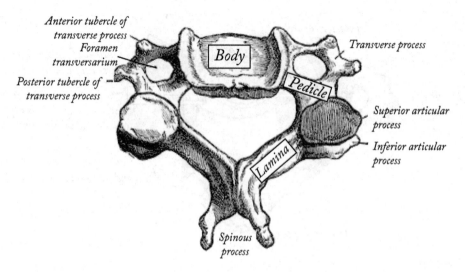

Figure 7.27 Typical cervical vertebra.

C3–7: The remaining cervical vertebrae are similar to one another. They contain bodies, pedicles, lamina, transverse processes with foramen, articulating facets, and bifid spinous processes (Fig. 7.27).

T1–12: The thoracic vertebrae contain bodies, pedicles, articulating facets, transverse processes, and spinous processes. They are larger than the cervical vertebra and connect with the ribs (Fig. 7.28).

L1–5

The lumbar vertebrae contain bodies, pedicles, lamina, articulating facets, and mamillary processes. They are the largest vertebrae (Fig. 7.29).

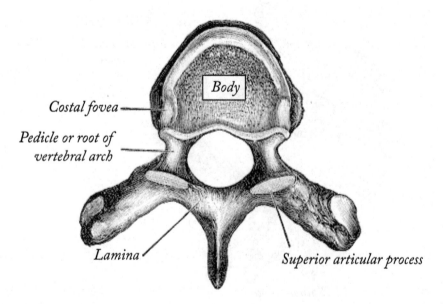

Figure 7.28 Typical thoracic vertebra.

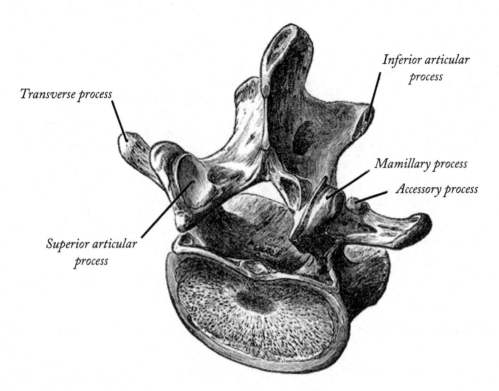

Figure 7.29 Typical lumbar vertebra.

Sacrum/Coccyx

The sacrum is a triangular curved bone located at the base of the spine. It is actually a series of 5 small vertebral bones that have fused. These bones begin to fuse at puberty and complete their fusion by around age 26. The sacrum articulates via articular processes with the 5th lumbar vertebra. The sacrum also articulates with the ilium of the coxal bones, forming the sacroiliac joints. Along the central posterior surface lies a ridge of bone from the fusion of the spinous processes of the sacral vertebrae. This ridge is called the medial sacral crest. There is also a series of eight paired holes called the sacral foramen. The sacrum is hollow forming a sacral canal that opens inferiorly with the sacral hiatus. The curvature is convex posteriorly and is more pronounced in males. The superior portion of the sacrum is called the base and contains a flat area called the sacral promontory (Fig. 7.30). The coccyx is another series of very small fused vertebral segments (3–5). These vertebrae do not completely fuse until late in adulthood.

The Ribcage

The ribcage consists of 12 pairs of ribs. There are true, false, and floating ribs. Usually ribs 1–7 are true ribs with ribs 8–10 being false ribs. Ribs 11 and 12 are floating ribs. The ribs attach to the vertebra in the back and the sternum in the front by way of cartilage connections (costochondral cartilage). True ribs connect directly to the sternum by way of their cartilage connections. False ribs connect to

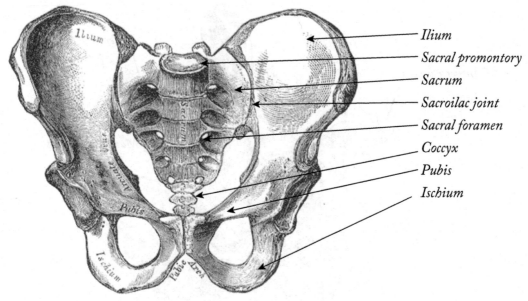

Figure 7.30 Pelvis and sacrum (anterior view).

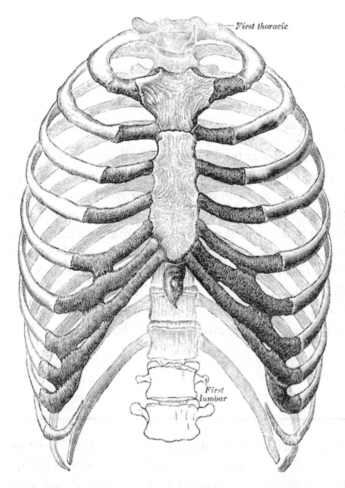

Figure 7.31. Ribcage.

the cartilage of true ribs and floating ribs connect only to the vertebrae in the back. There is no anterior connection to floating ribs (Fig. 7.31).

Sternum

The sternum has 3 parts. The most superior portion is called the manubrium. Just inferior to this is the body, and the most inferior portion is called the xiphoid process, which consists of cartilage. The sternum also articulates with the clavicle.

Hyoid Bone

The hyoid bone is located in the anterior region of the throat. It supports the larynx. A number of muscles that extend to the larynx, pharynx, and tongue attach to the hyoid bone (Fig. 7.32).

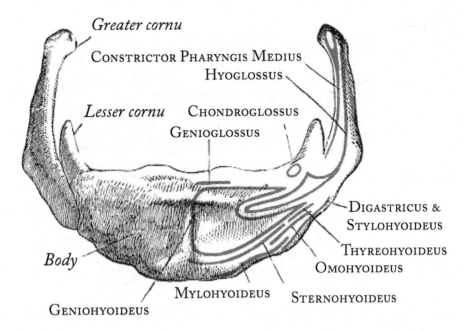

Figure 7.32 Hyoid bone.

The Appendicular Skeleton

The appendicular skeleton consists of the arms and legs (upper and lower extremities). The bones of the appendicular skeleton:

- Clavicle
- Scapula
- Humerus
- Radius
- Ulna
- Carpals
- Metacarpals
- Phalanges
- Coxal
- Femur
- Patella
- Tibia
- Fibula
- Calcaneus
- Talus
- Cuboid
- Navicular
- Cuneiforms
- Metatarsals
- Phalanges

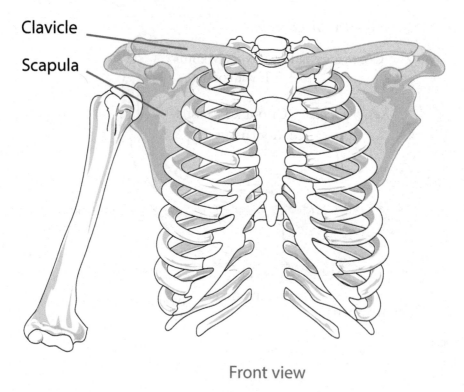

Figure 7.33 The pectoral girdle consists of the clavicle and scapula.

The upper extremity begins with what is called the pectoral girdle (shoulder girdle). This consists of the clavicle and scapula. The pectoral girdle acts as a support for the arms. The pectoral girdle attaches to the axial skeleton where the clavicle attaches to the sternum (sternoclavicular joint) (Fig. 7.33). This is the only direct attachment of the arm to the body. However, there are a number of muscles that also help to stabilize the connection.

Clavicles

The clavicles are located on the anterior portion of the thorax. They are the only S-shaped bones in the body. The clavicle articulates with the manubrium of the sternum and the acromion process of the scapula.

Scapula

The scapula is a triangular bone located in the posterior portion of the thoracic area. It articulates with the clavicle and the humerus. The borders of the scapula include superior, medial, and lateral borders. The glenoid cavity (fossa) is a concave surface on the lateral aspect of the scapula. It forms the "socket" of the ball-and-socket joint of the shoulder. The spine of the scapula is located on the posterior surface. It is a ridge of bone extending superiorly from medial to lateral. The acromion process is the terminal end of the spine of the scapula. It is a large process and articulates with the clavicle. The acromion process marks the highest point of the shoulder. The coracoid process is located on the anterior aspect of the scapula. This process is smaller than the acromion process and is located anterior and inferior to it. The supraspinous fossa is an indentation on the posterior portion of the scapula. It lies just above the spine. The infraspinous fossa is located just inferior to the spine of the scapula.

Bones of the Upper Extremity

The upper extremity consists of the arm, forearm, wrist, and hand. The bones of the upper extremity:

- Humerus
- Radius
- Ulna
- Carpals
- Metacarpals
- Phalanges

Humerus

The humerus is the proximal bone of the arm. It is a long tubular bone that articulates proximally with the scapula and distally with the radius and ulna. The head of the humerus is the proximal rounded end of the bone. The anatomical neck of the humerus is a small region that marks the end of the joint capsule between the humerus and the scapula. The surgical neck marks the beginning of the diaphysis. The greater tubercle is a rounded process on the lateral aspect of the proximal humerus. The lesser tubercle is a smaller rounded process on the medial aspect of the proximal humerus. The intertubercular groove (sulcus) is a groove between the greater and lesser tubercles. The deltoid tuberosity is a bump with a gradual slope on the lateral aspect of the humerus and is the site of attachment of the deltoid muscle. The lateral epicondyle is a widened area on the lateral aspect of the distal humerus. It is an important site of muscle attachments for the wrist extensor muscles. The medial epicondyle is a widened area on the medial aspect of the distal humerus. It is a site of attachment for wrist flexor muscles. The capitulum is a small rounded process at the distal end of the humerus on the lateral side. It articulates with the radius. The trochlea is a small spool-shaped process at the distal medial end of the humerus. It articulates with the ulna. The olecranon is an indentation on the posterior distal aspect of the humerus. The coronoid fossa is a small indentation on the anterior distal aspect.

Ulna

The ulna and radius both support the forearm (antebrachium). The ulna is on the medial side of the forearm. The bump on your elbow is actually the olecranon process of the ulna. The ulna articulates with the trochlea of the humerus and forms a hinge joint (Figs. 7.34, 7.35). The olecranon process is a rounded

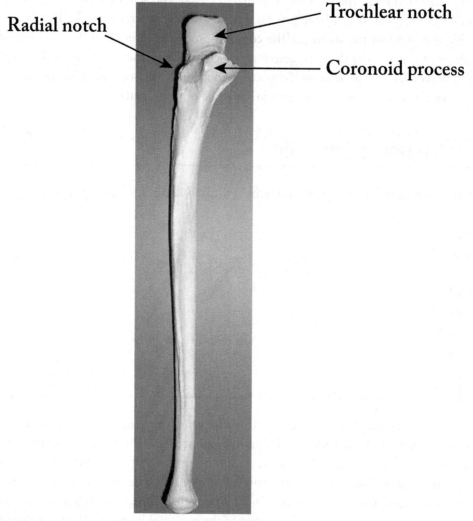

Figure 7.34 Anterior view of the ulna.

process on the proximal end of the ulna. The trochlear notch of the ulna articulates with the trochlea of the humerus. The radial notch is a flat spot that articulates with the radius. The styloid process of the ulna is a needle-like process at the distal end.

Radius

The radius is also located in the forearm. It articulates with the ulna and carpal bones. The radius allows for rotation of the forearm (Figs. 7.36, 7.37). The head of the radius articulates with the capitulum of the humerus. This joint can rotate. The radial tuberosity is a bump on the proximal aspect of the radius. The biceps muscle attaches there. The styloid process is a needle-like process on the distal aspect of the radius.

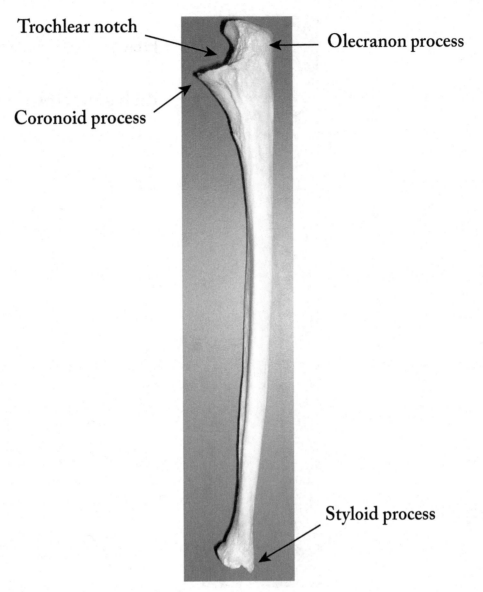

Figure 7.35 Medial view of the ulna.

Carpals

The carpal bones are located in the wrist. They consist of 8 bones that articulate with the radius and ulna proximally and the metacarpals distally (Fig. 7.38).

The 8 carpal bones:

- Scaphoid
- Lunate
- Triquetrum
- Pisiform
- Trapezium

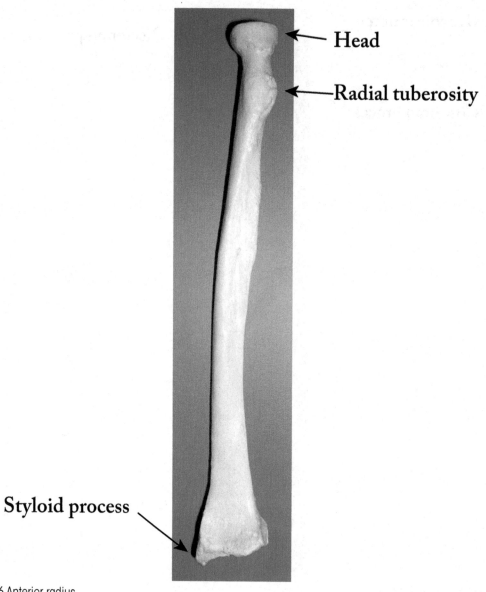

Figure 7.36 Anterior radius.

- Trapezoid
- Capitate
- Hamate

Metacarpals

The metacarpals are tubular-shaped bones that lie distal to the carpals. There are 5 metacarpals numbered accordingly from the thumb (1) to the little finger (5) (Fig. 7.39).

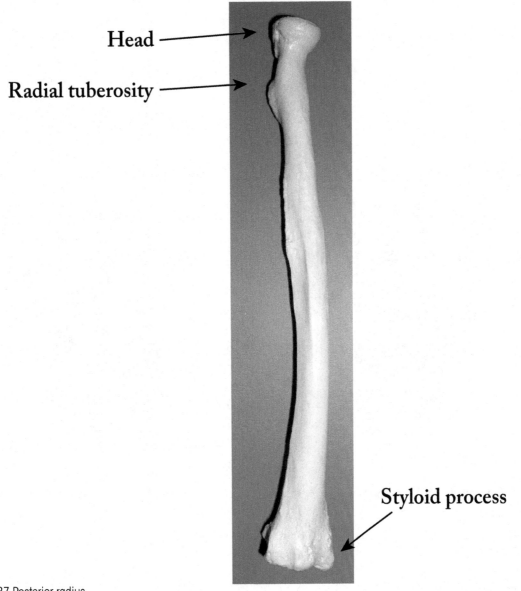

Figure 7.37 Posterior radius.

Phalanges

The phalanges comprise the fingers. They are numbered the same as the metacarpals and named for their location. The thumb has only a proximal and distal phalanx. The remaining fingers have proximal, middle, and distal phalanges (Fig. 7.40).

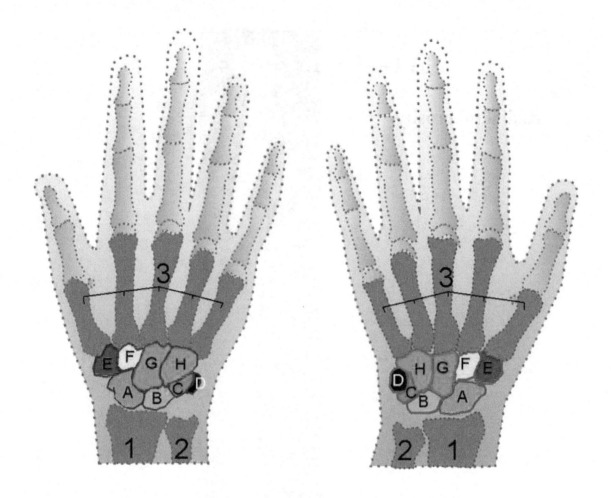

Figure 7.38 Carpals of the right hand.
A. Scaphoid
B. Lunate
C. Triquetrum
D. Pisiform
E. Trapezium
F. Trapezoid
G. Capitate
H. Hamate

Bones of the Lower Extremity

The lower extremity consists of the pelvis, leg, ankle, and foot. The bones of the lower extremity are as follows:

- Coxal
- Femur
- Patella

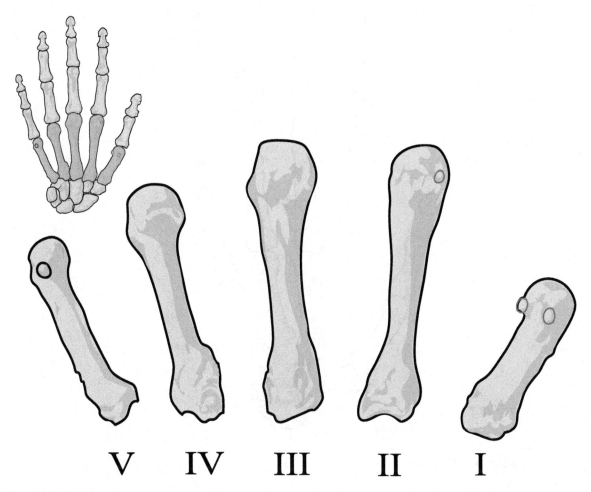

Figure 7.39 Metacarpals.

- Tibia
- Fibula
- Talus
- Calcaneus
- Tarsals
- Metatarsals
- Phalanges

The pelvic girdle consists of the 2 coxal bones.

Coxal Bone

The pelvis consists of the sacrum and 2 coxal bones. The coxal bones are actually 3 bones fused together. The 3 bones are the ilium, ischium, and pubis. The coxal bones articulate with the sacrum at the sacroiliac joints and the femurs at the hip joints (Fig. 7.41). The acetabulum is a socket-like concave structure

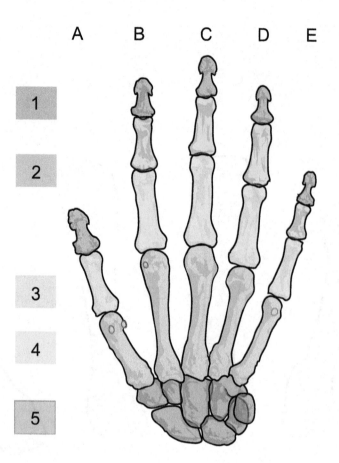

Figure 7.40 Bones of the hand.
1. Distal phalanx
2. Middle phalanx
3. Proximal phalanx
4. Metacarpals
5. Carpals
A. First
B. Second
C. Third
D. Fourth
E. Fifth

Notice the thumb has only proximal and distal phalanges. The remaining fingers have proximal, middle, and distal phalanges.

that articulates with the head of the femur to form the hip joint. The iliac crest is the most superior structure of the coxal bone. It is a ridge of bone that extends along the superior margin of the ilium. The anterior superior iliac spine is a bump on the anterior portion of the ilium. The iliac crest terminates here. This is an important site of muscle attachments. The posterior superior iliac spine is a bump on the posterior aspect of the ilium. The iliac crest terminates here posteriorly. The obturator foramen

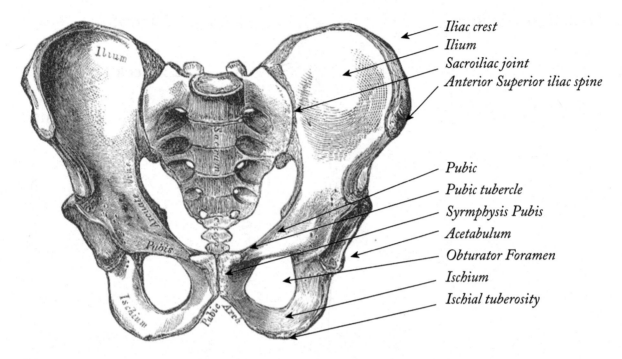

Figure 7.41. Coxal bones of the pelvis. The coxal bones are 3 fused bones consisting of the ilium, ischium, and pubis.

is a space that is formed by the pubis and ischium. The symphysis pubis is a fibrocartilaginous disc that forms a fibrous joint between the 2 pubic bones. The pubic tubercle resides on the anterior superior aspect of the pubis. The ischial tuberosity is a thickened area of bone located on the posterior aspect of the ischium. The hamstring muscles attach here. The pubic arch is the angle between the pubic bones.

Male and Female Differences

Generally the differences between male and female pelves are due to functions of childbirth. The pubic arch is greater in females and the ilia may be more flared. The sacrum tends to be more curved in males. The female pelvis is wider in all directions.

Femur

The femur is the longest bone in the body. It articulates with the acetabulum of the coxal bone proximally and with the patella and tibia distally (Figs. 7.42, 7.43). The head of the femur is a rounded process on the proximal end. The neck of the femur is the area that connects the head with the shaft. The greater

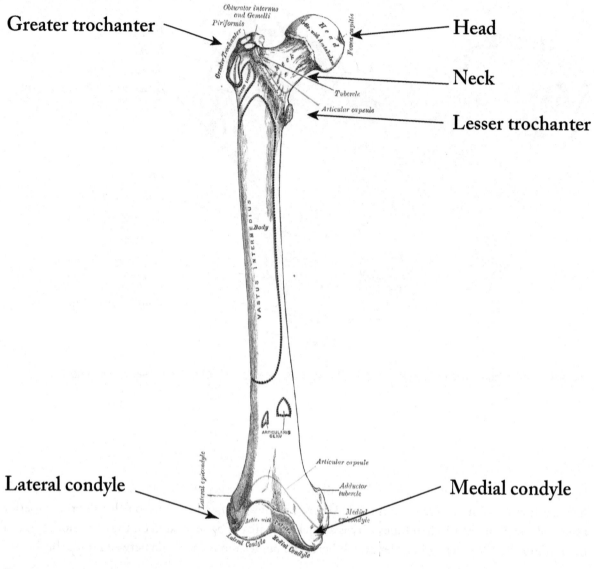

Figure 7.42 Anterior femur.

trochanter is a large process located on the proximal lateral aspect of the femur. The lesser trochanter is a smaller process located on the proximal medial aspect. There are 2 large rounded processes on the distal aspect of the femur called the medial and lateral condyles. The linea aspera is a roughened area on the posterior aspect. It is a site of muscle attachments.

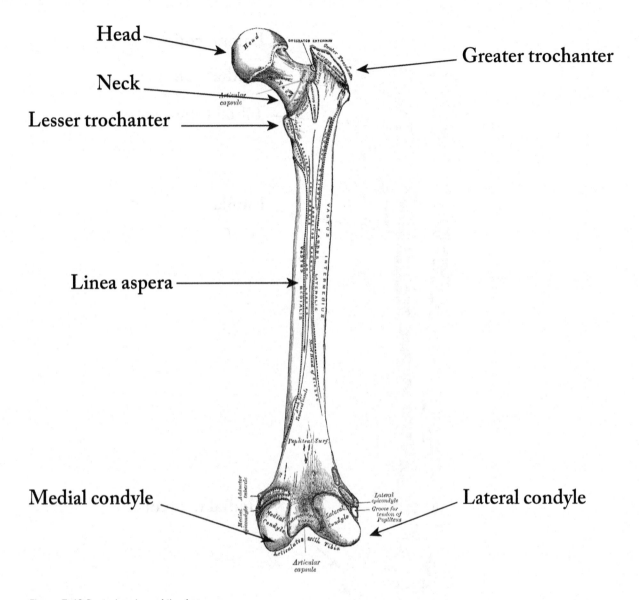

Figure 7.43 Posterior view of the femur.

Tibia

The tibia is the larger of 2 bones of the lower leg. It articulates with the femur, fibula, patella, and talus bones (Fig. 7.44). The tibial condyles are large rounded processes on the proximal aspect of the tibia. The 2 condyles are named medial and lateral. The tibial tuberosity is a broad bump on the anterior aspect of the tibia. The medial malleolus is a rounded process on the distal medial aspect of the tibia. It is the bump on the inside of the ankle.

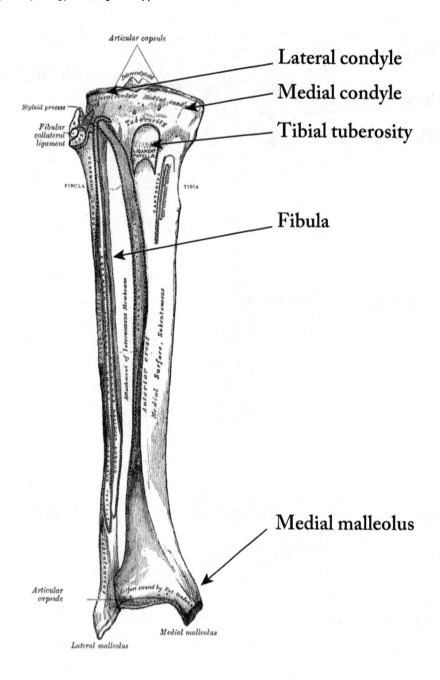

Figure 7.44 Anterior view of the tibia.

Fibula

The fibula is the lateral bone in the lower leg. It forms the lateral ankle and articulates with the tibia and talus bones (Fig. 7.45). The fibular head is a rounded process on the proximal end of the bone. The lateral malleolus is a rounded process on the distal end of the bone. It forms the lateral ankle.

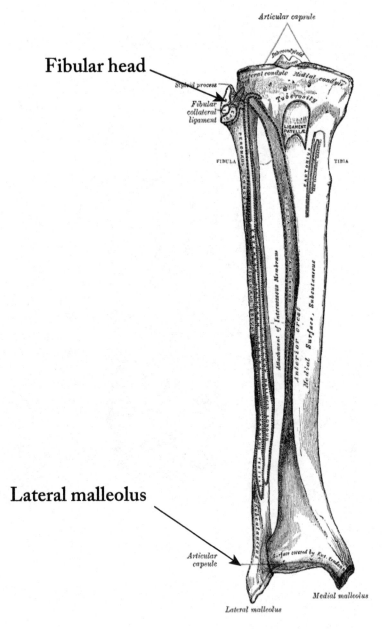

Figure 7.45 Anterior view of fibula.

The Ankle

The ankle and foot consist of the tarsals, metatarsals, and phalanges and has a similar construction to the wrist and hand (Fig. 7.46).

Tarsals:

- Calcaneus
- Talus

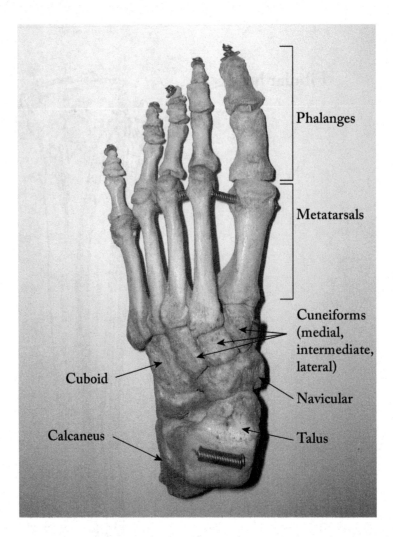

Figure 7.46 Foot and ankle.

- Navicular
- Cuboid
- Lateral cuneiform
- Intermediate cuneiform
- Medial cuneiform

Calcaneus and Talus

The calcaneus or heel bone is the largest of the tarsals. The talus forms the ankle joint with the tibia and fibula. These bones articulate with the navicular and cuboid bones. There are 3 cuneiform bones named for their position that articulate with the metatarsals.

Metatarsals and Phalanges

There are 5 tubular metatarsals that are named for their position (1–5). The phalanges are similar to those in the fingers. The big toe has only proximal and distal phalanges, while the remaining toes have proximal, middle, and distal phalanges.

CRITICAL-THINKING QUESTIONS

1. What are the structural and functional differences between a tarsal and a metatarsal?
2. What are the structural and functional differences between the femur and the patella?
3. If the articular cartilage at the end of one of your long bones were to degenerate, what symptoms do you think you would experience? Why?
4. What is the difference between closed reduction and open reduction? In what type of fracture would closed reduction most likely occur? In what type of fracture would open reduction most likely occur?
5. In terms of origin and composition, what are the differences between an internal callus and an external callus?
6. How do the structural differences of the pectoral and pelvic girdles reflect their functional differences?
7. Long-term bedridden patients sometimes develop decubitus ulcers (pressure sores). List three bony protuberances that might be sites for these ulcers.

Chapter Eight

Joints of the Skeletal System

Joints connect the bones of the body and allow it to move and grow. Joints are called articulations and can be classified according to the tissue that connects the bones. Joints can also be classified according to their degree of movement. Synarthrotic joints are immovable. Examples include the sutures of the skull. Amphiarthrotic joints are slightly moveable. Examples include interosseous ligaments that connect some of the long bones such as the radius and ulna.

Diarthrotic joints are freely moveable such as the shoulder and hip.

There are three basic categories of joints:

1. Fibrous
2. Cartilaginous
3. Synovial

Fibrous Joints

Fibrous joints are held together by dense connective tissue. There are three types of fibrous joints:

1. Syndesmosis
2. Suture
3. Gomphosis

A syndesmosis is a slightly movable joint formed by dense connective tissue between two bones. An example is an interosseous ligament that connects the radius and ulna together. An interosseous ligament also connects the tibia and fibula. A suture is a joint between the flat bones of the skull. The bones are united by a band of dense connective tissue called a sutural ligament. Sutures are considered synarthrotic or immovable. A gomphosis is a joint in which a cone-shaped process is united with a cone-shaped socket. A tooth is a good example of a gomphosis. The tooth unites with the bone via a periodontal ligament.

Cartilaginous Joints

In cartilaginous joints hyaline or fibrous cartilage unites the bones. There are two types of cartilaginous joints: symphysis and synchondrosis. A symphysis consists of areas of hyaline cartilage on the ends of the bones connected to a section of fibrocartilage. The intervertebral disc is an example of a symphysis. These are classified as amphiarthrotic or slightly movable. Another example is the symphysis pubis that connects the right and left pubic bones of the pelvis. A synchondrosis consists of hyaline cartilage uniting bones. An example of a synchondrosis is the cartilage between the ribs and the sternum, often referred to as the costochondral cartilage. Another example is the epiphyseal plate located in the epiphysis of a long bone.

Synovial Joints

Most of the joints in the body are synovial joints. Synovial joints are complex joints that consist of a number of parts. Synovial joints are freely movable or diarthrotic. Synovial joints are encapsulated by a synovial membrane. They contain fluid (synovial fluid) and cartilage on the ends of the bones. Strong bands of dense connective tissue called ligaments connect the bones together. Some synovial joints contain discs of fibrocartilage called menisci that act as small cushions to help dissipate force from the bones. Small sacs called bursa contain synovial fluid that helps to cushion the area around the joints and reduce friction.

There are a number of types of synovial joints named for their shape. The shape of the joint determines its movement (Fig. 8.1). Ball-and-socket joints consist of a rounded process and rounded socket. These include the hip and shoulder and allow for a variety of movements. Hinge joints consist of a convex surface and concave socket. Examples include the joint between the humerus and ulna as well as in some of the phalanges. Hinge joints move in only one plane. Condyloid joints consist of oval processes fitting into elliptical sockets. An example of this joint is the metacarpal phalangeal joint. Gliding joints consist of flattened surfaces connected together. Examples include the carpal bones of the wrist. Pivot joints consist of a cylinder fitting into a ring of bone. Examples include the joint between the atlas and axis of the spine and the joint between the radius and humerus. Saddle joints consist of two bones having both concave and convex surfaces. An example is the carpal-metacarpal joint of the hand.

Joint Movements

Joints move according to their shapes. The following movements are organized according to their respective joints.

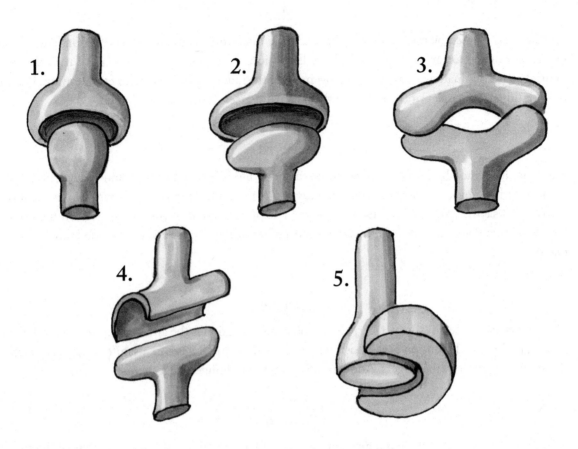

Figure 8.1 Types of synovial joints.
1. Ball and socket
2. Condyloid
3. Saddle
4. Hinge
5. Pivot

Shoulder

The shoulder joint consists of the scapula and humerus. Flexion consists of the humerus moving anterior in a sagittal plane. Extension consists of the humerus moving posterior in a sagittal plane. Abduction is the movement of the humerus away from the body in a coronal plane. Adduction is moving the humerus toward the body in a coronal plane. Internal rotation is moving the humerus along its long axis toward the body. External rotation is moving the humerus along its long axis away from the body.

Elbow

The elbow is formed by the connection between the humerus, radius, and ulna. Flexion is the anterior movement of the forearm in a sagittal plane. Extension is the posterior movement of the forearm in a sagittal plane. Supination is a rotational movement of the radius so that the palm faces upward. Pronation is a rotational movement of the radius so that the palm faces downward.

Wrist

The wrist is formed by the connection between the radius, ulna, and carpals and metacarpals. Flexion is the anterior movement of the carpals in a sagittal plane. Extension is the posterior movement of the carpals in a sagittal plane. Ulnar deviation is the lateral movement of the carpals toward the body in a coronal plane. Radial deviation is the lateral movement of the carpals away from the body in a coronal plane.

Fingers

The fingers are formed by the metacarpals and phalanges. Flexion is the anterior movement of the fingers in a sagittal plane. Extension is the posterior movement of the fingers in a sagittal plane. Abduction is the spreading apart of the fingers. Adduction is bringing the fingers together.

Hip

The hip joint consists of the coxal and femur bones. Flexion is the anterior movement of the femur in a sagittal plane. Extension is the posterior movement of the femur in a sagittal plane. Abduction is the lateral movement of the femur in a coronal plane away from the body. Adduction is the lateral movement of the femur in a coronal plane toward the body. Internal rotation is the movement of the femur along its long axis toward the body. External rotation is the movement of the femur along its long axis away from the body. Circumduction is the movement of the femur in a circular motion so that its distal end traces a circle.

Knee

The knee joint movements occur at the femur and tibia. Flexion is the anterior movement of the tibia in a sagittal plane. Extension is the posterior movement of the tibia in a sagittal plane.

Ankle

The ankle if formed by the tibia, fibula, talus, and metatarsals. Dorsiflexion is the upward movement of the foot (as if walking on the heels) in a sagittal plane. Plantarflexion is the downward movement of the foot (as if walking on the toes) in a sagittal plane. Inversion is the movement of the foot so the sole of the foot points medially. Eversion is the movement of the foot so the sole of the foot points laterally.

Spine

The spine is divided into cervical, thoracic, and lumbar areas. The movements are the same in all of these areas. Flexion is the anterior movement of the spine in a sagittal plane. Extension is the posterior movement of the spine in a sagittal plane. Right lateral flexion is lateral bending of the spine toward the right side. Left lateral flexion is the lateral bending of the spine toward the left side. Right rotation is the twisting of the spine toward the right. Left rotation is the twisting of the spine toward the left.

Other movements

Protraction is the forward movement of a part along a transverse plane. Retraction is the backward movement of a part along a transverse plane. Elevation is the upward movement of a part. Depression is the downward movement of a part.

Joint Examples

Shoulder

The shoulder consists of the scapula, humerus, and clavicle (Fig. 8.2). The joint between the scapula and humerus is a synovial ball-and-socket joint. As in all joints, the shoulder joint is held together by ligaments. Some of the important ligaments:

(1) Glenohumeral

These ligaments exist as three bands extending from the anterior wall of the glenoid fossa and attaching to the anatomical neck and lesser tubercle of the humerus.

(2) Coracohumeral

This ligament extends from the coracoid process of the scapula to the greater tubercle of the humerus.

(3) Transverse humeral

This ligament forms a band of connective tissue between the greater and lesser tubercles of the humerus. The long head of the biceps brachii is found in this groove.

(4) Glenoid Labrum

This is a rim of fibrocartilage that attaches to the glenoid fossa.

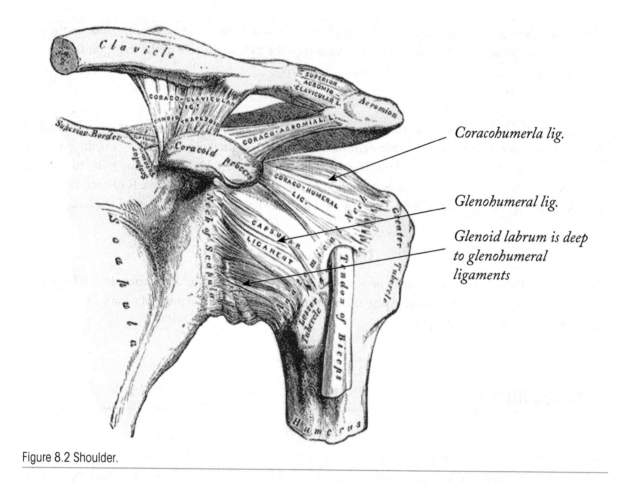

Figure 8.2 Shoulder.

Shoulder Separation and Dislocation

Shoulder separation occurs when the ligaments between the clavicle and scapula are torn. The lateral end of the clavicle often moves superiorly and protrudes (Fig. 8.3). A shoulder dislocation occurs between the scapula and humerus. Here the ligaments holding the humerus in the glenoid fossa are torn and the humerus comes out of the fossa.

Elbow

The elbow contains two articulations. One involves the humerus and ulna. The other involves the humerus and radius. The humeroulnar joint is formed by the trochlea of the humerus and the proximal portion of the ulna (Figs. 8.4, 8.5). This joint can only flex and extend. The humeroradial joint is formed by the capitulum of the humerus and the radial head. This joint can rotate. Some of the important ligaments are the ulnar and radial collaterals.

The ulnar collaterals connect the medial aspect of the medial epicondyle to the medial aspect of the coronoid process of the ulna. The radial collaterals connect the lateral epicondyle to the annular ligament

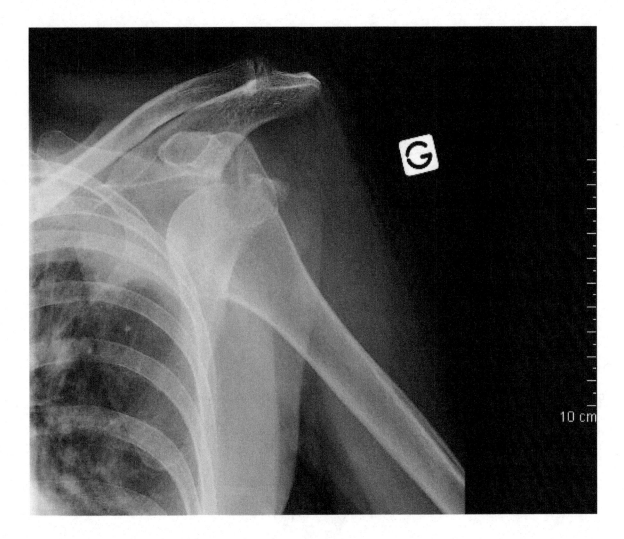

Figure 8.3. Dislocated shoulder. Notice the misalignment between the head of the humerus and the glenoid fossa.

of the radius. The annular ligament encircles the radial head and attaches to the trochlear notch of the ulna. This ligament can be prone to dislocation in children.

Hip

The hip joint consists of the femur and coxal bones. The head of the femur fits into the acetabulum of the coxal bone (Figs. 8.6, 8.7). The hip joint has the same motions as the shoulder. The ligaments:

The iliofemoral ligament extends from the ilium to the greater and lesser trochanters of the femur. It is Y-shaped and is considered the strongest ligament in the body. The ischiofemoral ligament extends from the ischium to the joint capsule of the femur. The pubofemoral ligament extends from the pubis to the joint capsule of the femur.

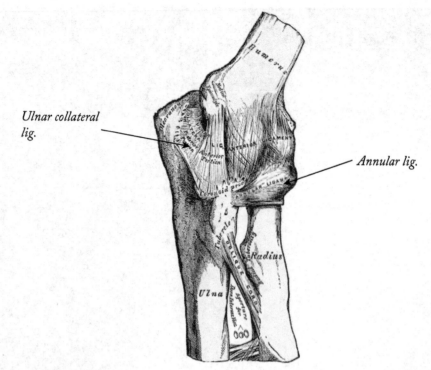

Figure 8.4 Elbow (medial projection). Notice how the annular ligament wraps around the head of the radius.

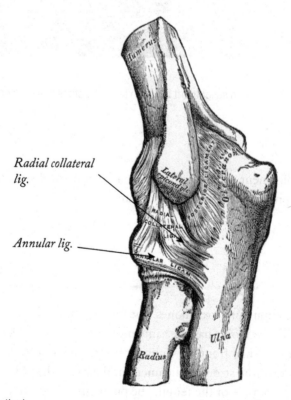

Figure 8.5 Elbow (lateral projection).

Joints of the Skeletal System | 139

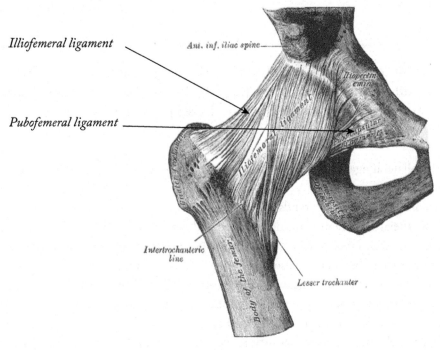

Figure 8.6 Hip (anterior view).

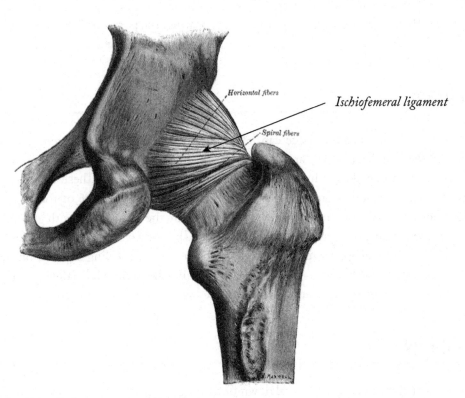

Figure 8.7 Hip (posterior view).

Knee

The knee is the most complex joint in the body. It is also the largest. It consists of the condyles of the tibia articulating with the condyles of the tibia (Fig. 8.8). The patella also articulates with the femur. The knee flexes and extends as well as rotates. It forms a locked position when extended. Some of the knee ligaments:

(1) The medial collateral ligament extends from the medial condyle of the femur to the medial condyle of the tibia.
(2) The lateral collateral ligament extends from the lateral condyle of the femur to the head of the fibula.
(3) The anterior cruciate ligament is inside the knee and extends from the posterior femur to the anterior tibia. The anterior cruciate (ACL) works to stop forward translation of the tibia on the femur. The posterior cruciate ligament is also inside the knee and extends from the anterior femur to the posterior tibia. It works to stop backward translation of the tibia on the femur.
(4) The arcuate popliteal ligament is on the posterior aspect of the knee. It is Y-shaped and extends from the lateral condyle of the femur to the fibular head. The patellar ligament extends from the inferior aspect of the patella to the tibial tuberosity. It is an extension of the common quadriceps tendon. The oblique popliteal is located in the posterior aspect of the knee and extends from the lateral condyle of the femur to the head of the fibula.

The knee also contains two fibrocartilage pads called menisci that help to cushion the joint. The medial and lateral menisci are located on top of the tibial condyles. Occasionally a meniscus can become injured and tear. The medial meniscus is more prone to tearing. In sports the "terrible triad" is known as tears to the medial collateral ligament, medial meniscus, and anterior cruciate ligament.

Joint Injuries

When the force exceeds what the tissue can handle, the tissue becomes damaged. Common injuries include tears to the muscles (strains) and tears to the ligaments (sprains). Both sprains and strains are graded 1, 2, or 3. In a first-degree injury 0 to 25% of the fibers are torn. These injuries typically take 1–2 weeks to heal. A second-degree injury is characterized by 25% to 50% of the fibers torn. These usually take from 2 to 4 weeks to heal. Third-degree injuries are the most severe with greater than 50% of the fibers torn. These injuries take at least 12 weeks to heal. The body reacts to these injuries by producing inflammation. The joint will appear red and swollen and cause pain. Healing depends on the severity of the injury as well as the health of the subject. In severe sprains the joint will become unstable due to the torn ligaments. In some cases the joint must be stabilized with splints, supports, or casts.

Osteoarthritis

Osteoarthritis is characterized by the breakdown of cartilage. It is the most common form of arthritis and tends to affect people in middle age and beyond. Osteoarthritis commonly affects the hands, knees, hips, and spine. In osteoarthritis the normal cartilage repair mechanisms malfunction and the cartilage

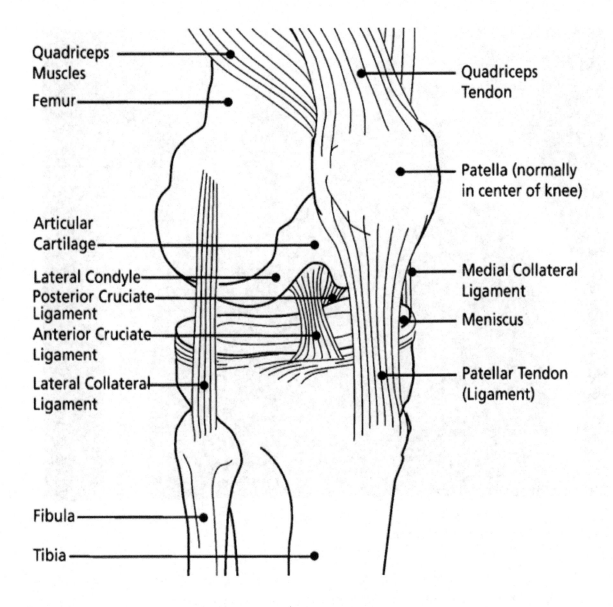

Figure 8.8 Knee ligaments.

begins to wear out. The joint space will become smaller and may progress to the point of bone rubbing on bone. The joint surfaces become roughened and cause pain and inflammation. There is no cure for osteoarthritis; however, severe cases are treated with joint replacement (Fig. 8.9).

CRITICAL-THINKING QUESTIONS

1. Define how joints are classified based on function. Describe and give an example for each functional type of joint.
2. Explain the reasons joints differ in their degree of mobility.

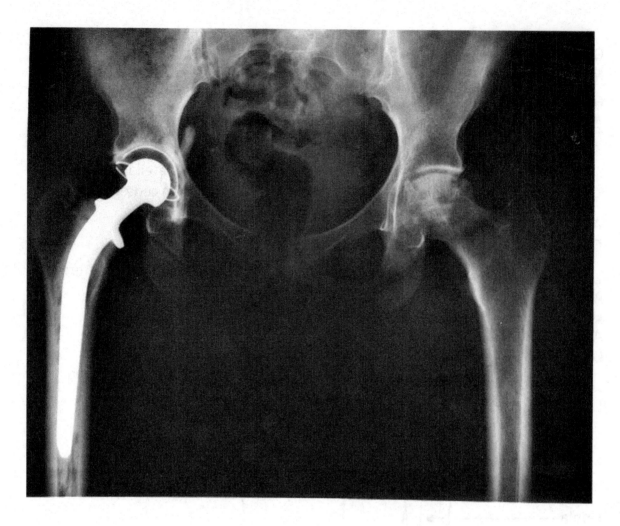

Figure 8.9 Osteoarthritis and joint replacement.

3. Distinguish between a narrow and wide fibrous joint and give an example of each.
4. Describe the two types of cartilaginous joints and give examples of each.
5. Both functional and structural classifications can be used to describe an individual joint. Define the first sternocostal joint and the pubic symphysis using both functional and structural characteristics.
6. Describe the characteristic structures found at all synovial joints.
7. Describe the structures that provide direct and indirect support for a synovial joint.
8. Briefly define the types of joint movements available at a ball-and-socket joint.
9. Discuss the joints involved and movements required for you to cross your arms together in front of your chest.
10. Discuss the structures that contribute to support of the shoulder joint.
11. Describe the sequence of injuries that may occur if the extended, weight-bearing knee receives a very strong blow to the lateral side of the knee.
12. What combination of movements is required for a person to clap their hands?

Chapter Nine

Muscular System

We have examined the supportive structures of the body, such as bones and cartilage, as well as how these structures are connected by ligaments, forming joints. We will now investigate how these supportive structures move. Muscles move bones by contracting and relaxing. Muscles are also important in keeping us alive. The heart is largely composed of muscles and the blood vessels contain a layer of muscles. The diaphragm is also a muscle. There are muscles that move our eyes and tongue, and food through our digestive tract. We will begin by looking at some general information that applies to all muscles then we will examine skeletal muscles in detail. In later chapters we will cover the muscles associated with various organs.

Types of Muscle Tissue

Muscle tissue consists largely of protein. Tiny protein filaments are bundled together in muscle. These filaments slide past each other, causing muscles to contract. Muscles contract in response to a signal from the nervous system. There are three basic types of muscle tissue. Skeletal muscle is characterized by densely packed protein filaments. Cardiac muscle is found only in the heart and also has densely packed protein filaments. Skeletal and cardiac muscles appear striated because of these filaments. Smooth muscle is found in the walls of the arteries and digestive system. It also consists of protein filaments but these are not as dense as skeletal or cardiac muscle. Smooth muscle does not appear striated because it is less organized.

When we describe the locations of skeletal muscles, we use the terms "origin" and "insertion." The origin of a muscle is the less mobile end of a joint. The insertion is the more mobile end of a joint. Think of how the body is structured. Joints need to be anchored on one end and more mobile on the other. Generally, there is more mobility at the distal ends of joints. Muscles connect to bones through dense connective tissue structures called tendons. Sometimes the tendons are broad and flattened. These are called aponeuroses. An example of an aponeurosis is a flat tendon on the lateral aspect of the thigh known as the iliotibial band.

Think of how muscles move joints. In order to move a joint in one direction you have to have at least one muscle on that side of the joint. To bring the joint back to its original position, you need to have at least one muscle on the opposite side of the joint. When the first muscle contracts, the other relaxes. The first muscle that produced the movement is called the agonist. The second muscle on the opposite side of the joint that opposes the movement is called the antagonist. Let's look at an example to illustrate this. The elbow can move into flexion or extension. The elbow has a muscle on the anterior side called the biceps. It also has a muscle on the posterior side called the triceps. Elbow flexion (bending the elbow) is caused by contraction of the biceps muscle. In this case we can say that the biceps muscle is the agonist. Since the triceps muscle opposes this movement, it is called the antagonist. If we straighten the elbow, the muscle that produces this movement is the triceps. So now the triceps is considered the agonist and since the biceps opposes this movement, it is considered the antagonist. So, when determining agonist and antagonist muscles, we first have to consider the specific movement of the joint. Muscles also tend to work together. However, there is usually one muscle that is most responsible for the movement. This muscle is called the prime mover. For example, there are a number of abdominal muscles. The rectus abdominus (known as the six-pack) runs right down the middle of the abdominals. The external obliques are located on the sides of the abdomen. During a sit-up (or crunch), the rectus abdominus muscle is most responsible for producing the movement. The rectus then is called the prime mover. The external obliques help out, so they are known as synergists. Some muscles are involved in holding bones in place. These muscles are called fixators or stabilizers. For example, when you move your shoulder there are muscles attached to your scapula that hold it in place.

Muscle Shapes

The shape of a muscle helps to determine how forcefully it can contract. There are three basic muscle shapes (Fig. 9.1). In some muscles the fibers are in a feather-like arrangement. These muscles are called pennate or bipennate. Some pennate muscles have all of their fibers arranged on one side. These are called unipennate. If the fibers are arranged at various places around a central tendon, the muscle is called multipennate. In straight muscles fibers are arranged along or parallel to the long axis of the muscle. Circular muscles are called orbicular muscles.

Muscles run over bones that act as pulleys and levers. There are three types of levers that involve muscle contraction. Muscles exert a force called a pull on a weight. In class 1 levers, the fulcrum is located between the pull and the weight. In class 2 levers, the weight is located between the fulcrum and the pull. In class 3 levers, the pull is located between the fulcrum and the weight. An example of a class 1 lever is the atlanto-occipital joint in the spine. The joint acts as a fulcrum while the posterior back muscles exert a pull on the skull. The joint lies between the muscles and the skull. An example of a class 2 lever is the temporomandibular joint. When the mouth opens, the weight or mandible is located between the fulcrum (TMJ) and the pull from muscles in the throat. Most muscles are arranged in a class 3 lever system. Our example involving the biceps muscle is a class 3 lever.

Muscle Contractions

There are three types of muscle contractions. All three are used in treating injuries in rehabilitation and physical therapy settings. In isotonic contractions (iso = equal, tonic = tone) the force remains the same but the length of the muscle changes. An example of an isotonic contraction is the classic

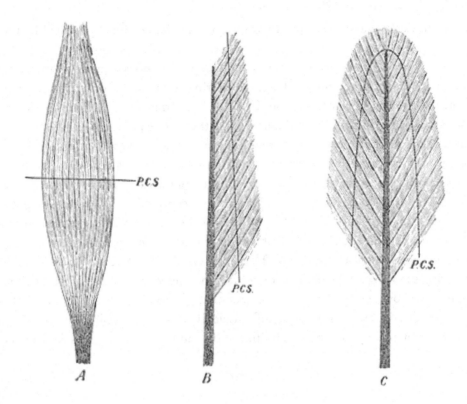

Figure 9.1 Muscle shapes.
A. Straight
B. Unipennate
C. Bipennate

biceps curl with a barbell. The force exhibited by the barbell does not change. However, the length of the bicep muscle can change by shortening during elbow flexion and lengthening during extension. Isotonic exercises are used in many gym settings in which participants use barbells and selectorized weight equipment. In isometric contractions (iso = equal, metric = length), the force can change but the length of the muscle remains the same. In isometric contractions there is no movement of the joint since the muscle length does not change. An example of an isometric contraction is to push against an object that cannot be moved, such as a wall. The participant can push with a little amount of force or a lot of force (force can change), but there is no movement of the joint. Isometric exercises are used in rehabilitation settings for the strengthening of damaged muscle tissue. They are relatively safe because the damaged area can be omitted during the exercises. For example, let's say an athlete injured her shoulder. Upon examination she was able to abduct her arm about 30 degrees before she experienced severe pain. Isometric exercises could then be used up to about 30 degrees of abduction. She would begin with using low amounts of force and then progress to higher amounts of force until the tissue healed. In isokinetic contractions (iso = equal, kinetic = motion) both the force and length of the muscle can vary but the contraction happens at a fixed speed. Isokinetic exercises are used primarily in rehabilitation settings. Sophisticated machines are used to control the speed of the exercise while allowing varying resistance. However, a simple treadmill is a good example of isokinetic exercise.

The participant can exercise at a fixed speed with varying degrees of force provided by the different incline angles of the treadmill.

Many exercises consist of two phases. There is a phase in which the muscle shortens during contraction and a phase in which the muscle lengthens during relaxation. During the shortening phase the muscle performs a concentric contraction. Concentric contractions are characterized by muscles shortening against a load. During the lengthening phase the muscle performs an eccentric contraction. Eccentric contractions are characterized by muscles lengthening against a load. Concentric contractions are used primarily to move a load, while eccentric contractions are used to slow down or stop a load. An example is the biceps curl exercise. During the flexion phase of the exercise, the biceps muscle shortens against the load. The biceps is said to perform a concentric contraction. When the weight is lowered during elbow extension, the biceps is lengthening against the load. Now the biceps is performing an eccentric contraction. Muscles are generally not well suited for eccentric contractions and can be more prone to injury during eccentric contractions. An example of this is the tear of rotator cuff muscles in the shoulder during throwing a baseball. The rotator cuff muscles work to internally rotate the shoulder. The throwing motion, however, consists of external rotation. The rotator cuff muscles work to decelerate the arm at the end of the thrown. The rotator cuff muscles do this by performing eccentric contractions. If the force is too great and exceeds the capabilities of the muscles, then the muscles can become torn or strained.

Overview of the Muscular System (Figs. 9.2, 9.3)

Note: There are many muscles in the human body. This work will describe only a sample of these muscles.

Muscles of the Head and Neck

The muscles of the head and neck move the face, larynx, and tongue. A sample of muscles of facial expression follows.

Muscles of facial expression (Figs. 9.4, 9.5)

Located on top of the head is a broad flat tendon called the epicranial aponeurosis. There is one muscle with two parts attached to the anterior and posterior sections of this tendon. The muscle is called the occipitofrontalis. The anterior portion lifts the eyebrows. The posterior is a weak head extensor and can cause headaches. There are two circular muscles called sphincters. The orbicularis oculi encircles the eye. It compresses the lacrimal gland and closes the eye. The orbicularis oris encircles the mouth. It causes the lips to pucker. The buccinator is located in the cheek. It compresses the cheek against the teeth. The zygomaticus muscle has major and minor divisions and attaches to the orbicularis oris and zygomatic bone. It raises the lateral ends of the mouth when smiling. The platysma is a very thin and superficial muscle located under the chin. It causes the action of frowning when contracted.

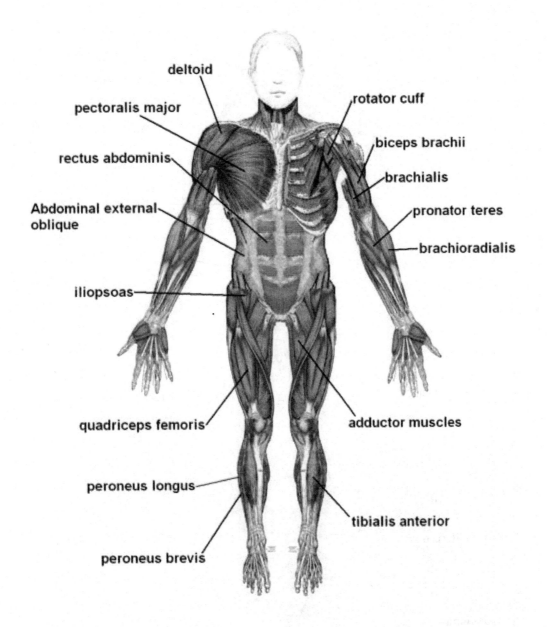

Figure 9.2 Overview of anterior muscles.

http://commons.wikimedia.org/wiki/Image:Muscles_anterior_labeled.png

Muscles of Mastication

The muscles of mastication (chewing) include the masseter, temporalis, and medial and lateral pterygoids (Figs. 9.6, 9.7). The masseter muscle attaches to the mandible and allows for closing the jaw. The temporalis is located in the lateral skull and attaches to the temporal bone. The temporalis aids in closing the jaw. In fact you can feel your temporalis muscle contract when touching the sides of your head when clenching your jaw. The medial and lateral pterygoids are deep muscles in the jaw. These can elevate, depress, protract, and cause lateral movement of the mandible. These muscles are often involved in temporomandibular joint (TMJ) disorder.

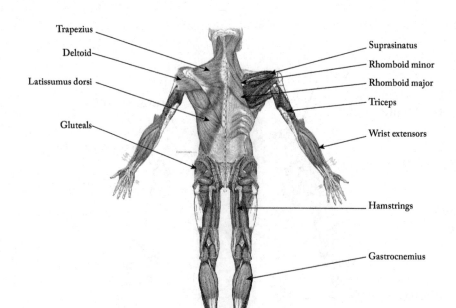

Figure 9.3 Overview of posterior muscles.

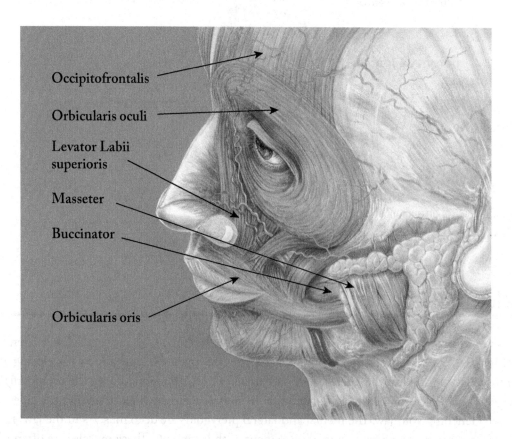

Figure 9.4 Muscles of facial expression.

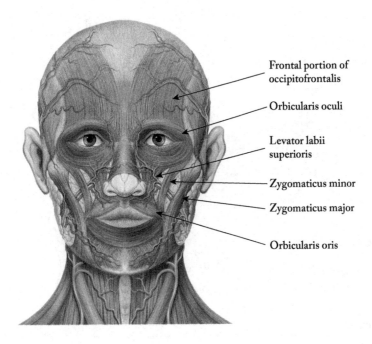

Figure 9.5 Facial muscles.

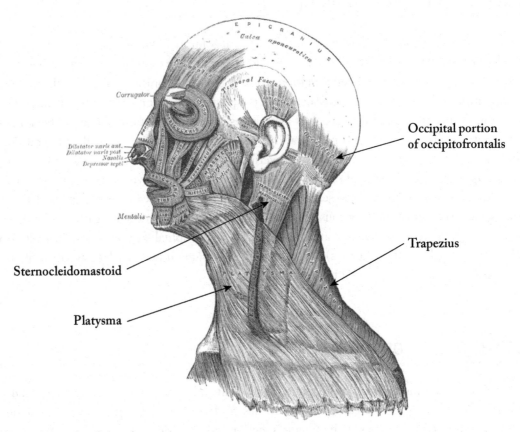

Figure 9.6 Lateral view of neck muscles.

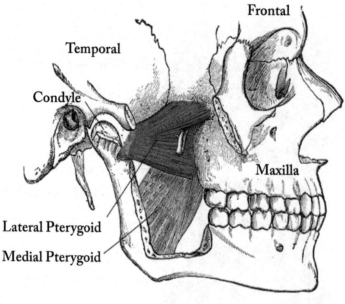

Figure 9.7 Medial and lateral pterygoids.

Head and Vertebral Column

There are a number of muscles that attach to the vertebral column and move the head (Figs. 9.8, 9.9). The sternocleidomastoid (SCM) attaches to the mastoid process of the temporal bone as well as the clavicle and sternum. It produces contralateral rotation when one muscle contracts and neck flexion when both muscles contract. The splenius capitis is located in the posterior portion of the neck. It helps bring head into an upright position (head extension). It also causes ipsilateral rotation and lateral flexion when one muscle contracts. The semispinalus capitis also produces head extension as well as lateral flexion and rotation. It connects to the occipital bone and vertebra of the cervical and thoracic spines. The erector spinae group of muscles consists of several muscles running up and down the spine. These consist of the spinalis, longissimus, iliocostalis, and semispinalis muscles. They are located in the cervical, thoracic, and lumbar spines. The names of these muscles give you a good clue as to their locations. For example, the spinalis muscles are located medially attaching directly to the spinal segments. The iliocostalis muscles attach to the ribs (iliocostalis thoracis) (costal = ribs). The longissimus muscles have long fibers and the semispinalis muscles run just lateral to the spinal segments.

Muscles of the Tongue

The muscles of the tongue include the genioglossus that pulls the tongue to one side when one side contracts and protrudes the tongue when both sides contract (Fig. 9.9). The hyoglossus depresses the tongue, while the styloglossus pulls the tongue superior and posterior.

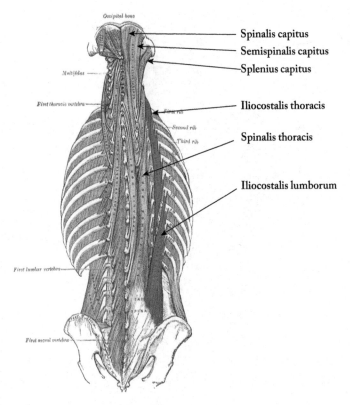

Figure 9.8 Erector spinae muscles.

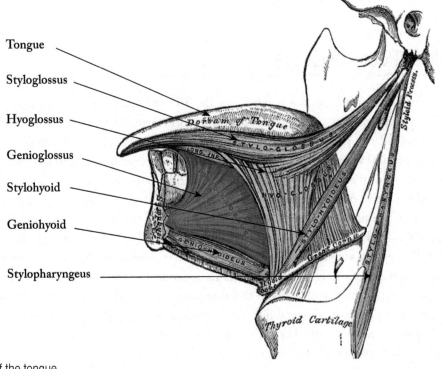

Figure 9.9 Muscles of the tongue.

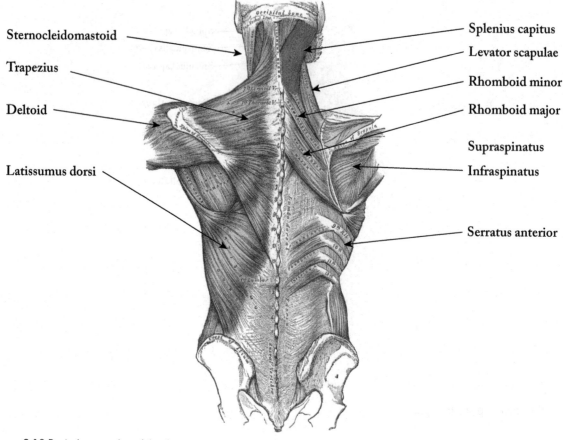

Figure 9.10 Posterior muscles of the thorax.

Shoulder/Pectoral Girdle

The shoulder is anchored by the pectoral girdle. The arm and scapula work together to allow the arm to move. The muscles of the pectoral girdle work to move the scapula in concert with the arm (Figs. 9.10, 9.11, 9.12). The trapezius has upper, middle, and lower divisions. The trapezius attaches to the thoracic and cervical vertebrae and extends upward to the occipital bone and laterally to the scapula. The upper portion raises the shoulder and scapula. The middle portion pulls the scapula toward the vertebral column and the lower portion pulls the scapula downward. The divisions of the scapula are evidenced by the direction of the fibers. There are two rhomboid muscles that pull the scapula upward and medially. The larger rhomboid major is the inferior muscle of the two. The smaller rhomboid minor is superior to the major. The levator scapula is a long thin muscle that attaches to the superior border of the scapula and extends upward to the occipital bone. As its name states, the levator scapula works to elevate the scapula. The serratus anterior attaches to the anterior surface of the scapula and extends to the ribs. The serratus anterior works to hold or stabilize the scapula against the ribcage. The pectoralis minor muscle is located deep to the major. It attaches to the upper ribs and extends to the coracoid process of the scapula. It works to pull the scapula anterior and inferior. The pectoralis minor is also an accessory muscle of inspiration. The deltoid is located on top of the humeral head. It attaches to the spine of the scapula and

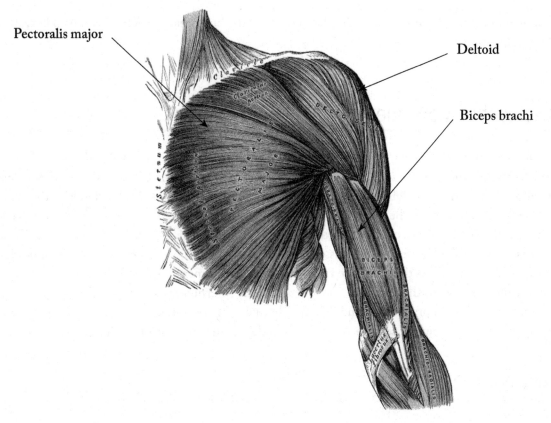

Figure 9.11 Muscles of anterior thorax and arm.

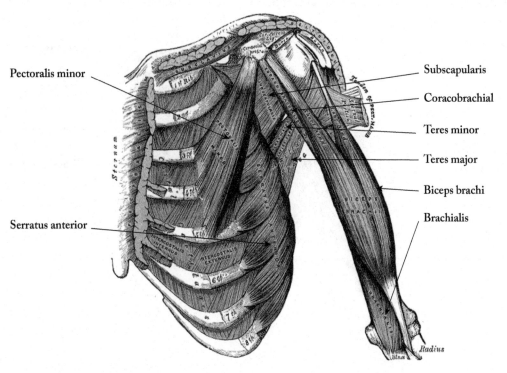

Figure 9.12 Deep muscles of the thorax and shoulder.

acromion process and extends to the deltoid tuberosity of the humerus. The deltoid works to flex, abduct, and extend the arm.

Muscles that Move the Arm

The arm can move into flexion, extension, adduction and abduction, and internal and external rotation, as well as combinations of these movements. The flexors include the coracobrachialis, pectoralis major, and deltoid. The coracobrachialis attaches to the coracoid process of scapula and extends to the shaft of the humerus. It runs deep to the deltoid and biceps muscles. The pectoralis major attaches to the clavicle, sternum, and costal cartilages of ribs and extends to the intertubercular groove of the humerus. The arm extensors include the teres major and latissimus dorsi. The teres major attaches to the lateral border of the scapula and extends to the intertubercular groove of the humerus. The latissimus dorsi attaches to the lower thoracic area to the iliac crest and extends to the intertubercular groove of humerus. The arm abductors include the supraspinatus and deltoid. The supraspinatus attaches to the posterior surface of the scapula above the spine of scapula and extends to the greater tubercle of the humerus.

The Rotator Cuff

The rotator cuff consists of four muscles, three of which are external rotators and one is an internal rotator. The first letter of each muscle can be taken to spell the acronym SITS, which stands for supraspinatus, infraspinatus, teres minor, and subscapularis (Fig. 9.13). The supraspinatus attaches to the superior

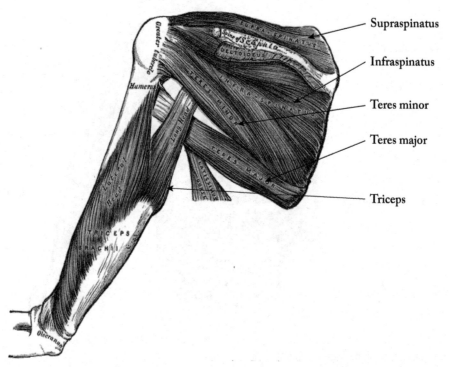

Figure 9.13 Posterior muscles of the arm.

portion of the scapula at the suprascapular fossa and extends to the greater tubercle of the humerus. It abducts as well as externally rotates the arm. The infraspinatus attaches to the posterior portion of the scapula at the subscapular fossa and extends to the greater tubercle of the humerus. It externally rotates the arm. The subscapularis attaches on the anterior surface of the scapula and extends to the lesser tubercle of the humerus. It is the only rotator cuff muscle that provides internal rotation. The teres minor attaches to the lateral border of the scapula and extends to the greater tubercle of the humerus. It externally rotates the arm.

Muscles that Move the Lower Arm

The forearm or antebrachium can move into flexion, extension, and rotation. The flexors include the biceps brachii, brachioradialis, and brachialis. The biceps brachii attaches to the scapula and extends to the radial tuberosity. This muscle has two heads at its proximal region. It works to flex the elbow. The brachialis lies deep to the biceps brachii and extends to the ulna. The brachioradialis attaches to the humerus and extends to the radius. There is only one muscle that functions in elbow extension. This muscle is the triceps brachii. This muscle is a large three-headed muscle that attaches to the scapula and humerus and extends to the ulna. It is the only muscle on the posterior side of the arm. The rotators of the forearm include the supinator, pronator teres, and pronator quadratus. The supinator attaches to the ulna and extends to the lateral aspect of the humerus. It works to move the wrist into supination. The pronator teres attaches to the humerus and ulna and extends to the radius. It works to move the wrist into pronation. The pronator quadratus attaches to the distal ulna and radius. It also works to pronate the wrist.

Hand/Wrist Muscles

The flexors of the hand and wrist include the flexor carpi radialis longus, flexor carpi ulnaris, palmaris longus, flexor digitorum superficialis, and flexor digitorum profundus. The wrist area contains a large number of tendons from muscles that move the wrist and hand. There is a large flat tendon on the palmar aspect of the wrist, known as the flexor retinaculum. A few muscles travel through this structure on their way to the metacarpals and phalanges of the hand (Fig. 9.14). The flexor carpi radialis longus attaches to the medial epicondyle and extends to the metacarpals. The flexor carpi ulnaris also attaches to the medial epicondyle and extends to the metacarpals. The palmaris longus muscle lies between the flexor carpi radialis longus and flexor carpi ulnaris, and extends to the flexor retinaculum of the wrist. The group of wrist flexors attach on the medial epicondyle. The flexor digitorum superficialis lies deep to the flexor carpi radialis longus and flexor carpi ulnaris, and extends to the proximal phalanges. The flexor digitorum profundus lies deep to the flexor digitorum superficialis and extends to the distal phalanges (Fig. 9.15). Some muscles extend through the flexor retinaculum and are prone to carpal tunnel syndrome, which is an inflammation of the median nerve. These muscles:

- Flexor carpi radialis longus and brevis
- Flexor digitorum profundus
- Flexor digitorum superficialis

Figure 9.14 Anterior forearm muscles.

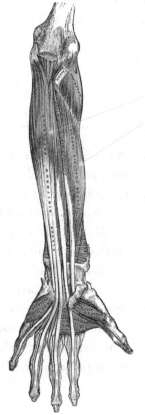

Figure 9.15 Deep anterior forearm muscles.

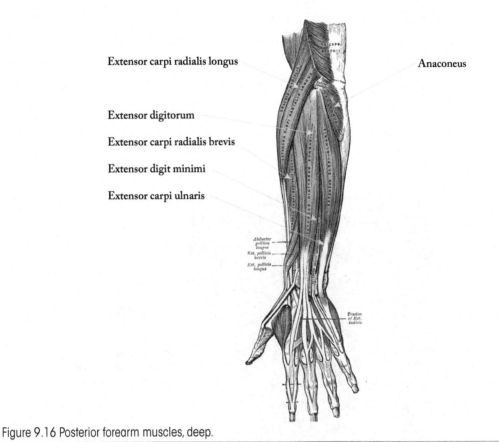

Figure 9.16 Posterior forearm muscles, deep.

The wrist and hand extensors are located on the posterior portion of the forearm (Figs. 9.16, 9.17). The wrist extensors have a common origin on the lateral epicondyle. Wrist extensor tendonitis, known as tennis elbow or lateral epicondylitis, can develop here. The wrist and hand extensors include the extensor carpi radialis longus, extensor carpi radialis brevis, extensor carpi ulnaris, extensor digitorum, and extensor digiti minimi.

Abdominal Wall Muscles

The contents of the abdomen are protected by a band of muscles (Figs. 9.18, 9.19, 9.20). There are three layers of abdominal muscles that include four muscles. The first layer consists of the rectus abdominus, which lies in the anterior and medial aspect of the abdomen, and the external oblique, which is located on the sides of the abdomen. The second layer consists of the internal obliques, which lie deep to the external obliques, and the third layer consists of the transverse abdominus. Some of the abdominal muscles attach to a broad dense band of connective tissue known as the linea alba. The linea alba extends from the xiphoid process to the symphysis pubis. The abdominal muscles aid in trunk flexion. They also compress the contents of the abdominal cavity, increase intra-abdominal pressure, and help to transmit force through the trunk to protect the spine and contents of the abdominal cavity. The transverse

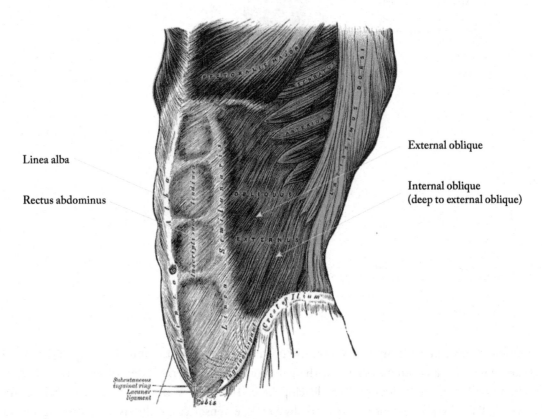

Figure 9.17 Superficial abdominal muscles.

Muscular System | 159

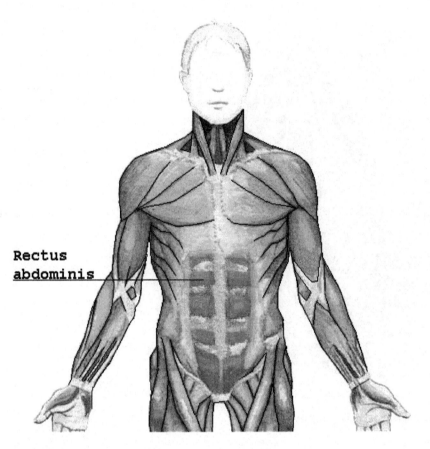

Figure 9.18 Rectus abdominus.

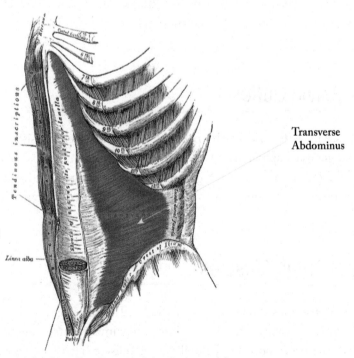

Figure 9.19 The transverse abdominus (the deepest abdominal muscle).

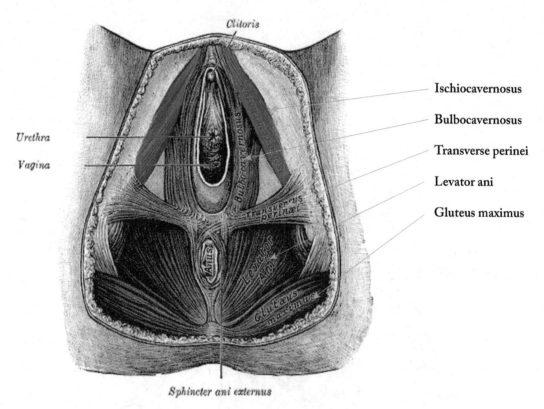

Figure 9.20 Muscles of the pelvic outlet (female).

abdominus muscle is a very important muscle in rehabilitation of low back injuries. This muscle acts as a natural back brace since its fibers run in a transverse plane.

Muscles of the Pelvic Outlet

The floor of the pelvic cavity is formed by the pelvic diaphragm, which consists of a layer of muscles (Figs. 9.21, 9.22). The urogenital diaphragm lies superficial to this layer and forms a second layer. The pelvic diaphragm consists of the levator ani and coccygeus muscles. The urogenital diaphragm consists of the superficial transverses perinea, bulbospongiosus (males only), ischiocavernosus, and the sphincter urethrae.

Pelvic and Upper Thigh Muscles

Muscles that move the thigh connect to the pelvis and femur. The anterior muscles include the iliopsoas and iliacus and the posterior muscles include the gluteus maximus, gluteus medius, gluteus minimus, and tensor fasciae latae. The psoas portion of the iliopsoas muscle has two divisions. The psoas major attaches to the lower lumbar vertebra and extends to the lesser trochanter of the femur. The psoas

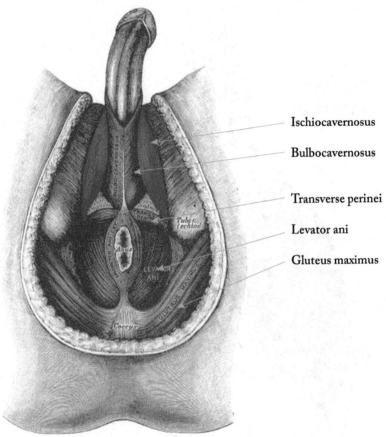

Figure 9.21 Pelvic outlet muscles (male).

minor muscle is smaller and inserts on the pubic bone. The iliacus muscle attaches to the ilium and also extends to the lesser trochanter of the femur. Since both the psoas major and iliacus share a common insertion point, they are often referred to as the iliopsoas. The iliopsoas works to flex the hip. The gluteus maximus is one of the strongest and largest muscles of the body (Fig. 9.23). It attaches to the iliac crest, sacrum, coccyx, and the aponeurosis of the sacrospinalis. It extends to the linea aspera of the femur and the iliotibial band. It works to produce hip extension. The gluteus medius lies deep to the gluteus maximus. It attaches to the ilium and extends to the greater trochanter of the femur. It works to produce hip abduction and extension. The gluteus minimus lies deep to the gluteus medius. It is the smallest gluteal muscle. The tensor fasciae latae is located on the lateral aspect of the thigh. It attaches to the iliac crest and extends to a band of dense connective tissue called the iliotibial tract or band. The iliotibial band extends down the lateral aspect of the femur to the tibia. It is a flat tendon or aponeurosis. Tendonitis can develop in this tendon in a condition known as iliotibial band syndrome. Deep muscles in the posterior pelvic area include the piriformis, obturator internus, obturator externus, superior and inferior gemellus, and quadratus femoris muscles. All of these muscles work to externally rotate and abduct the hip (Figs. 9.24, 9.25). Muscles on the proximal medial aspect of the thigh include the adductor longus, adductor brevis, adductor magnus, pectineus, and gracilis. These muscles attach to the pubic bone and extend down the thigh to various insertion points on the femur. They work to adduct the hip.

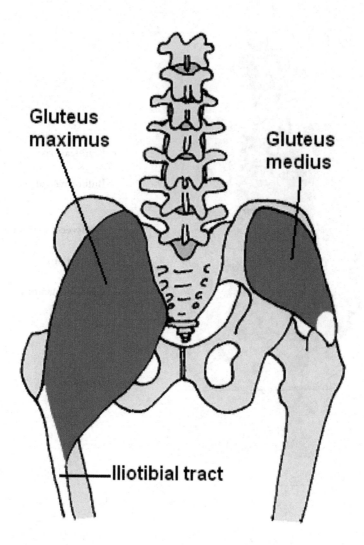

Figure 9.22 Gluteal muscles. The gluteus medius is deep to the gluteus maximus.

Anterior Thigh Muscles

The large muscles on the anterior portion of the thigh include the sartorius and quadriceps group (Fig. 9.26). The sartorius (tailor's muscle) attaches to the anterior superior iliac spine and extends from lateral to medial across the thigh to insert on the medial aspect of the upper tibia. This muscle has multiple actions including flexion, abduction, and external rotation of the hip. The quadriceps group consists of the rectus femorus, vastus medialis, vastus lateralis, and vastus intermedius. The quadriceps muscles work together to produce knee extension. The rectus femorus is located in the middle of the thigh. It attaches to the anterior superior iliac spine and extends inferiorly to the patella. The vastus medialis is located in the medial aspect of the thigh. It attaches to the linea aspera of the femur and extends to the patella. The vastus lateralis is located in the lateral aspect of the thigh. It attaches to the greater trochanter of the femur and extends to the patella. The vastus intermedius lies deep to the rectus femorus. It attaches to the femur and extends to the patella. All of the quadriceps muscles

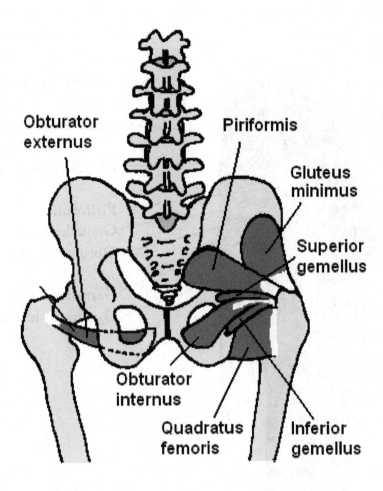

Figure 9.23 Deep muscles of the posterior pelvis.

have a common insertion point on the patellar ligament. The patellar ligament inserts on the tibial tuberosity.

Posterior Thigh Muscles

The posterior thigh muscles include the hamstring group (Fig. 9.27). The hamstrings consist of three muscles: the biceps femorus, semimembranosus, and semitendinosus. The hamstrings work to produce knee flexion. The biceps femorus is a two-headed muscle. The long head attaches to the ischial tuberosity and the short head attaches to the linea aspera and lateral supracondylar line of the femur. The muscle then extends inferiorly to attach to the head of the fibula. The semimembranosus attaches to the ischial tuberosity and extends inferiorly to attach to the medial condyle of the tibia and lateral condyle of the femur. The semitendinosus attaches to the ischial tuberosity and extends inferiorly to attach to the medial aspect of the upper tibia.

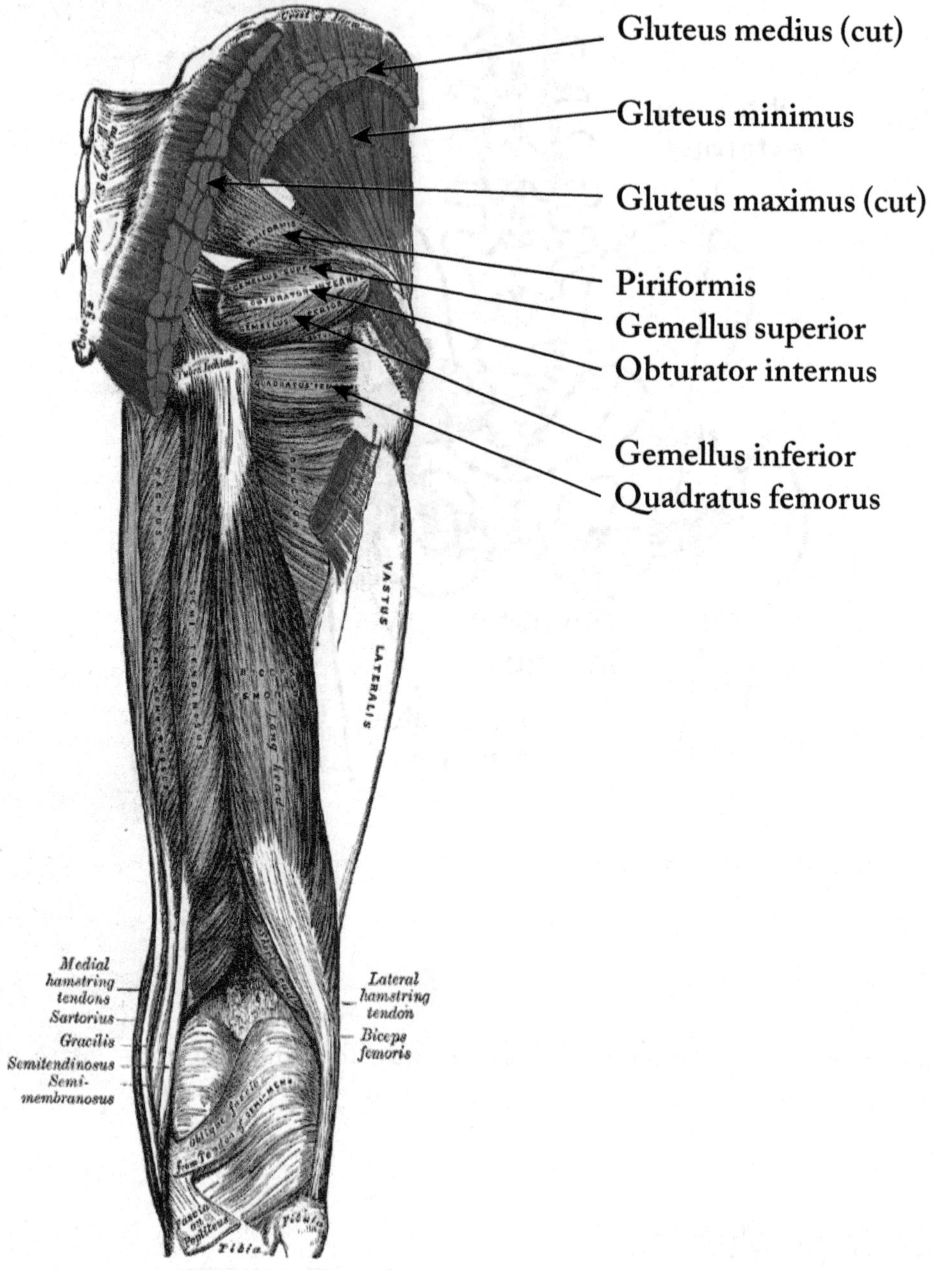

Figure 9.24 Deep muscles of the posterior pelvis.

Posterior Knee

Located in the posterior portion of the knee is the popliteus muscle. If the femur is fixed, the popliteus works to internally rotate the tibia. If the tibia is fixed, it works to externally rotate the femur.

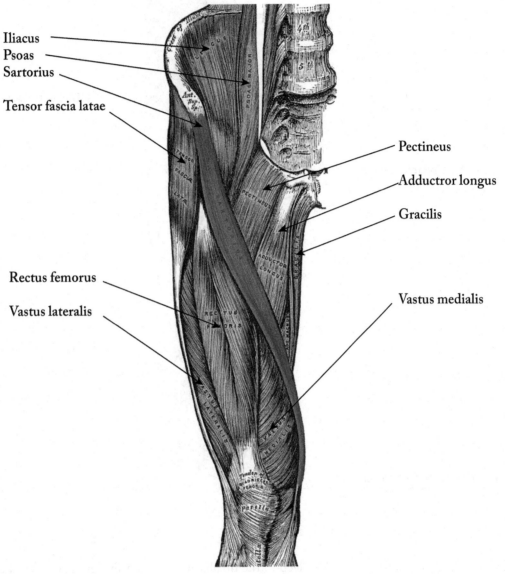

Figure 9.25 Muscles of the anterior thigh.

Muscles of the Anterior Leg

The muscles of the anterior portion of the leg work to dorsiflex the foot (Fig. 9.28). These include the tibialis anterior, extensor hallucis longus, extensor digitorum longus, and peroneus tertius. The tibialis anterior is located just lateral to the tibia. It attaches to the lateral condyle of the tibia, the lateral aspect of the proximal portion of the tibia, and the interosseous membrane that connects the tibia and fibula. It extends downward to attach to the medial cuneiform and first metatarsal. The tibialis anterior is involved in shin splints. The extensor hallucis longus lies deep to the tibialis anterior. It attaches to the anterior aspect of the fibula and interosseous membrane and extends downward to attach to the first distal phalanx. Besides being a synergist for dorsiflexion of the foot, it also extends the big toe.

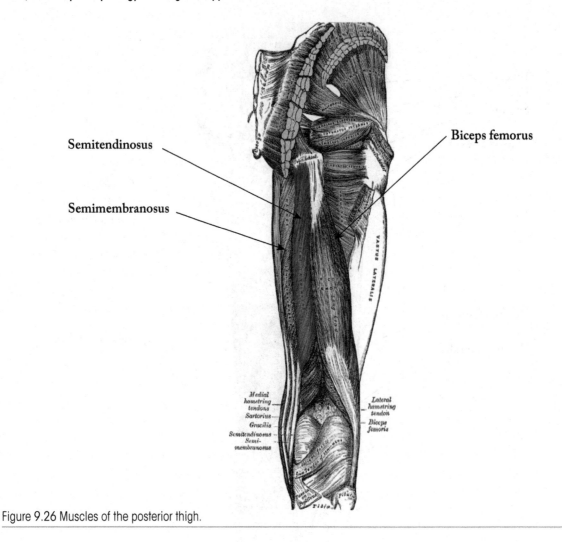

Figure 9.26 Muscles of the posterior thigh.

The extensor digitorum longus also lies deep to the tibialis anterior. It attaches to the lateral condyle of the tibia, shaft of the fibula, and interosseous membrane. It works as a synergist in dorsiflexion of the foot and extends the toes. It also works to tighten the plantar aponeurosis. The peroneus tertius is part of the peroneal group that includes the peroneus longus and peroneus brevis. This muscle works to dorsiflex and evert the foot. It attaches to the medial surface of the lower portion of the fibula and extends to the fifth metatarsal. The peroneal group works together to evert the foot.

Muscles of the Posterior and Lateral Leg

The muscles of the posterior leg work to plantarflex the foot (Fig. 9.29). These include the gastrocnemius, soleus, flexor digitorum longus, and tibialis posterior. The gastrocnemius is a two-headed muscle that crosses both the knee and ankle joints. Its action in the knee is to help with knee flexion. It also works to produce ankle plantarflexion. It attaches to the femoral condyles and posterior surface of the distal femur and extends downward to attach to the calcaneus. The soleus lies deep to the gastrocnemius. It attaches

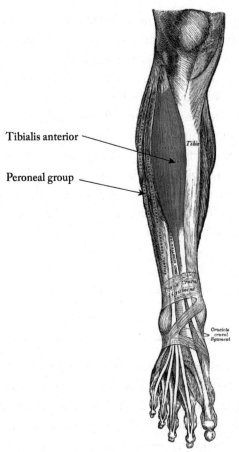

Figure 9.27 Anterior lower leg muscles.

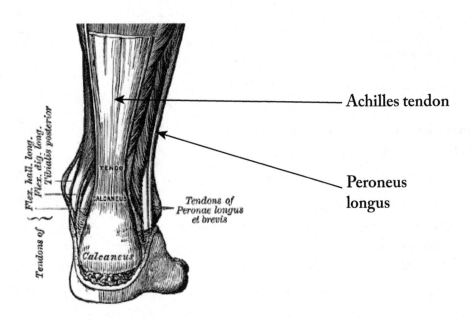

Figure 9.28 The gastrocnemius and soleus attach to the Achilles tendon. The peroneus longus is on the lateral aspect of the leg.

to the posterior aspect of the proximal fibula and tibia and extends downward to attach to the calcaneus. The soleus crosses only the ankle joint and produces ankle plantarflexion. The gastrocnemius and soleus both insert on the large Achilles (calcaneal) tendon and are known collectively as the triceps surae. The tibialis posterior is also a deep muscle of the posterior leg. It attaches to the posterior proximal surface of the tibia and fibula and extends downward to attach to the navicular, medial cuneiform, and second through fourth metatarsals. It works to produce plantarflexion and also helps to control pronation of the foot while walking. The flexor digitorum longus is a deep muscle of the posterior leg. It attaches to the posterior surface of the tibia and extends downward to attach to the second through fifth distal phalanges. It works to flex the toes and stabilizes the metatarsal heads. The peroneus longus is located on the lateral aspect of the lower leg. It attaches to the tibia and fibula and extends to the medial cuneiform and first metatarsal. The flexor hallucis longus is a deep muscle on the lateral aspect of the leg. It attaches to the distal portion of the fibula and interosseous membrane and extends to attach to the big toe. It works to flex the big toe.

Muscles of the Foot

The top of the foot is known as the dorsum of the foot (Figs. 9.30, 9.31). The only muscle located exclusively on the dorsum of the foot is the extensor digitorum brevis. This muscle attaches to the calcaneus and extensor retinaculum of the ankle and extends to the big toe and tendons of the extensor digitorum longus. It works to produce extension of the toes. The bottom, or sole, of the foot is known as the plantar region of the foot. This area contains four layers of muscles.

The layers from superficial to deep:

Layer 1:

- Flexor digitorum brevis
- Abductor hallucis
- Abductor digiti minimi

Layer 2:

- Quadratus plantus
- Lumbricales

Layer 3:

- Adductor hallucis
- Flexor digiti minimi brevis
- Flexor hallucis brevis

Layer 4:

- Dorsal interossei
- Plantar interossei

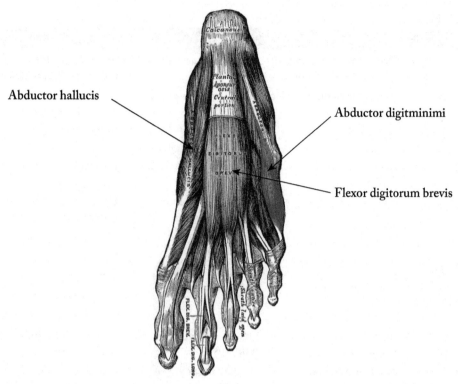

Figure 9.29 Dorsum of foot.

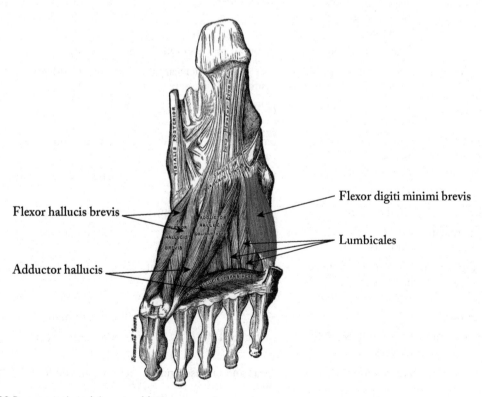

Figure 9.30 Deep muscles of dorsum of foot.

Table of Select Muscles

Muscle	Origin	Insertion	Action
Trapezius	Occipital bone, spines of C7 and all T vertebrae	spine and acromion of scapulaula	Extends head; elevates, depresses, rotates and retracts scapulaula
latissimus dorsi	lower vertebrae, iliac crest	intertubercular groove of humerus	Extends, adducts and medially rotates arm
serratus anterior	upper 8 ribs	anterior aspect of medial border of scapulaula	Protracts and rotates scapulaula
rhomboideus	spinous process of C1–T5	medial border of scapulaula	Retracts and rotates scapulaula
pectoralis major	clavicle, sternum, costal cartilages	greater tubercle of humerus	Flexes, adducts and medially rotates arm
pectoralis minor	ribs 3,4,5	coracoid process of scapulaula	Draws scapulaula anteriorly and inferiorly
acromiodeltoideus, spinodeltoideus and clavodeltoideus (deltoids)	clavicle, acromion and scapulaular spine	deltoid tuberosity of humerus	abducts arm
supraspinatus	supraspinous fossa of scapulaula	greater tubercle of the humerus	abducts and stabilizes humerus
infraspinatus	infraspinous fossa	greater tubercle of the humerus	lateral rotation of humerus
triceps brachii	axillary border of scapulaula, posterior humerus	olecranon process of the ulna	extends forearm, stabilizes shoulder
biceps brachii	coracoid process, intertubercular groove of the humerus	radial tuberosity	flexes arm and forearm, supinates hand
brachialis	distal anterior humerus	coronoid process of ulna	flexes forearm
flexor carpi ulnaris	medial epicondyle of humerus, olecranon process	base of 5th metacarpal, pisiform and hamate	flexes wrist, adducts hand
palmaris longus	medial epicondyle of humerus	palmar aponeurosis	flexes wrist
flexor carpi radialis	medial epicondyle of humerus	base of 2nd and 3d metacarpals	flexes wrist, abducts hand
pronator teres	medial epicondyle of humerus	lateral radius	pronates and flexes forearm
brachioradialis	distal humerus	styloid process of radius	flexes forearm
extensor carpi radialis	lateral epicondyle of humerus	metacarpals II and III	extends and abducts wrist
extensor digitorum communis	lateral epicondyle of the humerus	posterior surfaces of distal phalanges of digits 2–5	Extends fingers and wrist, adducts fingers
extensor digitorum	--	--	extends 5th digit
extensor carpi ulnaris	lateral epicondyle of humerus	metacarpal V	extends and adducts wrist
masseter	zygomatic arch	angle and ramus of mandible	elevates mandible
mylohyoideus	mandible	Hyoid	elevates hyoid

Muscle	Origin	Insertion	Action
digastricus	mandible and mastoid process	hyoid bone	elevates hyoid and depress mandible (open mouth)
sternohyoideus	manubrium and clavicle	hyoid bone	depresses hyoid and larynx
sternomastoideus	manubrium, clavicle	mastoid process	singly, rotates head to opposite shoulder; together, flexes head
oblique	lower 8 ribs	iliac crest and linea alba	flexion and rotation at waist
internal oblique	lumbodorsal fascia	lower 4 ribs	flexion and rotation at waist
transverse abdominis	iliac crest, cartilages of lowest ribs	linea alba and pubic crest	compresses abdominal wall
rectus abdominis	pubic crest and pubic symphysis	ribs 5–7 and xiphoid process	flexion at waist
tensor fascia latae	iliac crest and anterior superior iliac spine	iliotibial tract	flexes, abducts and medially rotates thigh
gluteus medius	ilium	greater trochanter of the femur	abduction and medial rotation of thigh
gluteus maximus	ilium, sacrum, coccyx	iliotibial tract, gluteal tuberosity of femur	extension and lateral rotation of thigh
sartorius	anterior superior iliac spine	tibia	flexes, abducts and laterally rotates thigh; flexes lower leg
gracilis	pubis	medial tibia	adducts thigh, flexes and medially rotates leg
adductor femoris	ischium and pubis	linea aspera of femur	adducts, flexes and laterally rotates thigh
biceps femoris	ischial tuberosity and femur	tibia and fibula	extends thigh and flexes lower leg
semitendinosus	ischial tuberosity	medial aspect of proximal tibia	extends thigh, flexes lower leg
semimembranosus	ischial tuberosity	medial condyle of tibia	extends thigh, flexes lower leg
vastus lateralis	linea aspera	patella and tibial tuberosity	extends lower leg, stabilizes knee
vastus medialis	linea aspera	patella and tibial	extends lower leg
vastus intermedius	proximal femur	patella and tibial tuberosity	extends lower leg
rectus femoris	anterior inferior iliac spine	patella and tibial tuberosity	extends knee, flexes thigh
gastrocnemius	medial and lateral condyles of femur	via Achilles tendon onto calcaneal tendon	flexes lower leg, plantarflexes foot
tibialis anterior	lateral condyle and tibial shaft	first cuneiform and first metatarsals	dorsiflexes and inverts foot
soleus	head of fibula and tibia	calcaneal tendon onto calcaneus	plantarflexes foot
fibularis longus	head of fibula	first metatarsal	plantar flexion
extensor digitorum longus	posterior tibia	distal phalanges of toes 2–5	Extend toes 2–5 and dorsiflexes ankle

Physiology of the Muscular System

How does a muscle contract? In order to answer this question we must first examine what tells a muscle to contract. Let's say you are sitting here reading and want to pick up a cup of coffee. In order to do so, you must send a command to the muscles in your arm. The command comes from a thought generated in your nervous system. The command travels from your brain to your spinal cord to a nerve that attaches to a muscle in your arm. The command tells your muscle to contract and your arm responds by moving closer to the coffee.

Muscles are made of protein. If we were to examine a skeletal muscle under a microscope, we would see that it is composed of tiny protein fibers or filaments. When a muscle receives a command from the nervous system to contract, the protein filaments slide past each other. In fact one of the filaments connects to the other and drags it along. Think of thousands of overlapping filaments sliding past each other as the muscle contracts. The command to contract must somehow get from the outside of the muscle to the inside. Tiny messengers called neurotransmitters bring the message from the nerve to the muscle. The message is then passed on by other chemical messengers that tell the protein filaments to contract. Muscles need energy to contract. Muscles must have some sort of power source in order to power the sliding filaments. The energy comes from ATP. ATP connects to one type of filament and extracts the energy so that it can pull the other filament along.

Muscle Structure

Before we can get into the details of how muscles contract, we must examine the microscopic structure of muscles. If we look at muscle tissue under a microscope, we will see that it consists of long cells and has light and dark areas. We say that the muscle has a striated appearance. The striations denote contractile units (Fig. 9.32).

If we look at the muscle down its long axis, we see that it consists of bundles within bundles. The outermost layer consists of connective tissue called fascia. The fascia continues along muscle to become tendons. The tendons connect the muscle to the bone. Deep to the fascia we see a layer of dense connective tissue covering the entire muscle. This layer is called the epimysium. Deep to the epimysium we see structures that look like bundles. These bundles are known as fascicles. Each fascicle consists of an outer connective tissue layer called the perimysium. Inside the perimysium are even smaller bundles of muscle fibers. Surrounding each muscle fiber is a layer of connective tissue called the endomysium. The muscle fibers consist of smaller protein filaments surrounded by plasma membrane, known as a sarcolemma (Fig. 9.33).

Muscle Fiber

Skeletal muscle cells are surrounded by a membrane called the sarcolemma and contain many nuclei. Inside the membrane are myofibrils packed with protein filaments. The myofibril extends along the entire length of the muscle. The myofibrils contain two types of myofilaments known as actin and myosin that overlap each other (Fig. 9.34). The actin or thin filament consists of a core fibrous protein called F actin

Muscular System | 173

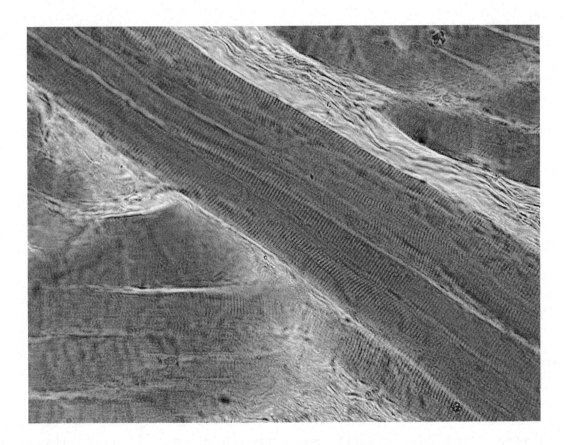

Figure 9.31 Skeletal muscle is characterized by long cells and has a striated appearance.

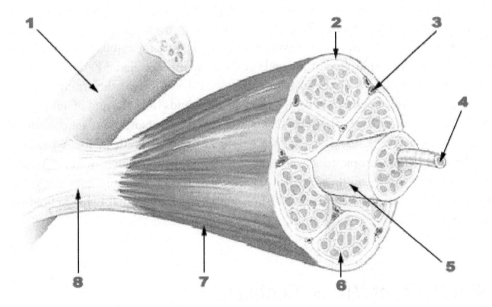

1. Bone
2. Perimysium
3. Blood vessel
4. Muscle fiber
5. Fascicle
6. Endomysium
7. Epimysium
8. Tendon

Figure 9.32 Structure of skeletal muscle.

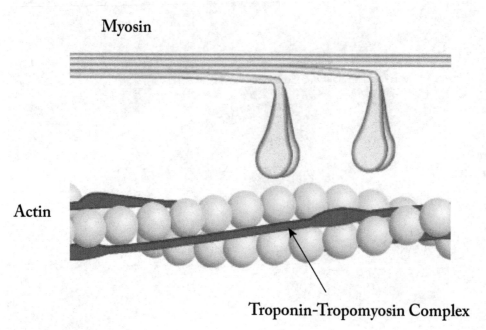

Figure 9.33 Actin and myosin.

twisted into a double-helix arrangement. The actin contains a binding site consisting of a polymer called G actin for the other protein filament called myosin. Surrounding each actin molecule is a complex of troponin and tropomyosin molecules. The tropomyosin covers the myosin-binding sites on the actin. Myosin molecules are known as thick filaments. Myosin contains a double-helix shaft portion and two globular protein heads. The heads can attach to the myosin-binding sites on the actin as well as use ATP. The myosin heads have APTase activity and can liberate the phosphate from ATP to release energy. The myosin heads also have a region that acts like a hinge, allowing myosin to bend.

The contractile unit in muscle is known as the sarcomere (Fig. 9.35). Protein connects the actin and myosin filaments together. This protein located at the ends of the filaments is called a Z-disc. The actin filaments directly attach to the Z-disc, while the myosin attaches via titin protein. The arrangement of overlapping actin and myosin creates a number of bands. The I-band extends from the end of one myosin filament to the other, including the Z-disc. The I-bands are also called light bands and are considered isotropic (equal in all directions). The A-bands extend the length of the myosin filaments. A-bands or dark bands are considered to be anisotropic (unequal in all directions). In the center of each A-band lies an area consisting only of myosin. This area is called the H-zone. The H-zone also contains a dark line running down the middle called the M-line. The M-line consists of protein that helps to hold the myosin filaments in place (Fig. 9.36).

Sliding Filament Model of Muscle Contraction

The essence of the sliding filament model of muscle contraction is the action of actin and myosin sliding past each other. When this happens, the sarcomere shortens and the muscle contracts. The process begins when a command or impulse is sent down a neuron that connects to muscle called a motor neuron.

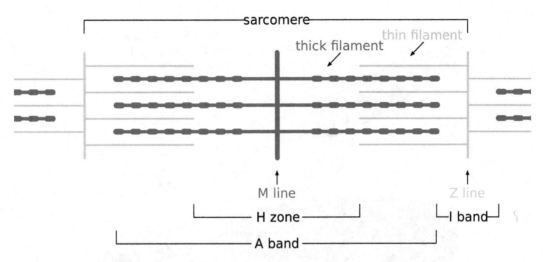

Figure 9.34 Diagram of a sarcomere.

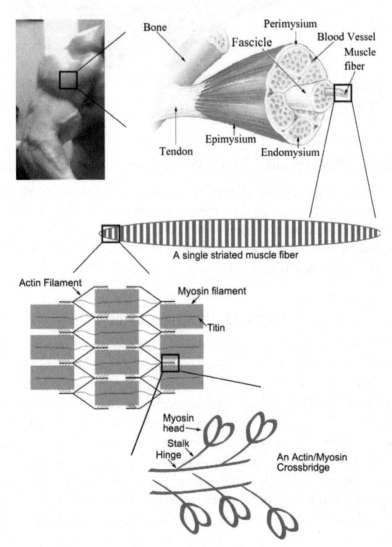

Figure 9.35 Skeletal muscle structure.

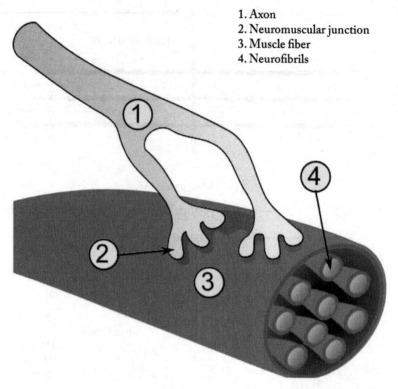

Figure 9.36 Motor neuron and muscle.

1. Axon
2. Neuromuscular junction
3. Muscle fiber
4. Neurofibrils

Step 1: Motor Neuron Sends Message to Muscle to Contract

The motor neuron releases a message in the form of a neurotransmitter to the muscle to tell it to contract. The neurotransmitter floats across an area between the neuron and muscle called the synaptic cleft (Fig. 9.37). The muscle side of the synaptic cleft is called the motor end plate. The sarcolemma is enfolded at the motor end plate in order to increase the surface area. The neurotransmitter involved in skeletal muscle contraction is acetylcholine (Fig. 9.38).

Step 2: Muscle Depolarizes

Muscle cells exist at a negative membrane potential or voltage. This negative potential in muscle cells is called resting membrane potential. Resting membrane potential is established by the various concentration gradients of electrolytes. The resting membrane potential of muscle (and nerve) cells is between -70mV and -90mV (mV = millivolts). When the acetylcholine floats across the synaptic cleft to the motor end plate, it attaches to receptors on transport proteins on the motor end plate. The transport proteins are sodium channels that are controlled by the acetylcholine. These are called ligand-gated channels because when the ligand (acetylcholine) attaches to the channel, the channel responds by opening and letting sodium into the cell. Since there is more sodium outside the cell than inside, opening the channel causes sodium to rush into the cell. This causes the voltage to change since sodium is positively charged. The cell's potential changes and becomes less negative (more positive). We say the cell is depolarizing.

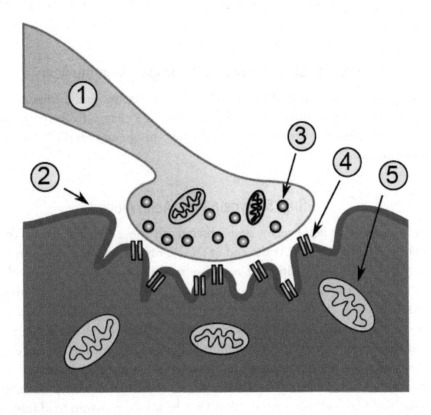

Figure 9.37 Neuromuscular junction.
1. Presynaptic terminal
2. Sarcolemma
3. Synaptic vesicles
4. Acetylcholine receptors
5. Mitchondrion

Step 3: Release of Calcium by the Sarcoplasmic Reticulum

The sarcolemma surrounding the muscle cell contains tube-like structures called T-tubules. The T-tubules reach into the muscle fiber and encircle the sarcomere. Since the T-tubule connects to the outside of the cell, it is filled with extracellular fluid. Between T-tubules lies a specialized type of endoplasmic reticulum called the sarcoplasmic reticulum. The sarcoplasmic reticulum is a network of membranous channels called cisternae. Cisternae near the T-tubules are wider and called terminal cisternae. A tubule and the two adjacent terminal cisternae are called a triad. The sarcoplasmic reticulum actively transports calcium, so it contains a high concentration of calcium. The concentration of calcium inside the sarcoplasmic reticulum is 2,000 times greater than inside the muscle cell. So a significant calcium gradient exists between the sarcoplasmic reticulum and the inside of the muscle cell. The sarcoplasmic reticulum responds to the depolarization of the muscle cell by opening calcium channels in the terminal cisternae of the sarcoplasmic reticulum. When these channels open, calcium rushes into the sarcoplasm of the muscle cell. This process is called excitation-contraction coupling.

Step 4: Calcium Binds to the Troponin on the Actin

Calcium rushes into the sarcoplasm of the muscle cell and attaches to the troponin on the troponin-tropomyosin complex wrapped around the actin. This causes a change in the position of the troponin that exposes the myosin binding site on the actin. The myosin can now bind with actin, forming what is known as a cross-bridge.

Step 5: Myosin Pulls Actin Along

Myosin can now move at its hinge region and subsequently move the actin along. This results in actin and myosin sliding past each other. At the end of a cycle of movement, the myosin must release from actin and return to its original position. It can now repeat the cycle and bind with another site on the actin. The cycle consists of cross-bridge formation, movement, release, and myosin's return to its original position. This cycle is called cross-bridge cycling. The energy needed for one cross-bridge cycle is provided by one ATP molecule. ATP binds to the myosin head, which has ATPase activity. The ATP decomposes into ADP and a phosphate. Once the calcium attaches to troponin and exposes the binding site, the myosin moves and binds to actin while releasing the phosphate and extracting the energy from the phosphate bond. ADP is released from the myosin head when myosin pulls actin along. Another ATP must again bind to the myosin head to allow for release of the myosin head from actin. ATP binds to the myosin head and decomposes into ADP and phosphate, which remain on the myosin head. The myosin head now releases from actin and resumes its resting position with the ADP and phosphate still on it. The energy from the ATP is stored in the myosin head. Movement of the myosin head while it is attached to actin is called the power stroke, while movement of the myosin head back to its original position is called the recovery stroke. Resting muscles store energy from ATP in the myosin heads while they wait for another contraction.

Muscle Twitch

A twitch occurs when one muscle fiber contracts in response to a command (stimulus) by the nervous system. The time between the activation of a motor neuron until the muscle contraction occurs is called the lag phase (sometimes called the latent phase). During the lag phase a signal called an action potential moves to the end of the motor neuron (axon terminal). This results in release of acetylcholine and depolarization of the motor end plate. The depolarization results in the release of calcium by the sarcoplasmic reticulum and subsequent binding of calcium to troponin, which causes the myosin binding site to be exposed. This is followed by the actual muscle contraction that develops tension in the muscle. This next phase is called the contraction phase. During the contraction phase the cross-bridges between actin and myosin form. Myosin moves actin, and releases and reforms cross-bridges many times as the sarcomere shortens and the muscle contracts. ATP is used during this phase and energy is released as heat. When the muscle relaxes, the tension decreases. This phase is called the relaxation phase. During this phase calcium is actively transported back into the sarcoplasmic reticulum using ATP.

The troponin moves back into position, blocking the myosin-binding site on the actin, and the muscle passively lengthens.

Muscle Stimulus and Contraction Strength

A skeletal muscle fiber will produce a given amount of force if the stimulus is strong enough to reach the threshold for muscle contraction. This is called the all-or-none law. Let's say that we are electrically stimulating a muscle fiber. We begin with a low amount of stimulation that does not reach the threshold to produce a contraction. The muscle fiber will respond by remaining relaxed; it will not contract. Now if we increase the stimulation so that enough is produced to reach the threshold, the muscle fiber will respond by contracting. Finally if we continue to increase the stimulus so that it well exceeds the threshold, the fiber will respond by contracting with the same force as when we just reached the stimulus. The muscle will not contract with greater force if the stimulus is greater. The muscle responds to stronger stimuli by producing the same force. In skeletal muscles a motor neuron can innervate many muscle fibers. This is called a motor unit. There are numerous motor units throughout skeletal muscles. Motor units act in a coordinated fashion. One stimulus will affect all of the muscle fibers innervated by a given motor unit. Whole muscles containing many motor units can contract with different amounts of force. More motor units are recruited to increase the force of contraction when needed. This phenomenon is called summation. In other words, increasing numbers of motor units are activated in order to increase the muscle's force of contraction.

Let's look at an example. Let's say that I am helping a friend move to a new house. I am holding an empty box while my friend fills it up with various items. The weight of the box or "load" is increasing. My biceps muscles must respond by increasing their force of contraction so that I will avoid dropping the box. As the load increases, more motor units are recruited, and the force of contraction increases to accommodate the load. Nerves contain many axons of neurons that innervate many motor units. If a nerve is stimulated to produce a stimulus that is below the threshold, no action potential is generated in the neurons and there is no muscle contraction. This is called a subthreshold stimulus. If the stimulus is strong enough to produce an action potential, we say that the stimulus is a threshold stimulus. As the stimulus increases, more motor units are recruited. We call this stimulus a submaximal stimulus. When the stimulus is strong enough to cause activation of all of the motor units associated with the nerve, we say that the stimulus is a maximal stimulus. A stimulus greater than a maximal stimulus (supramaximal stimulus) will not have any additional effect on the motor units. The ratio of neurons to muscle fibers differs in various muscles. Muscles involved in more precise movements, such as in the hands, have a smaller ratio of neurons to muscle fibers, whereas muscles involved in gross movements, such as the muscles in the thigh, have a higher number of fibers innervated by one neuron.

Muscle Contraction Frequency

When a muscle is stimulated by an action potential, it will contract. The time it takes for an action potential to occur is much shorter than the time it takes to contract a muscle. This means that another action potential can produce another contraction. As the frequency of action potentials increases, the frequency of muscle contraction also increases. There is a maximal frequency of action potentials that will cause a sustained contraction of a muscle. We call this phenomenon tetanus. Muscles in tetanus will not demonstrate even a partial relaxation. The tension produced by muscles increases along with the frequency of stimulation by action potentials. This phenomenon is known as multiple-wave summation.

During muscle contraction, calcium is released by the sarcoplasmic reticulum in response to depolarization of the sarcolemma. If a high frequency of action potentials is administered to the muscle, the calcium levels inside the cell remain high and the muscle responds by remaining in a contracted state. This allows for more cross-bridge formation and subsequent increase in tension of the muscle.

If a muscle is stimulated by an action potential and then allowed to relax, the next stimulus will produce a stronger contraction. This will continue for a few contractions, then the strength of contraction will level out. This phenomenon is called treppe.

Muscle Length-Tension Relationship

The length of a muscle is related to the tension generated by the muscle. Muscles will generate more force when stretched beyond their resting length to a point. Muscles stretched beyond this point will produce less tension. If the muscle is at its resting length, it will not produce maximal tension because the actin and myosin filaments excessively overlap. Myosin filaments can extend into the Z-discs and both filaments interfere with each other, limiting the number of cross-bridges that can form.

If the muscle is stretched, to a point, the tension will increase in the muscle. The actin and myosin filaments can now optimally overlap so that the greatest number of cross-bridges can form. If the muscle is overstretched, the tension will decrease. The actin and myosin filaments do not overlap causing a decrease in the number of cross-bridges that can form.

Types of Muscle Fibers

There are three major types of skeletal muscle fibers. These are called fast twitch, slow twitch and intermediate. Generally, fast-twitch fibers generate high force for brief periods of time. Slow-twitch fibers generate lower amounts of force but can do so for longer periods of time. Intermediate fibers have some characteristics of both fast- and slow-twitch fibers. Fast-twitch fibers are also called Type II fibers. Fast-twitch fibers are the predominant fibers in the body. They respond quickly to stimuli and can generate a good deal of force. They have a large diameter due to the large number of myofibrils. Their activity is fueled by ATP generated from anaerobic metabolism. Slow-twitch fibers respond much more slowly to stimuli than fast-twitch fibers do. They are smaller in diameter and contain a large number of mitochondria. They are capable of sustaining long contractions and obtain their ATP from aerobic metabolism. Slow-twitch fibers are surrounded by capillary networks that supply oxygenated blood for use in the aerobic energy systems. They also contain a red pigment called myoglobin. Myoglobin can bind oxygen (like hemoglobin) and provide a substantial oxygen reserve. Because of the reddish color of myoglobin, these fibers are often called red muscle fibers. Slow-twitch fibers are also called Type I fibers.

Intermediate fibers resemble fast-twitch fibers because they contain small amounts of myoglobin. They also have a capillary network around them and do not fatigue as readily as fast-twitch fibers.

They contain more mitochondria than fast-twitch but not as many as slow-twitch fibers. The speed of contraction and endurance also lie between fast- and slow-twitch fibers. Intermediate fibers are also called Type IIa fibers. Muscles that have a predominance of slow fibers are sometimes referred to as red muscles, such as in the back and areas of the legs. Likewise muscles that have a predominance of fast fibers are referred to as white muscles. It is interesting to note that there are no slow-twitch fibers in the eye muscles or muscles of the hands. The ratio of fast-slow-intermediate fibers is determined genetically. However, training can change the ratio of these fibers in skeletal muscles that contain all three types. For example, training for endurance can cause some fast-twitch fibers to become more like intermediate fibers.

Muscles Response to Exercise

There are three basic ways the muscular system responds to exercise. Let's look at this in the context of Jen who is beginning an exercise program. She has never been in a gym before and is excited to see the results of her efforts. Part of her program is weight lifting. Her trainer tests her on the first day and finds that she can lift 20 lbs. in a biceps curl. She then begins exercising three times per week. After about two weeks she finds that she can now lift 25 lbs. She is excited about her improvement in just two weeks of training. Jen asks her trainer to measure her biceps and they find that there is no difference in size. If the muscle size has not changed, then what is responsible for Jen's increase in strength?

One of the first ways muscles respond to training is to increase synchronous contraction of motor units. When motor units contract at different points in time (asynchronous contraction), then they cannot generate as much force as when they contract together. Training increases synchronous contraction so that the motor units work together to generate higher amounts of force. Jen continues her program and finds that after about 8–10 weeks, there is some increase in her biceps circumference. This is due primarily to hypertrophy or an increase in the cross-sectional diameter of muscle fibers. The number of muscle fibers does not change, but the size of the fibers increases. The number of protein filaments, mitochondria, and enzymes, and glycogen reserves increases. Jen may also experience some small amount of hyperplasia. Hyperplasia is an increase in the number of muscle fibers resulting from mitosis. The increase is slight, as most of the increase in size is attributed to hypertrophy.

Cardiac Muscle

Cardiac muscle is found only in the heart. Like skeletal muscle, it has a high concentration of myofilaments and is striated. However, there are also a number of structural differences between skeletal and cardiac muscles. Cardiac muscles are smaller and generally contain one nucleus, whereas skeletal muscles are multinucleated. They have a different arrangement of T-tubules and no triads. The sarcoplasmic reticulum does not have terminal cisternae. Cardiac muscle fibers are powered by aerobic metabolism and contain energy reserves in the form of glycogen and lipids. Cardiac muscle cells contain large numbers of mitochondria to utilize aerobic energy systems (Fig. 9.39).

Cardiac muscle cells also contain a specialized kind of cell junctions called intercalated discs that allow the flow of chemicals between cells and help to maintain the structure of the muscle. This allows for a greater transmission of electrical signals across large areas of cardiac muscles. The discs also allow

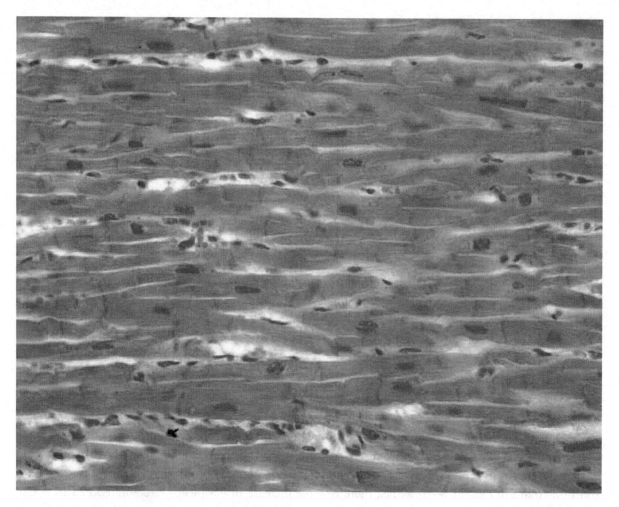
Figure 9.38 Histology of cardiac muscle.

adjacent fibers to pull together in a more coordinated contraction. Instead of motor units working separately in skeletal muscle, intercalated discs allow cardiac muscle to contract in large uniform segments. Cardiac muscle can also contract without a stimulus from the nervous system. Cardiac muscle contains self-generating action potential cells called pacemaker cells or nodes. The pacemaker cells, however, can respond to the nervous system by changing the rate and force of contraction of cardiac muscle cells. Cardiac muscle cannot undergo tetanic contractions due to the structure of the cell membrane.

Smooth Muscle

Smooth muscle cells are found throughout the body in organs, blood vessels, and tube-like structures. Smooth muscles contain actin and myosin and are long spindle-shaped cells. Actin and myosin are not arranged in sarcomeres, so smooth muscle is not striated. Instead the actin and myosin are scattered throughout the muscle. Smooth muscle has no T-tubules and the myosin has a larger number of globular protein heads. Smooth muscle contraction differs from skeletal or cardiac contraction in that when calcium is released by the sarcoplasmic reticulum, it binds with a calcium-binding protein called calmodulin that activates an enzyme called myosin light chain kinase. This enzyme allows for the formation of cross-bridges. Because of the structure of smooth muscle, length and tension are not related. When smooth muscle is stretched, it adapts to its new resting length and can continue to contract. Smooth muscle cells are classified as multiunit or visceral. Multiunit smooth muscle is organized into motor units that are innervated by the nervous system. However, each cell can be connected to more than one motor unit. Visceral cells do not connect directly with motor neurons and are arranged in layers. Gap junctions connect layers of smooth muscle so that one area can influence others when contracting. This can produce a wave-like contraction called peristalsis (Fig. 9.40).

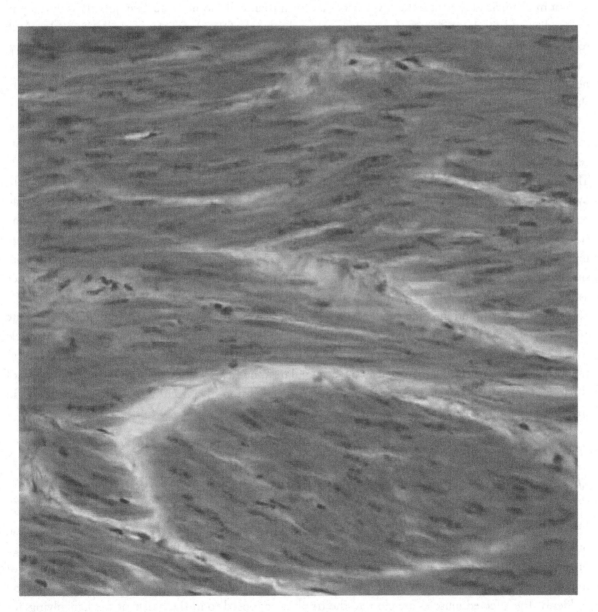

Figure 9.39 Histology of smooth muscle cells.

Development and Regeneration of Muscle Tissue

Most muscle tissue of the body arises from embryonic mesoderm. Paraxial mesodermal cells adjacent to the neural tube form blocks of cells called somites. Skeletal muscles, excluding those of the head and limbs, develop from mesodermal somites, whereas skeletal muscle in the head and limbs develop from general mesoderm. Somites give rise to myoblasts. A myoblast is a muscle-forming stem cell that migrates to different regions in the body and then fuse(s) to form a syncytium, or myotube. As a myotube is formed from many different myoblast cells, it contains many nuclei, but has a continuous cytoplasm. This is why skeletal muscle cells are multinucleate, as the nucleus of each contributing myoblast remains intact in the mature skeletal muscle cell. However, cardiac and smooth muscle cells are not multinucleate, because the myoblasts that form their cells do not fuse. Gap junctions develop in the cardiac and single-unit smooth muscle in the early stages of development. In skeletal muscles, ACh receptors are initially present along most of the surface of the myoblasts, but spinal nerve innervation causes the release of growth factors that stimulate the formation of motor end plates and NMJs. As neurons become active, electrical signals that are sent through the muscle influence the distribution of slow and fast fibers in the muscle.

Although the number of muscle cells is set during development, satellite cells help to repair skeletal muscle cells. A satellite cell is similar to a myoblast because it is a type of stem cell; however, satellite cells are incorporated into muscle cells and facilitate the protein synthesis required for repair and growth. These cells are located outside the sarcolemma and are stimulated to grow and fuse with muscle cells by growth factors that are released by muscle fibers under certain forms of stress. Satellite cells can regenerate muscle fibers to a very limited extent, but they help primarily to repair damage in living cells. If a cell is damaged to a greater extent than can be repaired by satellite cells, the muscle fibers are replaced by scar tissue in a process called fibrosis. Because scar tissue cannot contract, muscle that has sustained significant damage loses strength and cannot produce the same amount of power or endurance as it could before being damaged. Smooth muscle tissue can regenerate from a type of stem cell called a pericyte, which is found in some small blood vessels. Pericytes allow smooth muscle cells to regenerate and repair much more readily than skeletal and cardiac muscle tissues do. Similar to skeletal muscle tissue, cardiac muscle does not regenerate to a great extent. Dead cardiac muscle tissue is replaced by scar tissue, which cannot contract. As scar tissue accumulates, the heart loses its ability to pump because of the loss of contractile power. However, some minor regeneration may occur due to stem cells found in the blood that occasionally enter cardiac tissue.

Muscle Disorders: Dermatomyositis and Polymyositis

Dermatomyositis and polymyositis cause inflammation of the muscles. They are rare disorders, affecting only about one in 100,000 people per year. More women than men are affected. Although the peak age of onset is in the 50s, the disorders can occur at any age. Signs and symptoms: Patients complain of muscle weakness that usually worsens over several months, though in some cases symptoms come on suddenly. The affected muscles are close to the trunk (as opposed to in the wrists or feet), involving, for example, the hip, shoulder, or neck muscles. Muscles on both sides of the body are equally affected. In some cases, muscles are sore or tender. Some patients have involvement of the muscles of the pharynx (throat) or the esophagus (the tube leading from the throat to the stomach), causing problems with swallowing. In some cases, this leads to food being misdirected from the esophagus to the lungs, causing severe pneumonia. In dermatomyositis, there is a rash, though sometimes the rash resolves before muscle problems occur. A number of different types of rash can occur, including rashes on the fingers, the chest and shoulders, or on the upper eyelids. In rare cases, the rash of dermatomyositis appears but myopathy never develops. Other problems sometimes associated with these diseases include fever, weight loss, arthritis, cold-induced color changes in the fingers or toes (Raynaud phenomenon), and heart or lung problems.

Muscle Atrophy

The majority of muscle atrophy in the general population results from disuse. People with sedentary jobs and senior citizens with decreased activity can lose muscle tone and develop significant atrophy. This type of atrophy is reversible with vigorous exercise. Bedridden people can undergo significant muscle wasting. Astronauts, free of the gravitational pull of Earth, can develop decreased muscle tone and loss of calcium from their bones following just a few days of weightlessness. Muscle atrophy resulting from disease rather than disuse is generally one of two types: that resulting from damage to the nerves that supply the muscles, and disease of the muscle itself. Examples of diseases affecting the nerves that control muscles are poliomyelitis, amyotrophic lateral sclerosis (ALS or Lou Gehrig's disease), and Guillain-Barre syndrome. Examples of diseases affecting primarily the muscles are muscular dystrophy, myotonia congenita, and myotonic dystrophy, as well as other congenital, inflammatory, or metabolic myopathies. Even minor muscle atrophy usually results in some loss of mobility or power.

Common Causes:

- some atrophy that occurs normally with aging
- cerebrovascular accident (stroke)
- spinal cord injury
- peripheral nerve injury (peripheral neuropathy)
- other injury
- prolonged immobilization
- osteoarthritis
- rheumatoid arthritis
- prolonged corticosteroid therapy
- diabetes (diabetic neuropathy)
- burns
- poliomyelitis
- amyotrophic lateral sclerosis (ALS or Lou Gehrig's disease)
- Guillain-Barre syndrome
- muscular dystrophy
- myotonia congenita
- myotonic dystrophy
- myopathy

Muscular Dystrophy

Muscular dystrophy (MD) is a group of rare inherited muscle diseases in which muscle fibers are unusually susceptible to damage. Muscles, primarily voluntary muscles, become progressively weaker. In the late stages of muscular dystrophy, muscle fibers are often replaced by fat and connective tissue. In some types of muscular dystrophy, heart muscles, other involuntary muscles, and other organs are affected. The most common types of muscular dystrophy appear to be due to a genetic deficiency of the muscle protein dystrophin. There's no cure for muscular dystrophy, but medications and therapy can slow the course of the disease.

Medical Mystery: Sleep Twitches

The twitching phenomenon that happens in the early stage of sleep is called a hypnagogic massive jerk, or simply a hypnic jerk. It has also been referred to as a sleep start. There has been little research on this topic, but there have been some hypotheses put forth. When the body drifts off into sleep, it undergoes physiological changes related to body temperature, breathing rate, and muscular tone. Hypnic jerks may be the result of muscle changes. Another hypothesis suggests that the transition from the waking state to the sleeping state signals the body to relax. But the brain may interpret the relaxation as a sign of falling and then signal the arms and legs to wake up. Electroencephalogram studies have shown sleep starts affect almost 10 percent of the population regularly, 80 percent occasionally, and another 10 percent rarely. Muscle movement or twitching also may take place during the Rapid Eye Movement, or REM, phase of sleep. This also is the time when dreams occur. During the REM phase, all voluntary muscular activity stops with a drop in muscle tone, but some individuals may experience slight eyelid or ear twitching or slight jerks. Some people with REM behavioral disorder, or RBD, may experience more violent muscular twitching and full-fledged activity during sleep. This is because they do not achieve muscle paralysis, and as a result, act out their dreams. Researchers think that people with RBD lack neurological barriers that define the different stages of sleep. Research conducted by the Mayo Clinic shows that melatonin can help lessen RBD symptoms.

Microbiology

Clostridium tetani: Tetanus

Normally a nerve impulse initiates contraction of a muscle. At the same time, an opposing muscle receives the signal to relax so as not to oppose the contraction. Tetanus toxin blocks the relaxation, so both sets of muscle contract. The usual cause of tetany is lack of calcium, but excess of phosphate (high phosphate-to-calcium ratio) can also trigger the spasms.

Clostridium botulinum

Infant botulism (floppy baby syndrome) is the most common form of botulism in the U.S. of the four forms of botulism. If ingested, the toxin is absorbed in the intestine, goes to the blood, and on to the nervous system. It acts on the peripheral nervous system by blocking the impulse that is normally passed along to the nervous system; by blocking the impulse that is normally passed along to motor end plates so the muscle contraction can be released, resulting in paralysis.

CRITICAL-THINKING QUESTIONS

1. What effect does fascicle arrangement have on a muscle's action?
2. Movements of the body occur at joints. Describe how muscles are arranged around the joints of the body.
3. Explain the difference between axial and appendicular muscles.
4. Why are the muscles of the face different from typical skeletal muscle?
5. Describe the fascicle arrangement in the muscles of the abdominal wall. How do they relate to each other?
6. What are some similarities and differences between the diaphragm and the pelvic diaphragm?
7. The tendons of which muscles form the rotator cuff? Why is the rotator cuff important?
8. List the general muscle groups of the shoulders and upper limbs as well as their subgroups.
9. Which muscles form the hamstrings? How do they function together?
10. Which muscles form the quadriceps? How do they function together?
11. Why is elasticity an important quality of muscle tissue?
12. What are the five primary functions of skeletal muscle?
13. What are the opposite roles of voltage-gated sodium channels and voltage-gated potassium channels?
14. How would muscle contractions be affected if skeletal muscle fibers did not have T-tubules?
15. What causes the striated appearance of skeletal muscle tissue?
16. How would muscle contractions be affected if ATP were completely depleted in a muscle fiber?
17. What changes occur at the cellular level in response to endurance training?
18. What would be the drawback of cardiac contractions being the same duration as skeletal muscle contractions?
19. How are cardiac muscle cells similar to and different from skeletal muscle cells?
20. Why can smooth muscles contract over a wider range of resting lengths than skeletal and cardiac muscles can?
21. Which muscle type(s) (skeletal, smooth, or cardiac) can regenerate new muscle cells/fibers? Explain your answer.
22. Predict the consequences of administration of a compound that blocks the calcium channels on presynaptic terminals.
23. Predict the consequences of administration of a compound that binds to acetylcholine receptors.
24. Botulism food poisoning results from a toxin (poison) produced by bacteria. Death from botulism results from relaxation of the respiratory muscles (diaphragm). Assuming that this toxin affects the neuromuscular junction, propose a way in which this toxin could cause these muscles to relax and no longer contract.
25. The coroner was unable to remove a vial of pills from the clenched hand of a suicide victim. Explain the reason for this.

Section Three

Integration and Control Systems

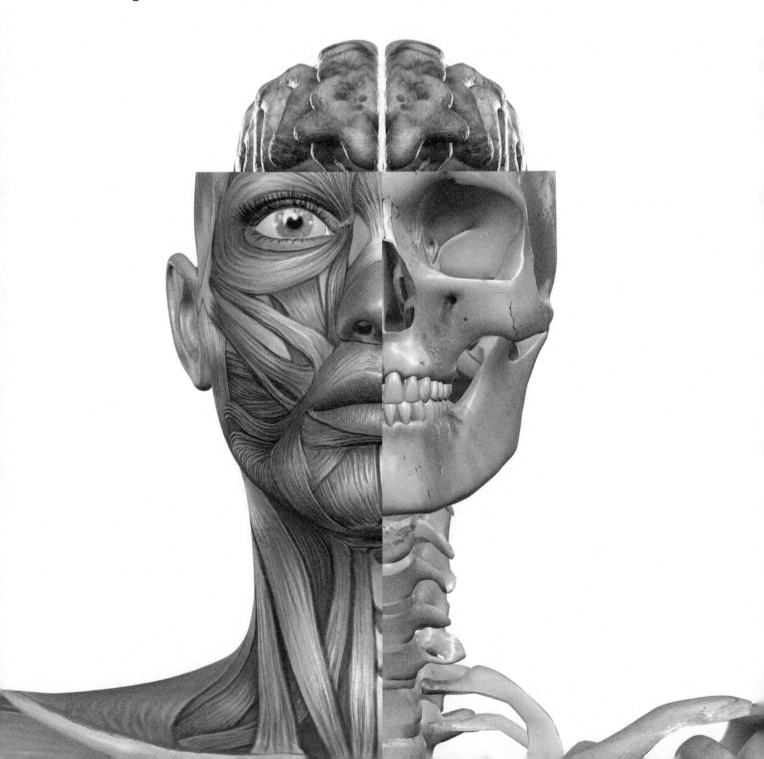

Chapter Ten

Nervous System I: Structure and Function

The central nervous system includes the brain and spinal cord. The brain and spinal cord are protected by bony structures, membranes, and fluid. The brain is held in the cranial cavity of the skull and consists of the cerebrum, cerebellum, and the brainstem. The nerves involved are cranial nerves and spinal nerves. The nervous system is divided into two components (Fig. 10.1). The central nervous system consists of the brain and spinal cord. The peripheral nervous system consists of nerves and a group of neurons known as the autonomic nervous system. We can begin to understand how the nervous system works by examining how information flows through it. Let's say that I have reached for a cup of hot tea and have moved it slightly, spilling tea on my hand. I must make the decision to let go of the cup and move my hand away. The sensation of touch, heat, and pain are first processed by sensory receptors located in my skin. All sensory receptors take information from the environment and convert it into a form that can be processed by the nervous system. The environment can be internal (inside the body) or external (outside the body). The information going into the receptor can be in many forms. For example, light rays enter the eye, sound waves enter the ear, and pressure is sensed by receptors that are deformed by either light or heavy pressure in the skin. Heat is also sensed by temperature receptors in the skin. The information coming out of the receptor is in the form of electrochemical impulses called action potentials (more about these later). The impulses from the sensory receptor then travel to the central nervous system via afferent pathways. These pathways generally consist of sensory nerves that attach to the receptors. The pathway continues to the spinal cord, which is part of the central nervous system. The impulse then travels upward toward the brain via a special pathway in the spinal cord called a spinal tract. Since the tract travels upward to the brain, it is called an ascending tract. The impulse then travels to the brain where the sensation of pressure and heat are processed. A decision is made in the brain to move the muscles of my arm and hand to let go of the cup. The impulse is now a motor impulse and it travels down the spinal cord following a spinal tract (this time a descending tract) and moves along an efferent (away from) pathway consisting of a motor nerve(s) to the muscles of my hand and arm. My hand lets go of the cup and moves away. We can think of the nervous system then in terms of stimulus (hot tea) and response (move hand). Many nervous system functions occur this way. But before we go deeper into how the nervous system works, we need to examine some structures.

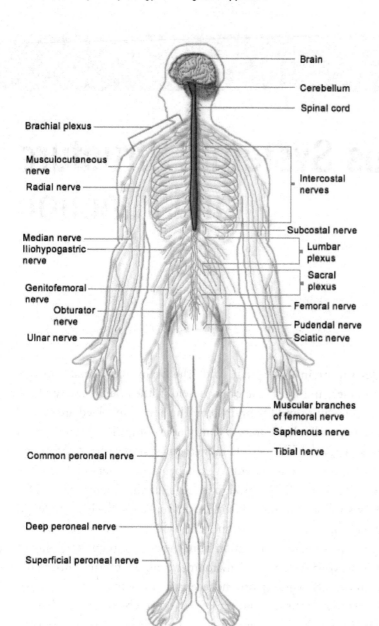

Figure 10.1 The nervous system.

Structure of the Spinal Cord

The spinal cord begins at the foramen magnum of the occipital bone and extends to the second lumbar vertebra. It ends in a cone-like structure called the conus medullaris (Fig. 10.2). A structure known as the cauda equina extends from the inferior end of the spinal cord. The cauda equina (horse's tail) consists of nerves that extend downward to exit the foramen of the lumbar and sacral vertebrae (Fig. 10.3). The spinal cord consists of cervical, thoracic, lumbar, and sacral segments. The spinal cord is also thicker in the cervical and lumbar areas. These areas are called the cervical and lumbar enlargements. The cord is thicker due to the larger numbers of nerves accommodating the upper and lower extremities. The spinal cord is anchored in place inferiorly by a thin ligament called the filum terminale that extends from the conus medullaris to the sacrum. The spinal cord is covered by a connective tissue covering called the meninges (Fig. 10.4). The meninges also cover the brain. The meninges consist of three layers. The dura mater is the most superficial layer. It forms a sac known as the thecal sac that encases the spinal cord. The thecal sac extends from the foramen magnum to the second sacral vertebra and is continuous with the brain. The space between the dura mater and the vertebrae is called the epidural space. Anesthetics are sometimes injected into this space (epidural injection). The middle layer of the meninges is known as the arachnoid mater. This is a thin layer consisting of simple squamous epithelium. The arachnoid mater adheres to the inner portion of the dura mater. The pia mater is the innermost membrane. It is closely attached to the spinal cord as a thin membrane. It continues inferiorly to produce the filum terminale. The pia mater also extends laterally to the dura mater at points along the spine to produce the dentate ligaments. These tiny ligaments work to anchor the cord in place. The space between the arachnoid and pia mater is known as the subarachnoid space. This space is filled with cerebral spinal fluid.

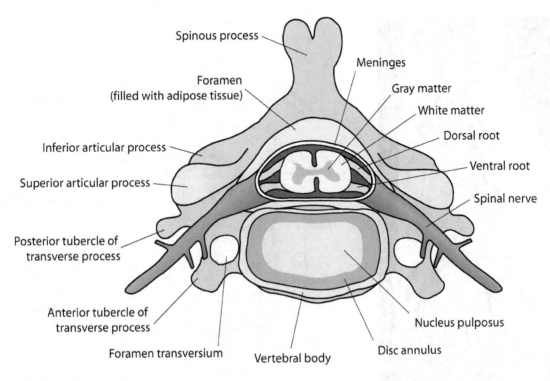

Figure 10.2 Spinal cord anatomy.

http://commons.wikimedia.org/wiki/File:Cervical_vertebra_english.png

The central structure of the spinal cord consists of an area of white matter surrounding a core of gray matter (Fig. 10.5). White matter consists of myelinated axons. Myelin is a lipid substance that helps to insulate the axons. The white matter consists of pathways called spinal tracts that carry information to and from the brain. The gray matter contains unmyelinated axons as well as other parts of neurons, giving it a darker color. In the center of the gray matter is the central canal. This long tubular structure carries CSF and is typically closed in many areas of the adult spinal cord. The white matter is divided into areas called funiculi. There are posterior, anterior, and two lateral funiculi. The funiculi contain the ascending and descending spinal tracts that carry information to and from the brain. The gray matter is divided into horns. There are anterior, posterior, and two small lateral horns. Like the white matter, the gray matter is symmetrically distributed on the right and left sides of the cord. The right and left sides of the gray matter are connected by the gray commissure.

Spinal Cord Tracts

One major function of the spinal cord is to carry information to and from the brain (Fig. 10.6). This information is carried by areas in the white matter called spinal tracts. Sensory information is carried to the brain by ascending tracts and motor information is carried from the brain by descending spinal tracts. Some tracts cross over (decussate—undergo decussation) to the contralateral side. The right side of the brain processes sensory information and sends motor information to the left side of the body and vice versa.

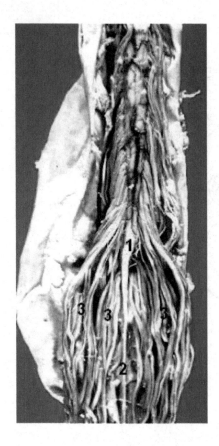

Figure 10.3 The cauda equina consists of nerves that travel inferiorly to exit at the lumbar and sacral vertebrae.

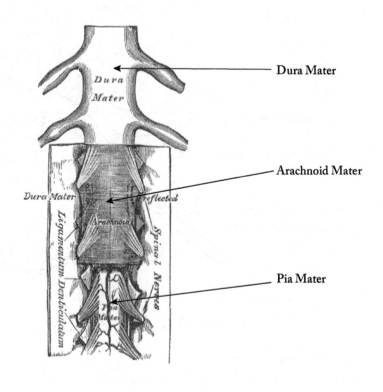

Figure 10.4 Meninges covering the spine.

Some important ascending tracts include the fasciculus gracilis, fasciculus cuneatus, spinothalamic, and spinocerebellar. There are generally three neurons that carry information from the stimulus to the brain. The first-order neuron carries information from the sensory receptor to the spinal cord. The second-order neuron carries the information to the thalamus and the third-order neuron carries the information to the cortex of the brain. The fasciculus gracilis is located in the posterior funiculus. This tract carries information related to discriminative touch, visceral pain, vibration, and proprioception. The tract carries this information from the middle thoracic and lower areas of the body. The fasciculus gracilis is part of the posterior spinal cord called the dorsal column. At the middle thoracic region (about T6) it combines with the fasciculus cuneatus. It contains first-order neurons that travel up the ipsilateral side of the cord and cross over at the brainstem in an area known as the medulla oblongata (gracile nucleus). The fasciculus cuneatus is also located in the posterior funiculus. It carries the same type of information as the fasciculus gracilis from the middle to upper areas of the body (T6 and above). It is also part of the dorsal column and its fibers cross over in the medulla (cuneate nucleus) as well.

The second-order fibers of the fasciculus gracilis and cuneatus combine to form the medial lemniscus from the medulla oblongata to the thalamus. The spinothalamic tract consists of two portions.

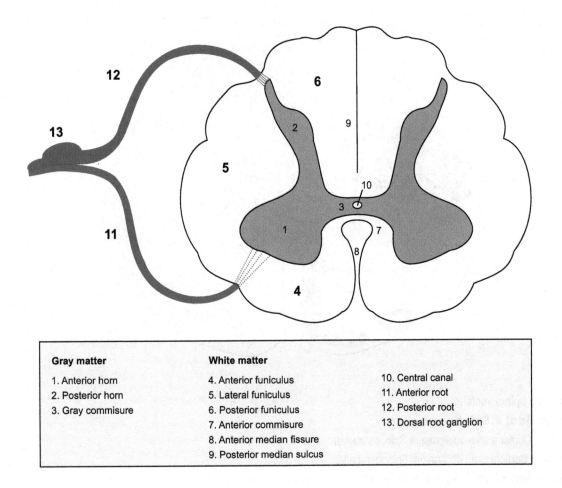

Figure 10.5 Transverse section of spinal cord.

The anterior spinothalamic and lateral spinothalamics are located in the anterior and lateral funiculi. The spinothalamics are sometimes referred to as the anterolateral system. The anterior spinothalamic tract carries information related to light touch and pain. Light touch is clinically defined as perceived sensation from stroking an area of the skin without hair. The fibers from the anterior spinothalamic tract cross at one or two segments above their entry point in the spine. The lateral spinothalamic tract is an important clinical tract because it carries information related to pain and temperature. Its fibers also cross in a similar way to the anterior spinothalamic tract. Lesions of the lateral spinothalamic tract will result in loss of pain and temperature. For example, in a Brown-Sequard lesion (sometimes called a hemisection of the spinal cord) there is a contralateral loss of pain and temperature below the level of the lesion, as well as a bilateral loss of pain and temperature at the segmental level of the lesion.

The spinocerebellar tract also consists of two portions. The anterior and posterior spinocerebellar tracts are both located in the lateral funiculus. The fibers in the posterior tracts do not cross, while the

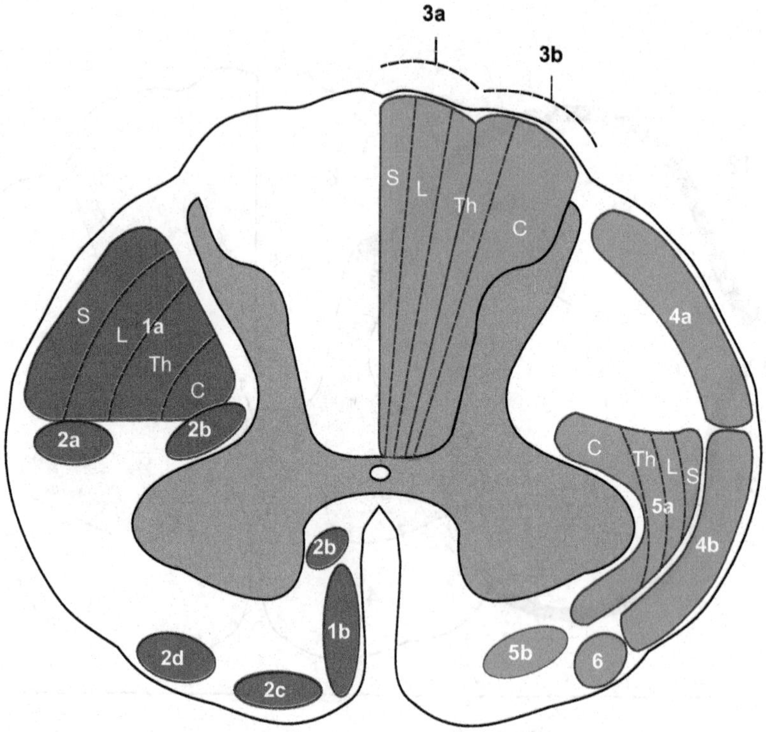

Figure 10.6 Spinal tracts.

1. Pyramidal **4. Spinocerebellars**
1a. Lateral Corticospinal 4a. Posterior Spinocerebellar
1b. Anterior Corticospinal 4b. Anterior Spinocerebellar
2. Extrapyramidal **5. Spinothalamics**
2a. Rubrospinal 5a. Lateral Spinothalamic
2b. Reticulospinal 5b. Anterior Spinothalamic
3. Dorsal Column
3a. Fasciculus Gracilis
3b. Fasciculus Cunneatus

anterior fibers cross at the medulla oblongata. The spinocerebellar tracts carry information related to coordination of muscles from the lower limbs and trunk to the cerebellum.

Important descending tracts include the corticospinal, reticulospinal, and rubrospinal. All these tracts carry motor information from the brain to the spinal cord. The corticospinal tract consists of anterior and lateral portions located in the anterior and lateral funiculi. These tracts are sometimes referred to as the pyramidal tracts. Fibers in the lateral tract cross over at the medulla oblongata. Fibers in the anterior portion cross at various levels in the spinal cord. Both tracts convey motor information to skeletal muscles. The rubrospinal tracts are located in the lateral funiculi. The fibers from these tracts cross over in the brain and descend through the lateral funiculi. The rubrospinal tracts also carry motor information to skeletal muscles. They also carry information about posture and coordination.

The reticulospinal tracts consist of anterior and lateral tracts. They are located in the anterior and lateral funiculi. Some of the fibers cross while others do not. These tracts carry information related to muscular tone and activity of sweat glands.

Nerves

Nerves are bundles of nerve fibers. It is important to realize that since nerves contain numerous fibers, some of these fibers can carry sensory information, while others carry motor information. Therefore one nerve can carry both sensory and motor information. This type of nerve is known as a mixed nerve. A nerve can carry sensory information only (sensory nerve) or motor information only (motor nerve) (Fig. 10.7). The outer layer of a nerve consists of the epineurium. The epineurium consists of dense connective tissue that surrounds and protects the nerve. Inside the nerve, the fibers are bundled in fascicles with each fascicle surrounded by a sheath called a perineurium. Inside the fascicles are bundles of neurons, each surrounded by a thin layer of loose connective tissue called the endoneurium.

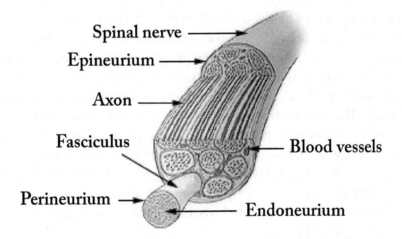

Figure 10.7 Nerve structure.

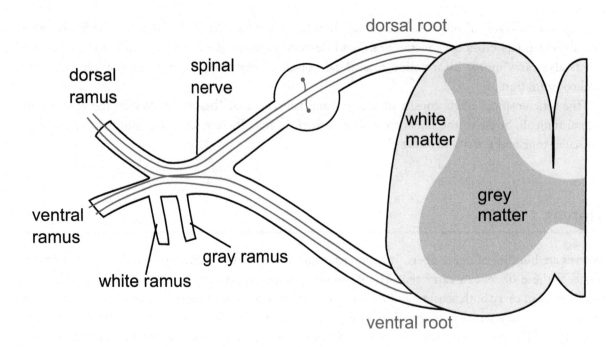

Figure 10.8 Spinal nerve.

Spinal Nerves

There are 31 pairs of spinal nerves. They are named after their attachment point in the spine. For example, cervical nerves are named C1–C8, thoracic T1–T12, lumbar L1–5, and sacral S1–S5. All spinal nerves are mixed nerves and carry both sensory and motor information (Fig. 10.8). Spinal nerves consist of two nerve roots that exit the spine. The dorsal root carries sensory afferent information and divides into eight small rootlets that enter the spine. Lateral to the rootlets lies a structure called the dorsal root ganglion. A ganglion is a collection of cell bodies. The ventral root consists of six to eight rootlets that exit the spine and combine. The ventral root carries motor information to muscles. After exiting the spinal canal, the spinal nerve forms several branches sometimes called rami. The posterior branch (ramus) innervates the back. It carries sensory information from the central region of the back as well as motor information to the muscles of the spine. The ventral branch (ramus) innervates the sides and anterior trunk. The ventral rami form intercostal nerves that run between the ribs. Some ventral rami form complicated networks of nerves called plexi (plexus). Nerves containing input from several spinal nerves exit a plexus and continue to the skin and muscles of a specific part of the body. The meningeal branch courses back into the spinal canal and innervates the vertebrae, meninges, and spinal ligaments. The visceral branch becomes part of the autonomic nervous system. Spinal nerves carry sensory information from the surface of the body. Each nerve carries sensation from a specific area of the body called a dermatome (Fig. 10.9).

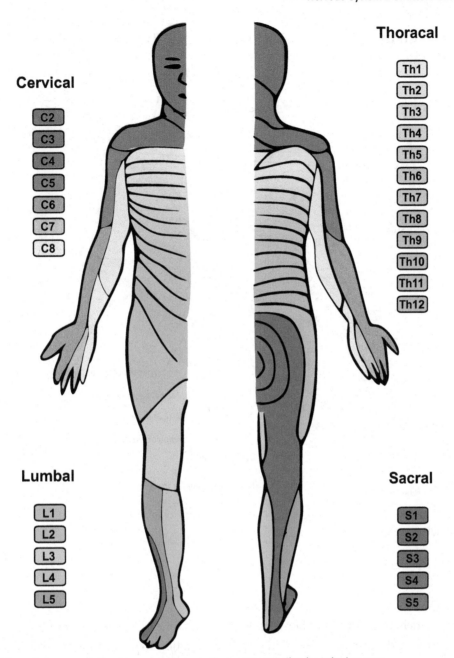

Figure 10.9 Dermatomes are defined areas of the body that carry sensation by spinal nerves.

Plexi

There are four major plexi in the human body. The cervical plexus (C1–C4) (Fig. 10.11) innervates the posterior head and skin of the neck. The phrenic nerve (C3, 4, 5) emerges from the cervical and brachial plexi and runs through the thorax to innervate the diaphragm. The brachial plexus (C5–T1) consists of the ventral rami from spinal nerves C5–T1. The rami form three trunks and the trunks become six divisions, which again join to form three cords. Five branches emerge from the three cords, which constitute the major nerves of the upper extremity. These include the axillary, radial, musculocutaneous, ulnar, and median nerves. The axillary nerve carries sensory information from the shoulder and motor

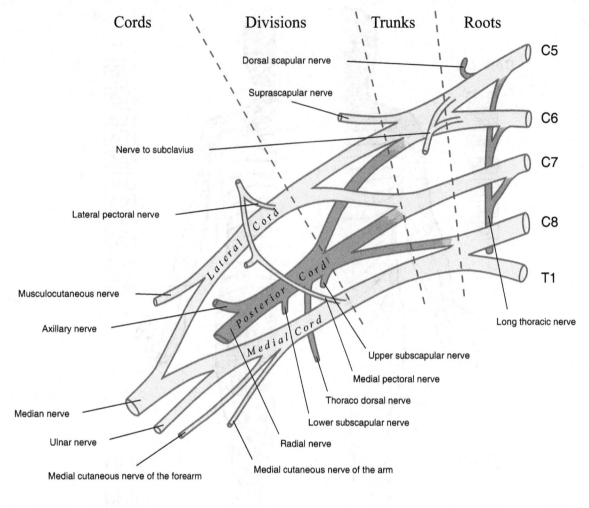

Figure 10.10 Brachial plexus.

information to the deltoid and teres minor muscles. The radial nerve carries sensory information from the posterior portion of the arm and hand, and motor information to the supinator, brachial, and extensor muscles of the upper extremity. The musculocutaneous nerve carries sensory information from the forearm and motor information to the anterior muscles of the upper extremity. The ulnar nerve carries sensory information from the medial two fingers and medial wrist as well as motor information to the hand muscles and the flexor carpi ulnaris and flexor digitorum profundus. The median nerve carries sensory information from the lateral fingers, thumb, and wrist, as well as motor information to the wrist flexor and thenar muscles. The lumbar plexus consists of the ventral rami from spinal nerves L1–L4 (Fig. 10.12). The sacral plexus consists of the ventral rami from spinal nerves L4–S4. Sometimes both plexi are referred to as the lumbosacral plexus. The major nerves exiting the lumbosacral plexus are the obturator, femoral, and sciatic. The obturator nerve carries sensory information from the medial thigh and motor information to the hip adductor muscles. The femoral nerve carries sensory information from the anterior and lateral thigh and motor information to the iliopsoas, Sartorius, and quadriceps muscles. The sciatic nerve is the largest nerve in the body. It runs down the posterior portion of the leg and splits into two divisions in the popliteal area (tibial and common peroneal). It carries sensory information from

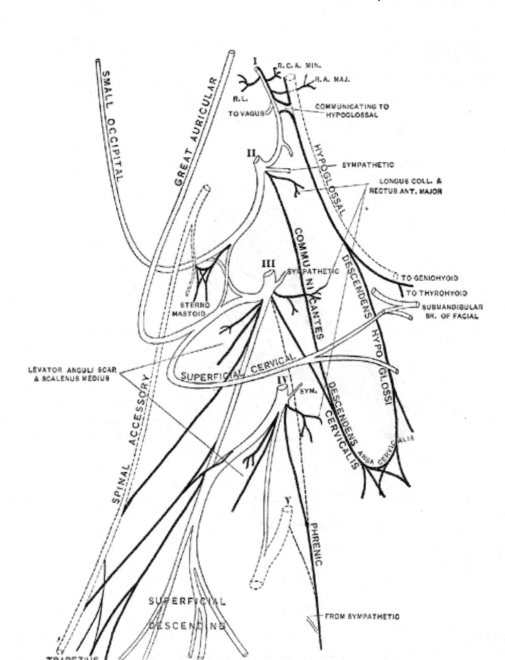

Figure 10.11 Cervical plexus.

the posterior portion of the leg as well as the anterior and lateral portions of the area below the knee. It carries motor information to the posterior thigh and leg muscles.

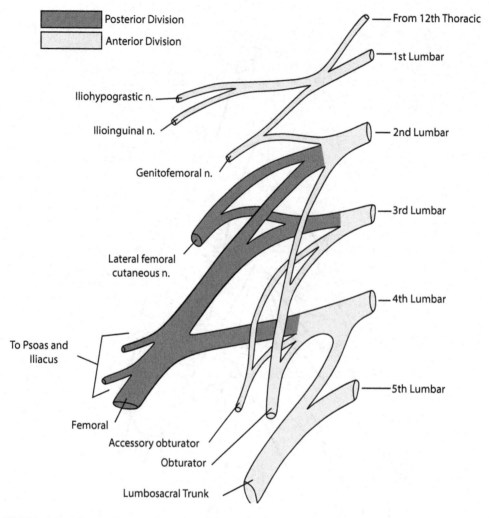

Figure 10.12 Lumbar plexus.

The Brain

The brain consists of four major structures: the cerebral cortex, diencephalon, brainstem, and cerebellum.

Fetal Development of the CNS

The central nervous system (CNS) develops from a flat tissue structure called the neural plate (Figs. 10.13, 10.14). The neural plate forms neural folds on the lateral sides. The neural folds contain elevated portions called neural crests. At the center of the neural plate lies the neural groove. During fetal development, the neural folds move toward each other and meet in the midline, forming a neural tube. The superior portion of the neural tube becomes the brain and the inferior portion becomes the spinal cord. The neural crest contains neural crest cells that eventually separate from the neural crest and

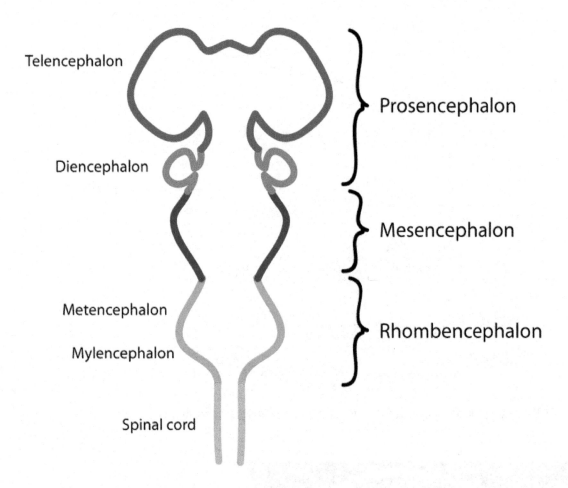

Figure 10.13 Developing CNS.

develop into the autonomic and sensory neurons of the peripheral nervous system. A series of pouch-like structures also develop from the anterior portion of the neural tube. The walls of these structures become parts of the brain, while the hollow areas become the ventricles. The developing brain can be divided into three main regions. These are the forebrain (prosencephalon), midbrain (mesencephalon), and hindbrain (rhombencephalon). The forebrain divides into the telencephalon and diencephalon. The midbrain remains as one structure and the hindbrain divides into the myelencephalon, which eventually becomes the medulla oblongata and the pons and cerebellum.

The Brainstem

The brainstem lies between the cerebral cortex and the spinal cord. It consists of the midbrain, pons, and medulla oblongata (Figs. 10.15, 10.16, 10.17, 10.18, 10.19, 10.20). The medulla oblongata is the most inferior portion of the brainstem and contains a number of centers for controlling heart rate,

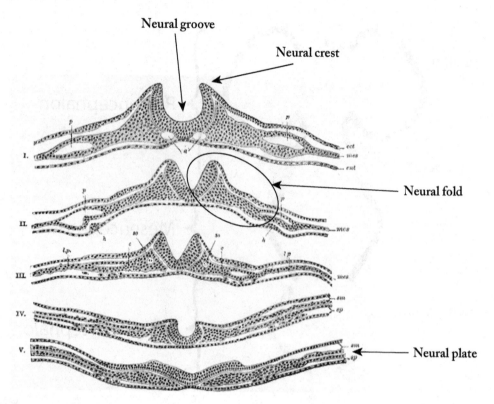

Figure 10.14 The neural plate develops folds that unite in the center to produce the neural tube.

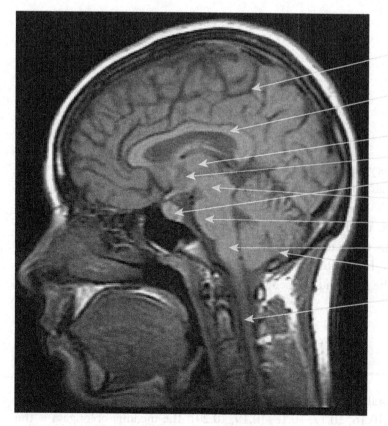

Figure 10.15 MRI of the brain.

Nervous System I: Structure and Function | 205

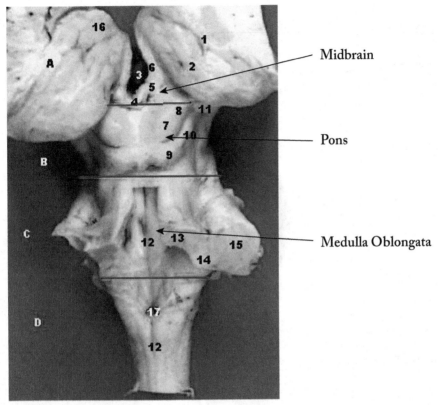

Figure 10.16 Brainstem, posterior view.

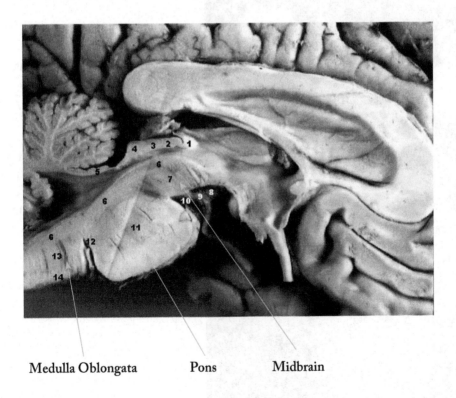

Figure 10.17 Brainstem, lateral view.

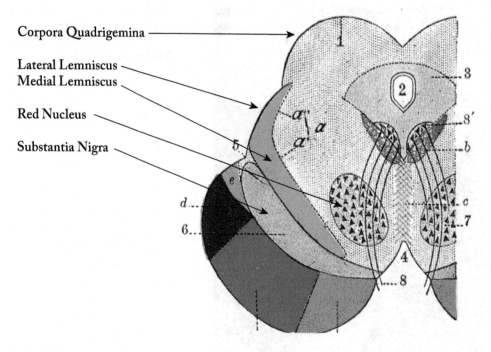

Figure 10.18 Brainstem, coronal view.

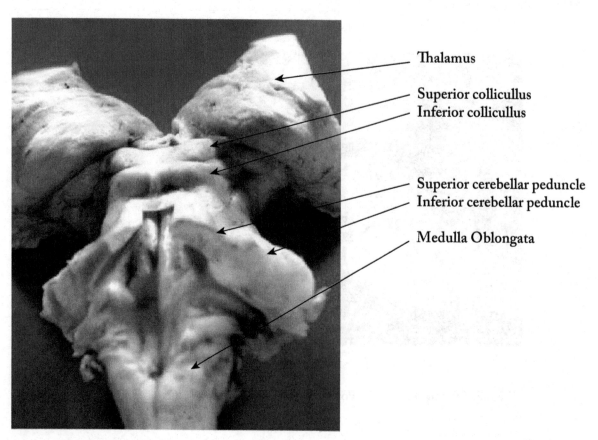

Figure 10.19 Brainstem, posterior view.

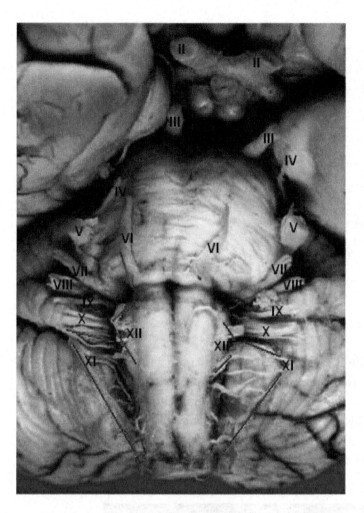

Figure 10.20 Brainstem, anterior view showing locations of cranial nerves.

respiration, swallowing, vomiting, and blood vessel diameter. These centers consist of nuclei that are clusters of neuron cell bodies. The spinal tracts also continue through the medulla, connecting the spinal cord with the brain. The medulla contains two rounded structures called olives, which consist of nuclei that help to control balance, coordination, and sound information. On the anterior surface of the medulla lie two enlargements called pyramids. The pyramids consist of the descending spinal cord tracts. The pons is the middle section of the brainstem. The pons also contains spinal cord tracts as well as nuclei that help to control respiration and sleep. A number of cranial nerve nuclei are located in the pons (CN V, VI, VII, VIII, IX). The midbrain is the most superior portion of the brainstem. It contains the nuclei of cranial nerves III, IV, and V. The roof or tectum of the midbrain contains the corpora quadrigemina, which consist of four nuclei. The two superior nuclei are called the superior colliculi, while the inferior are called the inferior colliculi. The superior colliculi help to control the movement of the head toward stimuli, including visual, auditory, and touch. The superior colliculi receive input from the eyes. The inferior colliculi help to process hearing and also receive input from the skin and cerebrum. The floor of the midbrain is called the tegmentum. It contains two reddish-colored structures called the red nuclei that process information for unconscious motor movements. The midbrain also contains the cerebral peduncles that carry motor information from the cerebrum to the spinal cord. The substantia nigra resides in the midbrain and processes information relating to tone and coordination of muscles. The reticular formation is located throughout the brainstem and is concerned primarily with regulating sleep-wake cycles.

The Cerebellum

The cerebellum is located posterior and inferior to the cerebrum (Fig. 10.21, 10.22, 10.23, 10.24). It is connected to the brainstem via three cerebellar peduncles (superior, middle, and inferior peduncles). The cerebellum contains both gray and white matter. The white matter branches much like a tree and is called

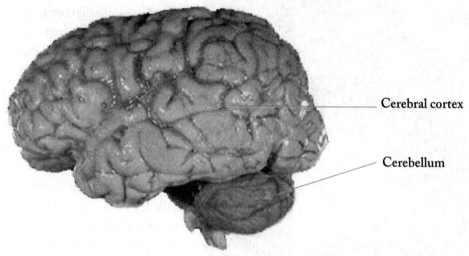

Figure 10.21 The cerebellum.

the arbor vitae. The cerebellum contains a number of different types of neurons, but one in particular—the Purkinje cell—is the largest cell in the brain. These cells have very intricate dendritic networks that can synapse with as many as 200,000 other fibers. Purkinje cells are inhibitory cells and function in processing motor information. The cerebellum can be divided into three parts. The flocculonodular lobe is the inferior portion. The vermis constitutes the middle portion and the two lateral hemispheres make up the remaining portion. The cerebellum functions in processing information related to complex

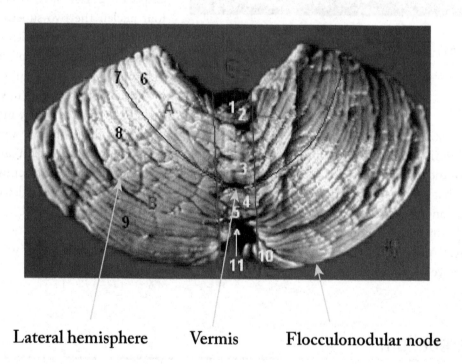

Figure 10.22 Cerebellum, inferior view.

Nervous System I: Structure and Function | 209

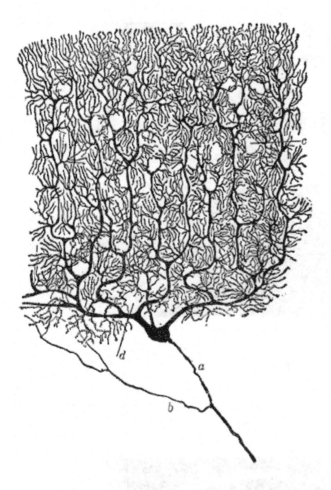

Figure 10.23 Highly branching Purkinje cells found in the cerebellum.

movements, coordination, and unconscious proprioception.

The Diencephalon

The diencephalon lies between the brainstem and cerebrum. It consists of the thalamus, hypothalamus, subthalamus, and epithalamus (Fig. 10.25). The thalamus is the largest part of the diencephalon (Figs. 10.26, 10.27, 10.28). It consists of two lateral portions connected by a stalk called the interthalamic adhesion, sometimes referred to as the intermediate mass. The thalamus carries all sensory information to the cerebral cortex with the exception of the sense of smell, which is carried directly to the frontal lobe of the cerebral cortex by the olfactory nerves. The thalamus is sometimes referred to as a relay station for sensory information. Examples of sensory information include auditory information that synapses in the medial geniculate nucleus; visual information that

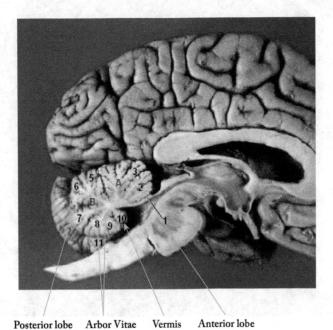

Figure 10.24 Cerebellum, sagittal view.

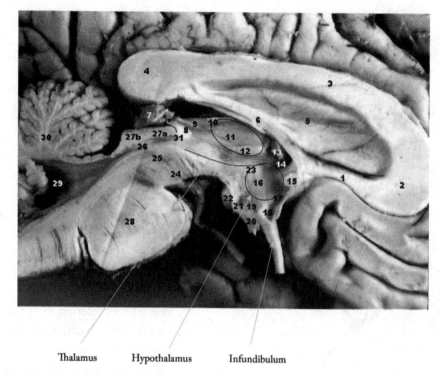

Figure 10.25 Diencephalon, thalamus, and hypothalamus.

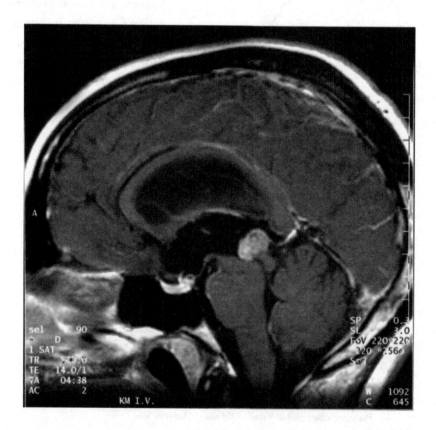

Figure 10.26 The pineal gland (highlighted area) is located in the posterior region of the diencephalon.

Nervous System I: Structure and Function | 211

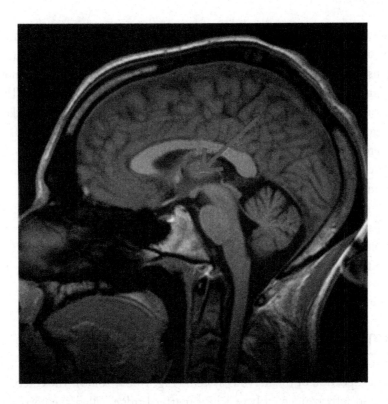

Figure 10.27 MRI showing location of thalamus.

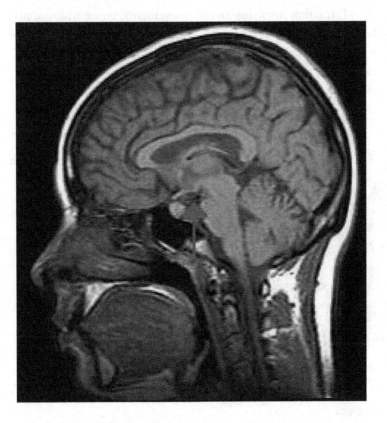

Figure 10.28 MRI showing location of hypothalamus.

synapses in the lateral geniculate nucleus; and motor information from the basal nuclei, motor cortex, and cerebellum, synapsing in the ventral anterior and lateral nuclei. The thalamus is also intimately involved in emotions due to its connections to the limbic system. The hypothalamus lies inferior and anterior to the thalamus. It contains the mamillary bodies on its anterior surface. The mamillary body processes information associated with the sense of smell and emotions. A stalk-like projection called the infundibulum projects anterior and inferior and connects to the pituitary gland. The hypothalamus is intimately connected with the endocrine system and helps to regulate hormones. The hypothalamus also regulates body temperature, thirst, hunger, and sexual drive, and is involved in processing emotions, mood, and sleep, along with the reticular activating system.

The epithalamus is located posterior and superior to the thalamus. It is a small area that works to process the sense of smell and emotional responses. The pineal body (gland) is also located in this area. It helps to regulate sleep-wake cycles by secreting the hormone melatonin (Fig. 10.26). The subthalamus is located inferior to the thalamus. It contains nuclei that are involved in controlling motor information.

The Cerebrum

The cerebrum is largest portion of the nervous system (Fig. 10.29, 10.30). The cerebrum consists of two hemispheres (right and left) connected by a white matter bridge called the corpus callosum. On the surface of the cerebrum are folds called gyri and grooves called sulci. Deep grooves are known as fissures. Each hemisphere is divided into lobes. The lobes are the frontal, parietal, temporal, and occipital. The frontal lobe processes information involving motor movements, concentration, planning, and problem solving, as well as the sense of smell and emotions. The parietal lobes process sensory information with

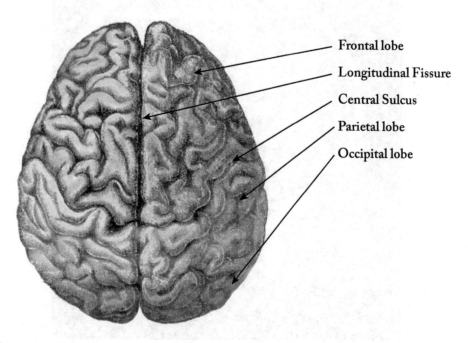

Figure 10.29 Cerebrum superior view.

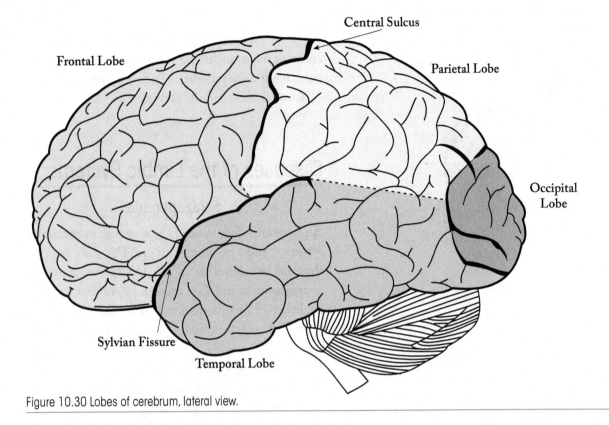

Figure 10.30 Lobes of cerebrum, lateral view.

the exception of hearing, smell, and vision. The temporal lobes process information related to hearing, smell, and memory, as well as abstract thought and making judgments. The occipital lobe processes visual information. Some lobes are divided by fissures. Along the superior aspect of the cerebrum lies the longitudinal fissure that divides the parietal lobes. The lateral fissure (Sylvian fissure) is located on the lateral aspect and separates the temporal from parietal lobes. One sulcus, called the central sulcus, is located midway on the lateral aspect of the cerebrum and extends from superior to inferior. The central sulcus separates the frontal from parietal lobes. Deep in the lateral fissure is the insula, which is often referred to as a fifth lobe of the cerebrum. The cerebrum also has a medulla that consists of white matter tracts. Association fibers connect regions of the cerebral cortex to other regions within the same hemisphere. Commissural fibers interconnect both hemispheres. The corpus callosum is the largest group of commissural fibers. Projection fibers connect the cerebrum to other portions of the brain and spinal cord and form the internal capsule. The basal nuclei are located in the inferior portion of the cerebrum as well as the diencephalon and midbrain. The cerebral basal nuclei consist of the caudate and lentiform nuclei. The lentiform nuclei divide into the putamen laterally and globus pallidus medially. These nuclei work to control motor information along with the subthalamic nuclei and substantia nigra.

Limbic System

The limbic system consists of portions of both the cerebrum and diencephalon and is involved in the emotions as well as reproduction and memory (Fig. 10.31). The limbic system contains the cingulate gyrus located just superior to the corpus callosum and the parahippocampal gyrus located on the medial aspect of the temporal lobe. The limbic system also contains nuclei including the dentate nucleus, amygdala, mammillary bodies of the hypothalamus, the olfactory cortex, and the fornix.

Diseases of the Limbic System:

Schizophrenia

An increased dopamine (DA) response in the limbic system results in schizophrenia. DA may be synthesized or secreted in excess, DA receptors may be supersensitive, and DA regulatory mechanism may be defective. Symptoms are decreased by drugs which block DA receptors. Symptoms of schizophrenia are:

1. Loss of touch with reality
2. Decreased ability to think and reason
3. Decreased ability to concentrate
4. Decreased memory
5. Regress in child-like behavior
6. Altered mood and impulsive behavior
7. Auditory hallucinations

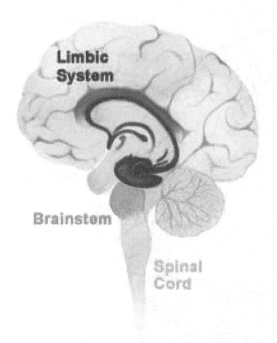

Figure 10.31 Limbic system.
http://commons.wikimedia.org/wiki/Image:Brain_limbicsystem.jpg

Symptoms may be so severe that the individual cannot function.

Depression

Depression is the most common major mental illness and is characterized by both emotional and physical symptoms. Symptoms of depression are:

1. Intense sadness and despair
2. Anxiety
3. Loss of ability to concentrate
4. Pessimism
5. Feelings of low self-esteem
6. Insomnia or hypersomnia
7. Increased or decreased appetite
8. Changes in body temperature and endocrine gland function

Ten to 15% of depressed individuals display suicidal behavior during their lifetime. The cause of depression and its symptoms are a mystery but we do understand that it is an illness associated with biochemical changes in the brain. A lot of research has been conducted to explain that it is associated with a lack of the amines serotonin and norepinephrine. Therefore, pharmacological treatment strategies often try to increase amine concentrations in the brain. One class of antidepressants is monoamine oxidase inhibitors. Mono amine oxidase is an enzyme that breaks down amines like norepinephrine and serotonin. Because the antidepressants inhibit their degradation, they will remain in the synaptic cleft for a longer period of time, making the effect just as if you had increased these types of neurotransmitters. A newer class of antidepressants is selective serotonin reuptake inhibitors (SSRIs). With SSRIs decreasing the uptake of serotonin back into the cell, that will increase the amount of serotonin present in the synaptic cleft. SSRIs are more specific than the monoamine oxidase inhibitors because they affect only serotonergic synapses. You may recognize these SSRIs by their trade name, Prozac and Paxil.

Bipolar Disorder

Another common form of depression is manic depression, now better known as bipolar disorder. Mania is an acute state characterized by:

1. Excessive elation and impaired judgment
2. Insomnia and irritability
3. Hyperactivity
4. Uncontrolled speech

People who have bipolar disorder display mood swings between mania and depression. The limbic system receptors are unregulated. Drugs used are unique mood stabilizers. The hippocampus is particularly vulnerable to several disease processes, including ischemia, which is any obstruction of blood flow or oxygen deprivation, Alzheimer's disease, and epilepsy. These diseases selectively attack CA1, which effectively cuts through the hippocampal circuit.

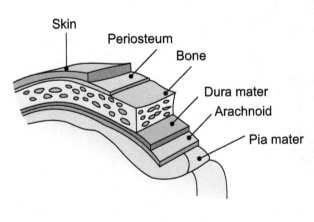

Figure 10.32 The brain has 3 layers of meninges.

The Cerebral Spinal Fluid System

We explored the meninges of the spinal cord earlier in this chapter. These coverings are consistent with the brain. The meninges cover both the brain and spinal cord (Fig. 10.32). However, there are some differences in how the membranes are structured in the brain. The dura mater in the brain adheres to the inner portions of the bones of the skull. The dura also produces folds that extend into some of the brain's fissures. The falx cerebri is a fold of dura mater that extends into the longitudinal fissure. It connects to the crista galli of the ethmoid bone. The tentorium cerebelli lies between the cerebrum and cerebellum in a transverse plane. The falx cerebelli lies between the cerebellar hemispheres.

The dura mater in the brain also forms sinuses, which are hollow areas that contain venous blood and cerebral spinal fluid. The superior sagittal sinus lies between falx cerebri and periosteum of the skull. The inferior sagittal sinus lies deep within the falx cerebri and superior to the corpus callosum. The arachnoid mater is the middle layer of meninges. The pia mater makes a very close connection to the surface of the brain. The subarachnoid space exists between the pia and arachnoid maters.

Cerebral spinal fluid (CSF) is derived from the plasma of the blood. It contains none of the large elements of the blood such as plasma proteins. It acts as a shock absorber and cushions the brain and spinal cord. CSF is produced by small vascular structures called choroid plexi. A choroid plexus contains ependymal cells that produce CSF. These cells actively transport sodium to the outside of the cells, creating a gradient that pulls fluid out of the blood vessels. The blood vessels in a choroid plexus form a blood–brain barrier between the blood and CSF. The capillaries inside of the brain also form a blood–brain barrier. These cells are surrounded by nervous system cells called astrocytes. These cells work to form tight junctions that help to regulate the substances passing into the brain. Examples of substances that can pass through the blood–brain barrier include lipid-soluble drugs and alcohol. Water-soluble substances can also enter the brain via transport proteins. CSF not only circulates in the subarachnoid space but also within hollow chambers located in the brain. These chambers are called ventricles (Fig. 10.33). There are two lateral ventricles separated by a fibrous membrane called the septum

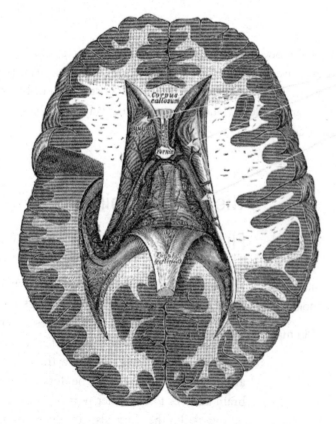

Lateral ventricles

Choroid plexus

Figure 10.33 Lateral ventricles of brain.

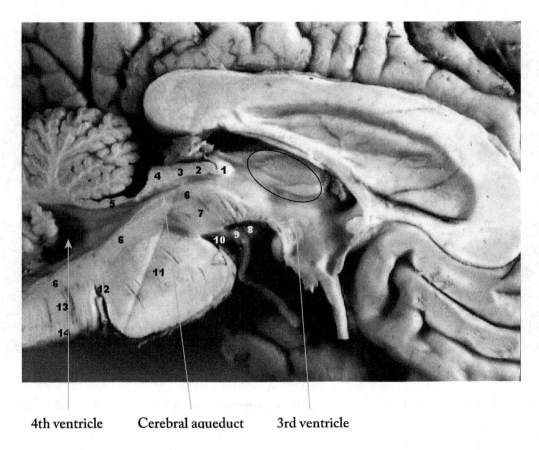

4th ventricle   Cerebral aqueduct   3rd ventricle

Figure 10.34 Third and fourth ventricles of the brain.

pellucidum, a third and fourth ventricle. The lateral ventricles are located within the cerebral hemispheres. The third ventricle lies between the two halves of the thalamus in the diencephalon. The fourth ventricle lies between the brainstem and the cerebellum (Figs. 10.34, 10.35). The ventricles are all connected via foramen (holes) or tubular passages. The lateral ventricles connect to the third ventricle via the interventricular foramen. The third ventricle connects to the fourth via a tube passing through the midbrain called the cerebral aqueduct (aqueduct of Sylvius). The fourth ventricle connects with the central canal of the spinal cord. The fourth ventricle also connects with the subarachnoid space via lateral and medial apertures. The median aperture is called the foramen of Magendie and the two lateral apertures are called the foramen of Luschka. CSF is produced by the choroid plexi that make about 500 ml/day. However, some of the CSF is absorbed so there is only about 140 ml in the system at any one time. This is due to the CSF being absorbed by arachnoid granulations. Arachnoid granulations are masses of arachnoid tissue located in the dural venous sinuses. CSF can move into the blood at these locations.

The Cranial Nerves

There are 12 pairs of cranial nerves. Eleven of these originate in the diencephalon or brainstem, while one pair originates in the frontal lobe of the brain. The cranial nerves can carry sensory information, motor information, or both. The sensory information consists of touch, pain, and vision. Motor information

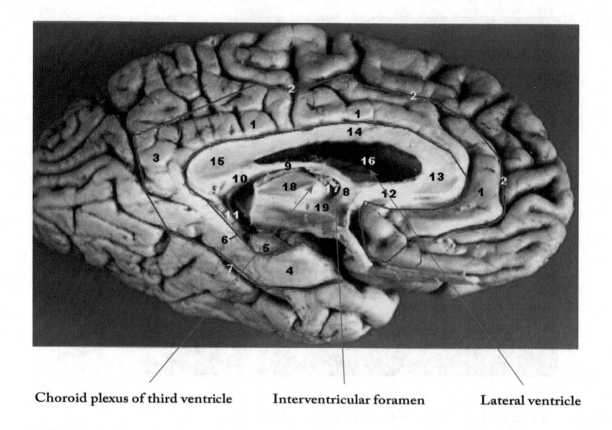

Figure 10.35 CSF circulatory structures.

controls skeletal muscles. Some cranial nerves also carry information for the parasympathetic nervous system. The cranial nerves are usually designated as Roman numerals (I–XII).

Cranial Nerve I Olfactory

The olfactory nerve is a sensory nerve. It carries the information for the sense of smell (Figs. 10.36, 10.37). The olfactory nerve is the only nerve that originates in the frontal lobe of the brain and its fibers pass through the cribriform plate of the ethmoid bone to reach the upper nasal passages. There its receptors collect sensory information in the form of changes in chemical concentrations of substances that are interpreted by the cerebral cortex as smell. The olfactory nerves enter the olfactory bulbs located near the crista galli of the ethmoid bone before entering the cerebrum.

Cranial Nerve II Optic

The optic nerves are sensory nerves. They carry information relating to vision from the retina of the eyes and pass through the optic canals of the sphenoid bone (Fig. 10.38). They then form the optic chiasm before entering the lateral geniculate nuclei of the thalamus. There they synapse with projection fibers that carry the information to the occipital lobe. At the optic chiasm the medial half of the fibers cross over to the opposite side of the brain. A few fibers bypass the lateral geniculate and synapse in the superior colliculus of the midbrain in the brainstem.

Nervous System I: Structure and Function | 219

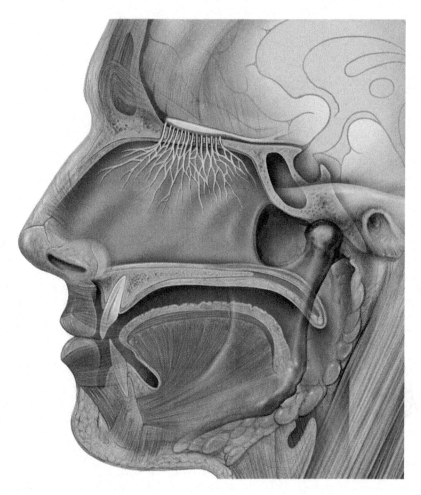

Figure 10.36 Olfactory nerve.

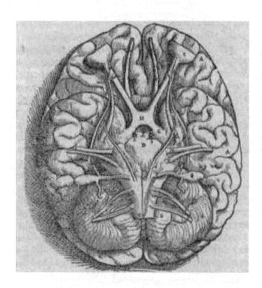

Figure 10.37 Olfactory bulb (highlighted in red). The olfactory fibers travel to the olfactory bulb.

Cranial Nerve III Oculomotor

The oculomotor nerves are motor nerves that innervate some of the muscles of the eye including the superior and inferior rectus, medial rectus, and inferior oblique (Fig. 10.39). They also innervate the levator palpebrae superioris muscles that move the eyelids. When these nerves are damaged, patients will experience an inability to track objects with their eyes (strabismus), which can lead to double vision (diplopia). The oculomotor nerves also carry information for the autonomic nervous system that changes the pupil size. These fibers synapse in the ciliary ganglion.

Cranial Nerve IV Trochlear

The trochlear nerves are motor nerves that innervate the superior oblique muscles of the eyes (Fig. 10.40). The nerves derive their names from their location

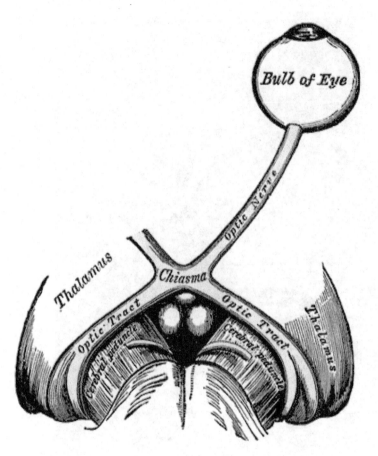

Figure 10.38 Optic nerve.

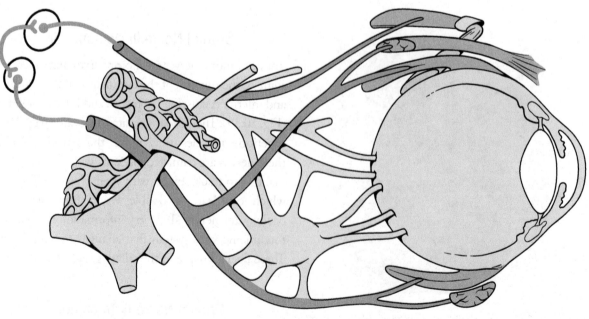

Figure 10.39 Oculomotor nerve.

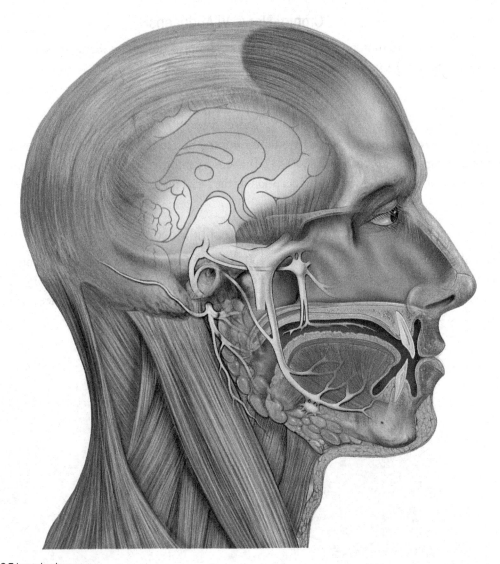

Figure 10.40 Trigeminal nerve.

near a ligamentous structure called the trochlea. The trochlea connects to the superior oblique muscle and acts as a pulley.

Cranial Nerve V Trigeminal

The trigeminal nerves are mixed nerves carrying both sensory and motor information. The trigeminal nerves originate in the pons. They form a large semilunar ganglion before splitting into three divisions. The superior ophthalmic branch carries sensory information from the upper portion of the face above the eyelids. The middle maxillary branch carries sensory information from the middle portion of the face from below the lower eyelid to the upper lip. The lower mandibular branch carries sensory information from the mandible. The mandibular branch also carries motor information to the muscles of mastication, including the masseter and temporalis.

Cranial Nerve VI Abducens

The abducens nerves are motor nerves carrying information to the lateral rectus muscles of the eyes. If the abducens nerve is damaged, the eye will move inward.

Cranial Nerve VII Facial

The facial nerves are mixed nerves. They carry motor information to the muscles of the face and are responsible for producing facial expressions (Fig. 10.41). The sensory information consists of taste from the anterior two thirds of the tongue along with proprioception of the facial muscles and deep pressure in the face.

Cranial Nerve VIII Vestibulocochlear

The vestibulocochlear nerves are sensory nerves. They carry sensory information regarding hearing, balance, and equilibrium from the inner ear. The nerves form two branches. A vestibular branch innervates the vestibule and semicircular canals of the ear and carries information related to balance and equilibrium.

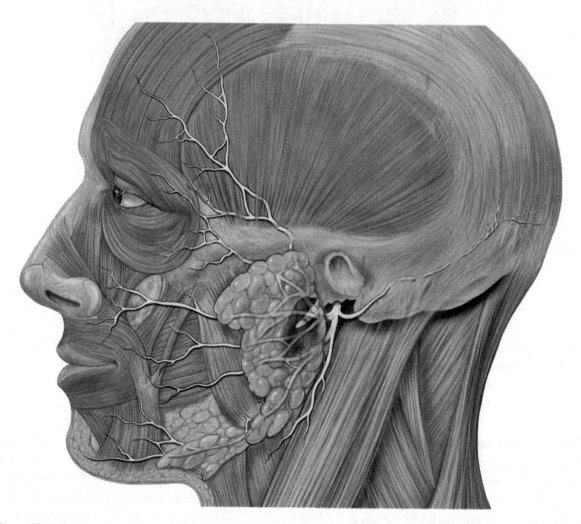

Figure 10.41 Facial nerve.

A cochlear branch carries hearing information from the cochlea of the inner ear.

Cranial Nerve IX Glossopharyngeal

The glossopharyngeal nerves are mixed nerves (Fig. 10.42). They carry sensory information regarding taste from the posterior one third of the tongue as well as motor information to the muscles in the pharynx for swallowing.

Cranial Nerve X Vagus

The vagus nerves are mixed nerves. They carry sensory information from the viscera of the esophagus, respiratory tract, and abdomen. They carry motor information to the heart, stomach, intestines, and gallbladder. The vagus nerves also carry information for coordination of swallowing. The vagus nerves are important autonomic nervous system nerves.

Cranial Nerve XI Spinal Accessory

The spinal accessory nerves are motor nerves. They carry information to the muscles of the neck and upper back, including the sternocleidomastoid and trapezius. What is unique about the spinal accessory nerves is that some of the motor fibers originate in the anterior gray horns of the first five cervical segments of the spinal cord. These fibers enter the foramen magnum and join fibers originating in the medulla oblongata. The combined fibers then exit the cranium at the jugular foramen and divide into two branches. The internal branch joins the vagus nerve and innervates the vocal cords, pharynx, and soft palate. The external branch controls the sternocleidomastoid and trapezius muscles.

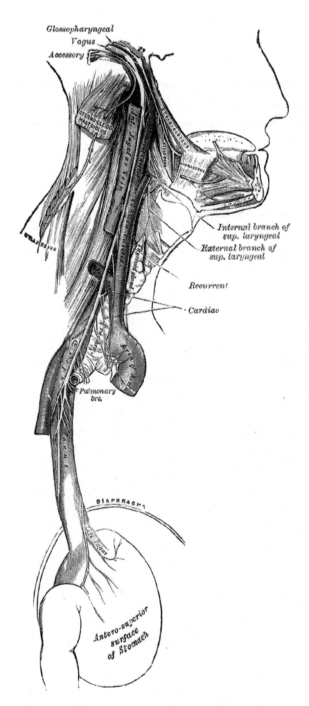

Figure 10.42 Glossopharyngeal nerve (yellow).

Cranial Nerve XII Hypoglossal

The hypoglossal nerves are motor nerves. They carry primarily motor information to the muscles that move the tongue. One way to check the hypoglossal nerves is to ask the patient to stick their tongue out. Deviation of the tongue from one side to the other indicates a problem with the hypoglossal nerve.

The Autonomic Nervous System

The autonomic nervous system can be thought of as an "automatic" system because it works to maintain homeostasis in the body even when it is in an unconscious state (Fig. 10.43). The autonomic nervous system (ANS) can control respiratory, cardiovascular, urinary, digestive, and reproductive functions.

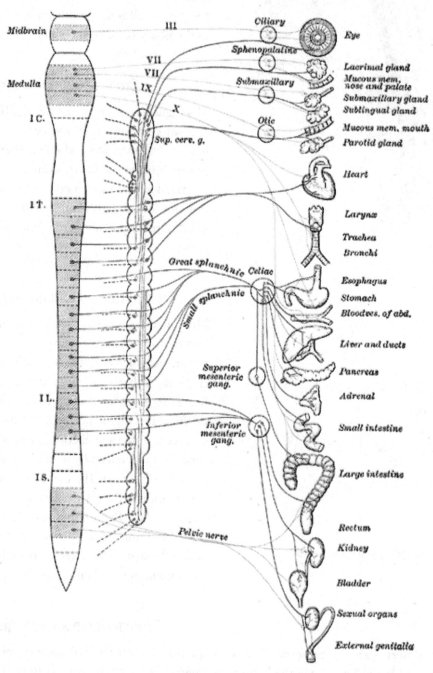

Figure 10.43 Autonomic nervous system. The sympathetic division is colored red, the parasympathetic division is colored blue.

It works to maintain balance of fluids, electrolytes, blood pressure, nutrients, and blood gases. The ANS does this by sending motor impulses to viscera, and cardiac and smooth muscles. Since it sends motor impulses to viscera, the ANS is also known as a visceral motor system. The ANS is divided into two subdivisions. The sympathetic is often referred to as the fight-or-flight system. It is located in the thoracic and lumbar spines and sends fibers to the viscera. The parasympathetic division begins in the cervical and lower lumbar spines and sends fibers to the same viscera as the sympathetic. The sympathetic and parasympathetic divisions typically have the opposite effect on organs and thus work to maintain balance based on the body's needs. For example, the sympathetic system can increase heart rate while the parasympathetic system decreases it.

Sympathetic Nervous System

The sympathetic nervous system works to increase heart rate, dilate air passages, increase activity of sweat glands, increase glucose levels in the blood, dilate the pupils, and decrease digestive activity. It can increase the amount of blood moving to the cardiac and skeletal systems while decreasing blood flow to the skin. It also decreases urinary activity. The sympathetic nervous system must send two neurons to the viscera. The first neuron has its cell body in the brainstem or spinal cord. Its axon extends from the ventral roots of the spinal nerves to the paravertebral ganglia (sometimes called the sympathetic chain) of the spinal cord. The sympathetic chains are located on each side of the spinal cord. Each sympathetic chain consists of 3 cervical, 10–12 thoracic, 4 or 5 lumbar, and 4 or 5 sacral ganglia. These fibers are known as preganglionic fibers. They synapse with fibers in the ganglia that are postganglionic. The preganglionic fibers then are short, while the postganglionic fibers are long.

The spinal nerves send two branches to the paravertebral ganglia. One branch contains fibers from the spinal nerve traveling to the ganglia. This branch is called the white communicating ramus. It is white due to the myelinated neurons. The other branch carries unmyelinated fibers from the ganglion to the viscera. This branch is called the gray communicating ramus. Once the preganglionic fibers reach the paravertebral ganglia, they synapse with a postganglionic neuron at that level, synapse with a postganglionic neuron at another level, or don't synapse with another neuron but exit the ganglia as splanchnic nerves. The splanchnic nerves synapse with postganglionic neurons at ganglia called the prevertebral ganglia. The prevertebral ganglia are located near the abdominal aorta and form a plexus there (Fig. 10.44). There are three ganglia associated with the abdominal aortic plexus, which include the superior mesenteric, inferior mesenteric, and celiac. The celiac ganglion is sometimes called the solar plexus. It is important to understand the role of the adrenal gland in the sympathetic nervous system. The adrenal glands are pyramid-shaped glands that sit on top of the kidneys. They consist of two parts; an outer cortex and an inner medulla. The adrenal medulla contains neurons that secrete epinephrine, norepinephrine, and dopamine. Preganglionic sympathetic fibers synapse with neurons in the adrenal medulla. Preganglionic fibers of the sympathetic nervous system are classified as cholinergic. This means that they respond to the neurotransmitter acetylcholine. Postganglionic sympathetic fibers are classified as adrenergic, which means they respond to adrenaline. There are exceptions, however, and these include the postganglionic fibers that innervate sweat glands and some superficial blood vessels. These structures respond to acetylcholine and are therefore considered cholinergic.

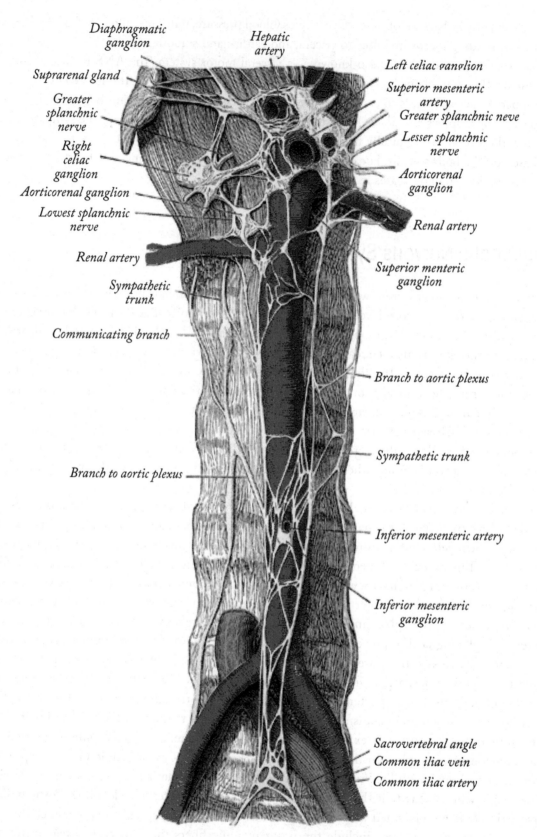

Figure 10.44 The abdominal aortic plexus contains the celiac and mesenteric ganglia.

Parasympathetic Nervous System

The parasympathetic division of the ANS originates in the brain and sacral region of the spinal cord. It is sometimes referred to as the craniosacral division of the ANS. The cell bodies lie in the brainstem (midbrain, pons, medulla oblongata) as well as in sacral segments 2–4 of the spinal cord. The preganglionic neurons are much longer than those in the sympathetic nervous system. They synapse at terminal ganglia near the target organs. The parasympathetic nervous system fibers in the brain exit via cranial nerves. The oculomotor nerve carries parasympathetic fibers that control the lens and pupil of the eye. Preganglionic fibers enter the ciliary ganglion (behind the eye) and synapse with postganglionic fibers that innervate the ciliary and pupillary constrictor muscles. The facial nerve also carries parasympathetic fibers. These fibers control the tear, salivary, and nasal glands. The glossopharyngeal nerve carries fibers that control salivation. The vagus nerve carries a large number of preganglionic fibers. These fibers travel to plexi including the cardiac, esophageal, and pulmonary plexi. The fibers emerge from the plexi and continue through the diaphragm to innervate organs in the abdominal cavity such as the pancreas, stomach, intestines, kidney, ureter, liver, and part of the colon. Parasympathetic fibers from the sacral segments form pelvic splanchnic nerves that continue to the inferior hypogastric plexus. Most of these become pelvic nerves that innervate the reproductive organs, rectum, urinary bladder, and the remainder of the colon. Preganglionic and postganglionic parasympathetic fibers are considered cholinergic.

Autonomic Neurotransmitters

There are two types of adrenergic receptors: alpha and beta. Both of these respond to adrenaline but can elicit different effects in different organs. Norepinephrine is broken down by monoamine oxidase via the process of reuptake. Alpha adrenergic receptors are generally excitatory while beta receptors are inhibitory. However, it is a good idea not to generalize, as there are exceptions due to subclasses of receptors such as alpha 1, alpha 2, beta 1, beta 2. For example, if norepinephrine (NE) stimulates alpha receptors on blood vessels, they vasoconstrict. However, if NE stimulates beta receptors on blood vessels in skeletal and cardiac muscle, they vasodilate. There are also two types of cholinergic receptors. Muscarinic receptors are located in the membranes of target tissue of postganglionic parasympathetic neurons. Nicotinic receptors are located between pre- and post-ganglionic neurons. Acetylcholine is degraded by acetylcholinesterase located in the post-synaptic membrane.

Organ	Autonomic Innervation	Type of Receptor	Action
Eye: pupil	sympathetic parasympathetic	alpha muscarinic	dilation of the pupil constriction of the pupil
Eye: ciliary muscle	sympathetic parasympathetic	beta muscarinic	allows far vision allows near vision
Tear glands	sympathetic parasympathetic	beta muscarinic	vasoconstriction secretion of tears
Salivary glands	sympathetic parasympathetic	alpha muscarinic	vasoconstriction and secretion of mucus with a low enzyme count secretion of watery saliva with a high enzyme count

Organ	Autonomic Innervation	Type of Receptor	Action
Heart	sympathetic parasympathetic	beta alpha muscarinic	dilation of coronary arteries, increased heart rate, increased force of contraction, increased rate of pacemaker conduction coronary artery constriction slows, heart rate, reduces contraction and conduction, constricts coronary arteries
Bronchii	sympathetic parasympathetic	beta muscarinic	dilation constriction and mucus secretion
Esophagus	sympathetic parasympathetic	alpha muscarinic	vasoconstriction peristalsis, secretion of mucus
Stomach and Intestines	sympathetic parasympathetic	beta alpha muscarinic	inhibition of peristalsis and secretion vasoconstriction, sphincter contraction peristalsis and secretion
Spleen	sympathetic	alpha	contraction
Adrenal medulla	sympathetic	-	adrenaline and noradrenaline secreted into the bloodstream
Liver	sympathetic	beta	break down of glycogen (glycogenolysis)
Gallbladder	sympathetic parasympathetic	beta muscarinic	relaxation contraction
Pancreas	sympathetic	alpha beta	inhibition of insulin secretion stimulation of insulin secretion
Descending colon	sympathetic parasympathetic	alpha beta muscarinic	vasoconstriction inhibition of peristalsis and secretion peristalsis and secretion
Sigmoid colon, rectum and anus	sympathetic parasympathetic	alpha beta muscarinic	constriction of sphincter muscles inhibition of peristalsis and secretion peristalsis and secretion
Bladder	sympathetic parasympathetic	alpha beta muscarinic	contraction of sphincter relaxation of detrusor muscle contraction of detrusor muscle
Penis	sympathetic parasympathetic	- muscarinic	ejaculation erection
Clitoris	parasympathetic	muscarinic	erection
Uterus	sympathetic	alpha beta	contraction relaxation
Blood vessels in:			
Skin	sympathetic	alpha	constriction
Muscle	sympathetic	cholinergic	dilation
Sweat glands except palm of hands	sympathetic	muscarinic	sweating
sweat glands on palms of hands	sympathetic	alpha	sweating
Arrector pili muscles at root of body hair	sympathetic	alpha	piloerection (making hair "stand on end")
Adipose tissue	sympathetic	beta	lipolysis (break-down of fat to release energy)

Drugs

A drug is, generally speaking, any substance that changes the way your body works. Some drugs have a medicinal effect, and some are used recreationally. They have diverse effects, depending on the drug. Drugs can do anything from diminish pain to preventing blood clots to helping a depressed person. Different drugs work in different ways, called the mechanism-of-action. The drugs covered here will all act on the nervous system via receptors on different neurons. There are also drugs that change how enzymes work, but that's not part of the nervous system (at least directly) and will not be discussed here. You've probably heard the terms stimulant (excitatory) and depressant (inhibitory). This is a broad way of classifying drugs that work on the CNS. Depressants slow down neural function, and stimulants speed it up. Most of the common depressants (including alcohol, benzodiazepines, barbiturates, and GHB) work on GABA receptors, although there are others. Opiates, for example, work on mu opioid receptors and also produce inhibitory effects, and some antipsychotics block serotonin. See the alcohol section below to see one way this can work. Stimulants work mostly with epinephrine, dopamine, or serotonin (or a combination of them). Many of them either mimic one, or stop them from leaving the synapse, causing more action potentials to be fired. Methamphetamine, discussed below, is a fairly typical stimulant drug.

Drug Abuse

Scientists have long accepted that there is a biological basis for drug addiction, though the exact mechanisms responsible are only now starting to be being identified. It is believed that addictive substances create dependence in the user by changing the brain's reward functions, located in the mesolimbic dopamine system—the part of the brain that reinforces certain behaviors such as eating, sexual intercourse, exercise, and social interaction. Addictive substances, through various means and to different degrees, cause the synapses of this system to flood with excessive amounts of dopamine, creating a brief rush of euphoria more commonly called a "high." Some say that abuse begins when the user begins shirking responsibility in order to afford drugs or to have enough time to use them. Some say it begins when a person uses "excessive" amounts, while others draw the line at the point of legality, and others believe it amounts to chronic use despite degenerating mental and physical activity. Below are some drugs that are abused frequently: acid/LSD, alcohol, various tryptamines and phenethylamines, cocaine, ecstasy/MDMA, heroin, inhalants, marijuana, methamphetamine, PCP/phencyclidine, prescription medications, smoking/nicotine, and steroids.

Alcohol

Alcohol is, and has been for thousands of years, one of the most commonly used drugs in the world. It is legal, with some restrictions and exceptions, nearly everywhere. It is a common misconception that somehow alcohol is "better" or "safer" than other recreational drugs. This is simply NOT the case. Alcohol is a depressant, and as such it has the potential to cause coma, respiratory depression/arrest, and possibly death. Compared with some other (illegal in most places) drugs of recreational value (such as marijuana and serotonin-based hallucinogens like LSD or psilocybin) alcohol is far more toxic and has more risk of overdose. That doesn't mean that moderate drinking will probably hurt you, though. Short-term effects from drinking (listed roughly as they appear, and as dosage goes up) are: decreased inhibitions and thusly judgment, flushing of the face, drowsiness, memory problems begin, severe motor impairment, blurry vision, dizziness, confusion, nausea, possible unconsciousness, coma, death (due to respiratory arrest or possibly aspiration on vomit). Alcohol produces these effects mainly via the GABA receptors in the brain. When GABA (or in this case alcohol) binds to its receptor, it lets either Cl^2 ions in or K^1 out. This is called hyperpolarization, or an inhibitory post-synaptic potential (IPSP). It makes it harder for the neuron to depolarize and hence harder for it to fire an action potential, slowing neural function. At higher doses, alcohol will start to block NMDA. NMDA is involved in memory (see the long-term potentiation section) so this is thought to account for memory blackouts.

Methamphetamine

In the U.S., medically prescribed methamphetamine is distributed in tablet form under the brand name Desoxyn®, generally for Attention Deficit Hyperactivity Disorder (ADHD) but also for narcolepsy or obesity. Illicit methamphetamine comes in a variety of forms. Most commonly it is found as a colorless crystalline solid, sold on the street under a variety of names, such as crystal meth or crystal. Methamphetamine may also be referred to as shards, rock, pony, crissie, crystal, glass, ice, Jib, critter, Tina, tweak, or crank. "Dope" may refer to methamphetamine or other drugs, especially heroin or marijuana. The term "speed" can denote any stimulant including other amphetamines (e.g., Adderall), cocaine, and methylphenidate (Ritalin). Methamphetamine can be injected (subcutaneous, intramuscular, or intravenous), smoked, snorted, swallowed, or used rectally or sublingually; the latter two being fairly uncommon. After administration, methamphetamine takes from a few seconds (smoked or injected IV) to around 30 minutes (oral) for effects to arise, lasting around eight hours depending on the route of administration. Effects/side effects are euphoria, anorexia, increased energy, clenching of the jaw/grinding of teeth (bruxism), weight loss, insomnia, tooth decay, and psychosis, among others. Methamphetamine is neurotoxic to at least some areas of the brain, and owes most of its effects to the neurotransmitters dopamine, norepinephrine, and serotonin it releases. It also blocks the reuptake of those neurotransmitters, causing them to stay in the synaptic cleft longer than normal.

Marijuana

Marijuana contains a myriad of chemicals, called cannabinoids, that have psychoactive and medicinal effects when consumed, the major one being tetrahydrocannabinol (THC). THC serves to mimic the endogenous neurotransmitter anandamide (also found in chocolate) at the CB1 receptors in the brain. Other cannabinoids include Cannabidiol (CBD), cannabinol (CBN), and tetrahydrocannabivarin (THCV). Although THC is found in all parts of the plant, the flower of the female plant has the highest concentration, commonly around eight percent. The flowers can be used, or they can be refined. Trichomes contain most of the THC on the flowers and can be removed by a few different methods. These removed trichomes are called kief. Kief can, in turn, be pressed into hashish. By far the most common way to consume any of these products is by smoking, but it can be taken orally as well.

Cannabis has a very long, very good safety record. Nobody on record has ever died because of marijuana, directly at least. It is estimated that it would take 1–1.8 kilograms of average-potency marijuana, taken orally, to have a fifty percent chance of killing a 68-kg human. Despite this, the possession, use, or

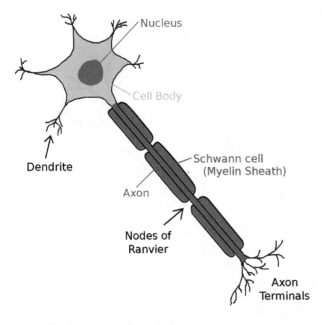

Figure 10.45 Structure of a typical neuron.

sale of psychoactive cannabis products became illegal in many parts of the world in the early 20th century. Since then, while some countries have intensified the enforcement of cannabis prohibition, others have reduced the priority of enforcement to the point of de facto legality. Cannabis remains illegal in the vast majority of the world's countries. The nature and intensity of the immediate effects of cannabis consumption vary according to the dose, the species or hybridization of the source plant, the method of consumption, the user's mental and physical characteristics (such as possible tolerance), and the environment of consumption. This is sometimes referred to as set and setting. Smoking the same cannabis either in a different frame of mind (set) or in a different location (setting) can alter the effects or perception of the effects by the individual. Effects of cannabis consumption may be loosely classified as cognitive and physical. Anecdotal evidence suggests that the Cannabis sativa species tends to produce more of the cognitive or perceptual effects, while Cannabis indica tends to produce more of the physical effects.

Nervous System Physiology

The primary cell that carries information in the nervous system is called the neuron (Fig. 10.45). The basic parts of a neuron include the cell body, dendrites, and axon. There are a number of different types of neurons. The primary neuron we will be concerned with is the multipolar neuron. This neuron has a number of processes extending from the cell body, with one being the axon. Other types of neurons include bipolar and unipolar neurons (Fig. 10.46). Bipolar neurons have two processes extending from their cell bodies. One is a dendrite and the other an axon. These are found in the special senses such as the eye, ear, and nose. Unipolar neurons have one process extending from their cell bodies. On one side, the process branches into dendrites. The other side enters the brain or spinal cord. The cell bodies sometimes reside in ganglia.

Neurons have a number of other parts. The cell body or perikaryon contains many of the cell organelles. These include mitochondria, microtubules, Golgi apparatus, and a granular cytoplasm. The cell body also contains Nissl Bodies, which are membranous packets of chromatophilic substance consisting of rough endoplasmic reticulum. Some neurons are myelinated (white matter). Their axons are surrounded by a covering of myelin. In the peripheral nervous system, a type of cell called a Schwann cell is responsible for producing the myelin sheath. Axons connect to the cell bodies of neurons via the axon hillock. The axon hillock is important in the process of producing an electrical stimulus called an action potential. The axon hillock has a large number of sodium- and potassium-gated channels. Axon terminals are located at the distal ends of axons. The axon terminals contain synaptic vesicles containing neurotransmitter.

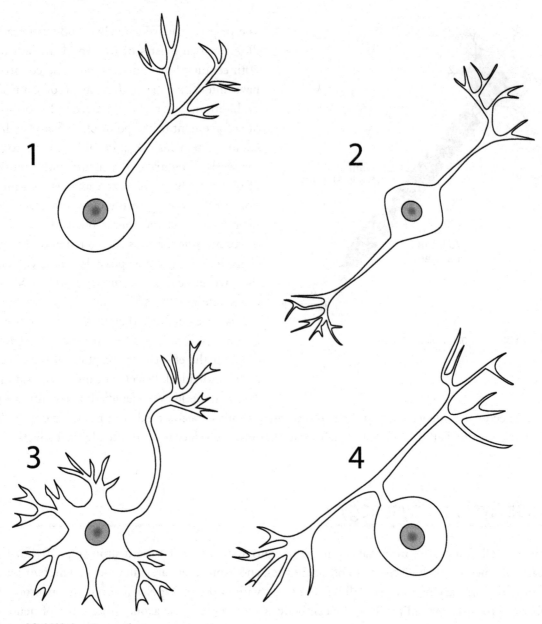

Figure 10.46 Various types of neurons.
1: Unipolar neuron
2: Bipolar neuron
3: Multipolar neuron
4: Pseudounipolar neuron

Neuroglia

The nervous system also contains cells that support neurons called neuroglia. There are a few different types of neuroglia that have a number of important functions. Astrocytes work to provide structural support and may also help in regulating electrolytes within the interstitium (Fig. 10.47). They are star

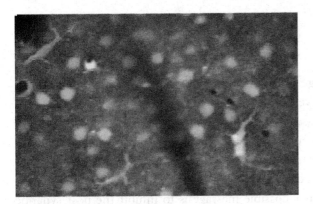

Figure 10.47 Astrocytes (green) in the cortex.

shaped and can be found between neurons and blood vessels. Astrocytes help to maintain the blood–brain barrier. They also help to repair damaged areas in the central nervous system by forming scar tissue. Oligodendrocytes produce the myelin that surrounds white matter axons in the brain and spinal cord. Ependymal cells form the lining of the central canal and ventricles in the spinal cord and brain. They are also found in the choroid plexi of the brain. They help to produce CSF by providing a porous membrane for blood plasma to pass through. Microglia are very small cells that are located throughout the central nervous system. They provide support and help to clean up debris through phagocytosis.

Neural Communication: Resting Membrane Potential

The nervous system is a vast information network. Neurons send messages to others that can either stop the message or pass it along. Next we will examine how neurons communicate. We will begin by learning about a concept known as the resting membrane potential. Neurons, like many other cells in the body, do not exist at equilibrium with their surroundings. In fact there is a net negative charge on the inside of the neuron with respect to the outside. This negative charge exists mostly because of differences in membrane permeability to different electrolytes. It turns out that the cell membranes of neurons are slightly permeable to sodium and potassium. Although they are permeable to both sodium and potassium, they are slightly more permeable to potassium. There are also a number of negatively charged ions inside of the neuron's cell membrane. These include phosphates, sulfates, ATP, RNA, and proteins. These negatively charged ions cannot leave the cell. So if potassium (which is positively charged) is allowed to move out of the cell, then the inside of the cell becomes more negative (due to the presence of the negative ions) than the outside of the cell. As this ionic gradient increases, some positive ions are attracted back into the cell. Eventually the cell reaches a steady state by which potassium diffuses out of the cell at the same rate that it moves into the cell via the ionic gradient.

There is much more sodium outside of the cell than inside. The neuron's cell membrane is not very permeable to sodium so just a little sodium moves into the cell via its concentration gradient. We also have the sodium-potassium pump working to maintain both sodium and potassium gradients by moving sodium out of the cell and potassium into the cell. Remember that the sodium-potassium pump requires energy in the form of ATP. The nervous system has a lot of these pumps in order to function. In fact about 70% of the energy used by the nervous system is used by the sodium-potassium pumps. So, if we put all of these effects together, we end up with a net negative charge on the inside of the cell with respect to the outside. This negative charge is approximately -70 millivolts (mV) and is called resting membrane potential.

Depolarization

Neurons communicate by sending chemical messages from one neuron to another. These chemicals are called neurotransmitters. The neurotransmitters move from one neuron to another across an area known as the synaptic cleft. The neuron sending the message is called the presynaptic neuron. The neuron receiving the message is called the post-synaptic neuron. Once the neurotransmitter floats across the synaptic cleft, it attaches to a receptor on the post-synaptic neuron. There are two possible messages carried by neurotransmitters. One is to trigger the post-synaptic neuron to send another message. This essentially moves the information forward. The other possible message is to inhibit the post-synaptic neuron (hold the information back). In order to trigger the post-synaptic neuron, the neurotransmitter will cause the opening of sodium gates on the post-synaptic neuron. In other words, the presynaptic neuron is said to be excitatory. When the sodium gates open, sodium rushes into the neuron. This changes the potential by making it less negative due to the positive sodium ions rushing into the cell. We say the cell is depolarizing. Remember that the resting membrane potential is negative (-70mV). In other words, the cell is polarized to begin with. Once the sodium gates open causing the cell to become less negative, there is less polarization. So the cell is depolarizing with the opening of sodium gates.

Threshold

If the stimulus is strong enough to cause enough of a change in potential to reach a certain level, the neuron will react by opening more sodium gates and depolarizing at a rapid rate. In neurons, the level is at about -55mV. In other words, if a stimulus is great enough to cause a neuron to depolarize to -55mV, then we say that it has reached the threshold. Once the neuron reaches the threshold it will continue to depolarize to about +30mV. The rapid change in potential from -55mV to +30mV is called an action potential. This is called the all-or-none principle, which means that once the threshold is reached, the neuron continues through the cycle of depolarization and repolarization to resting membrane potential (Fig. 10.48).

Action Potentials

Action potentials are generated at the axon hillock of neurons. There are a large number of sodium gates that react to changes in membrane potential. These sodium gates are called voltage-gated sodium channels because they open in response to a change in membrane potential. When a stimulus causes depolarization to the threshold, the voltage-gated sodium channels open, causing more voltage-gated channels to open, resulting in a large influx of sodium into the cell. The action of sodium channels causing more sodium channels to open is a positive feedback system. Voltage-gated potassium channels also open at the same time as the sodium channels. The potassium channels work more slowly than the sodium channels. The result is that some potassium diffuses out of the cell but much more sodium diffuses in. After the maximum depolarization is reached at about +30mV–+40mV, the sodium gates close and the potassium gates remain open, allowing potassium to diffuse out of the cell. This causes the

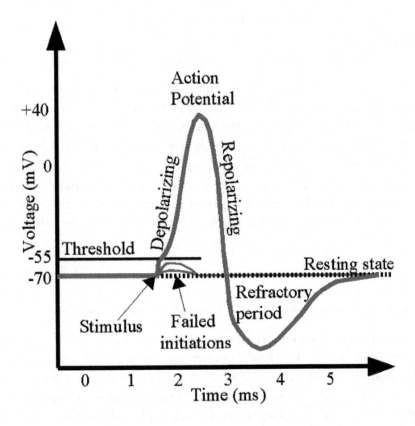

Figure 10.48 Events of an action potential.

membrane potential to become more negative. This occurs until the resting membrane potential is reached. Some neurons become slightly more negative for a brief time after an action potential. This is due to the voltage-gated potassium channels remaining open beyond the normal resting membrane potential. Once the potassium channels close, the cell returns to resting membrane potential. The sodium-potassium pump also helps to restore and maintain resting membrane potential.

Refractory Periods

Once an action potential is generated, the neuron will not respond to further stimuli for a period of time. This is known as the refractory period. There are two parts to this period. The first part is known as the absolute refractory period. During this time the neuron will not respond to any additional stimulus. The absolute refractory period occurs when the neuron is depolarizing due to the opening of sodium gates and continues to near the end of the repolarization phase. The second part is known as the relative refractory period. During this time the neuron will respond to a strong stimulus. Potassium channels are open during this time. The absolute refractory period ensures that neurons will not enter a state of continuous depolarization. It also sets a limit as to the frequency of action potentials generated.

Stimuli

Neurons are stimulated by neurotransmitters. The secretions of neurotransmitters can cause a graded response in membrane potentials. In other words, one neuron may send a certain amount of neurotransmitter to another neuron that is not strong enough to depolarize the second neuron to the threshold. This is called a subthreshold stimulus. If the same neuron continues to secrete more neurotransmitter, the second neuron may reach threshold and stimulate an action potential. This is called a threshold stimulus. The second neuron will respond to a threshold stimulus by generating one action potential. If the first neuron continues to secrete more neurotransmitter, the second neuron will continue to generate action potentials at a maximal frequency. This is called a maximal stimulus. If the stimulus is even stronger, the second neuron will respond by continuing to generate action potentials at the maximal frequency. This is known as a supramaximal stimulus. A stimulus between the threshold and maximal stimulus is known as a submaximal stimulus. The action potential frequency and strength of stimuli are thus related.

Propagation of Action Potentials

Once an action potential is generated at the axon hillock, it will be transmitted by the axon to the end of the axon terminals. We say the action potential propagates down the axon. This occurs much like a row of dominos falling over. One section of the axon stimulates the next, causing another action potential, which stimulates the next section, and so on. Another way to think of this process is as a wave of depolarization that moves down the axon. The process of propagation occurs because of local currents generated in adjacent sections of the axon. The outside of the axon's membrane becomes more negative as positive ions move inside the cell. At the same time the inside of the cell becomes more positive. The adjacent section has the opposite characteristics setting up a current that influences it. The action potential moves in only one direction because of the absolute refractory period. This method of propagation occurs in unmyelinated axons (Fig. 10.49).

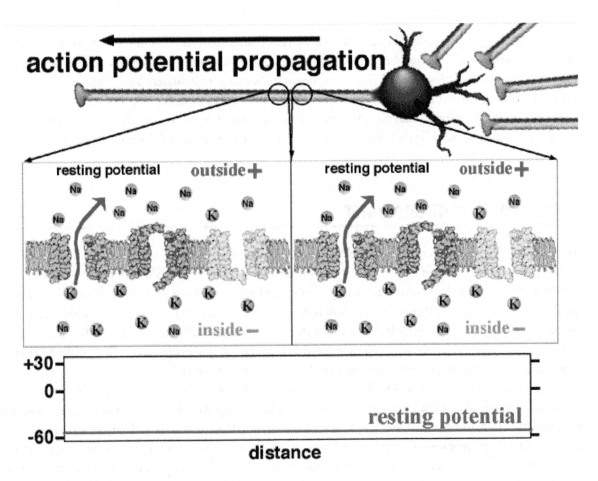

Figure 10.49 Propagation of an action potential.

Saltatory Conduction

Myelinated axons contain Schwann cells that produce myelin. The myelin is discontinuous with gaps called Nodes of Ranvier. Myelin is a lipid substance that acts as an insulator. This causes the action potential to move from one Node of Ranvier to the next. The nodes contain a high concentration of sodium gates. The action potential then appears to jump from node to node. This type of conduction is called saltatory conduction. Action potentials move much faster in myelinated versus unmyelinated axons. Let's illustrate this with an analogy. We have two rows of students with 12 students in each row. The student at the end of each row has a ball and has to get it to the student to the other end of the row as quickly as possible. The first row is given instructions to pass the ball from one student to the next until reaching the last student. The second row is told to throw the ball to the 4th student who then throws it to the 8th student and so on until reaching the 12th student at the end of the row. The instructor tells the students to begin at the same time. Which row will win the race? Obviously the second row wins because time is lost with the handling of the ball by every single student in the row versus every 4th student. The first row then represents an unmyelinated axon, while the second row represents a myelinated axon. The speed of conduction also relies on the thickness of the myelin sheath as well as the diameter of the axon. Greater myelination and larger diameter axons conduct action potentials faster. Nerve fibers are classified according to their size. Type A fibers are large-diameter myelinated fibers that quickly conduct action potentials (15–120 meters/second). Examples of type A fibers include sensory neurons and motor neurons to skeletal muscles. Type B fibers are medium-diameter myelinated fibers that conduct action potentials more slowly than Type A fibers. Type B fibers can conduct action potentials from 5–15 meters/second. Type C fibers have a small diameter and are unmyelinated fibers that conduct action potentials at 2 meters or less per second. Type B and C fibers are found in the autonomic nervous system.

Release of Neurotransmitter

The result of an action potential is the release of a neurotransmitter from a neuron. This occurs at the distal end of the axon at an area called the axon terminal. Axon terminals contain small packets of neurotransmitters called synaptic vesicles. Action potentials cause voltage-gated calcium channels to open, allowing calcium to diffuse into the axon terminal. Calcium causes the synaptic vesicles to attach to the cell membrane and release their neurotransmitters via exocytosis. The neurotransmitters are released into the synaptic cleft or space between the presynaptic and post-synaptic neurons. They diffuse into the cleft and attach to receptors on the membrane of the post-synaptic membrane.

Once neurotransmitters attach to the post-synaptic membrane, they can elicit one of two responses. They can either depolarize the post-synaptic membrane by causing sodium gates to open or hyperpolarize the membrane by causing potassium of chloride gates to open. After neurotransmitters attach to receptors on post-synaptic membranes, they are quickly degraded. There are two primary methods of degradation. One method involves enzymes in the post-synaptic membrane breaking down the neurotransmitter and allowing it to be recycled. For example, acetylcholine is broken down by acetylcholinesterase in the post-synaptic membrane into choline and acetic acid. Choline is transported back to the presynaptic membrane and combines with acetyl coenzyme A to form acetylcholine. Acetic acid is used to synthesize acetyl coenzyme A. The other method of degradation involves the neurotransmitter moving back to the presynaptic axon terminal where it is recycled. This method is called reuptake. Norepinephrine is transported back into the presynaptic terminal and is recycled or degraded by the enzyme monoamine oxidase (Fig. 10.50).

It is important not to generalize the effects of neurotransmitters. There are a variety of receptors that can either depolarize or hyperpolarize membranes for a given neurotransmitter. For example, norepinephrine can attach to one type of receptor that causes depolarization or a different type of receptor that causes hyperpolarization. Receptors can also exist on presynaptic membranes. For example, norepinephrine receptors on the presynaptic membrane can inhibit the release of more neurotransmitter. This allows some neurotransmitters like norepinephrine to control their own release.

Excitatory and Inhibitory Potentials

There are two responses a post-synaptic neuron can elicit from neurotransmitters: depolarization and hyperpolarization. Depolarization leads to the production of an action potential, so we say the post-synaptic potential is excitatory (excitatory post-synaptic potential EPSP). Likewise if the post-synaptic neuron becomes hyperpolarized in response to a neurotransmitter, there is a lesser likelihood

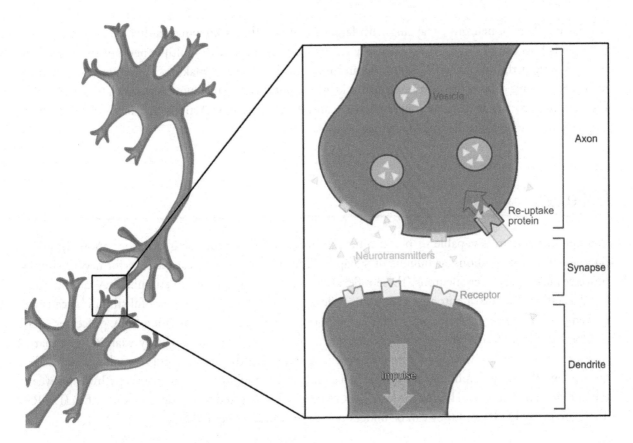

Figure 10.50 Neurotransmitter reuptake.

that an action potential will be generated. We say the potential is inhibitory (inhibitory post-synaptic potential IPSP). Many presynaptic neurons can synapse with one post-synaptic neuron. The effects of these multiple inputs sum to either facilitate or inhibit the production of an action potential. There are two ways in which summation occurs. Spatial summation occurs when multiple presynaptic neurons synapse with one post-synaptic neuron at the same time. Temporal summation occurs when multiple presynaptic neurons synapse with one post-synaptic neuron in a short amount of time. The first neurotransmitter may cause sodium gates to open on the post-synaptic membrane to allow it to depolarize. The second neuron then continues to stimulate the same post-synaptic neuron to depolarize.

Neuron Networks

The nervous system is immensely complex. One neuron can have input from thousands of other neurons to form complex networks. There are, however, three basic structures that are formed. One structure consists of many neurons synapsing with fewer and fewer neurons. This is known as a convergent network. Think of how many neurons it takes to make a decision to contract a muscle. The many neurons involved in making the decision converge to a few neurons that control the muscle. Another structure consists of smaller numbers of neurons synapsing with larger numbers. This is known as a divergent network. An example is how sensory input from a sensory receptor can synapse with multiple neurons in the central nervous system that produce the sensation and may involve decision making. Some networks involve feedback systems where the outputs feed back into inputs. These are known as oscillating networks. Oscillating networks help to prolong an action caused by a stimulus. Actions that are cyclical, such as respiration, may be controlled by oscillating circuits.

Reflexes

The nervous system is capable of performing extremely complex processing of information. Thought processes can take millions or billions of synapses. The nervous system can also perform very simple processes using just a few neurons. These reflexes are automatic responses to stimuli. They can be classified as either somatic or autonomic. Somatic reflexes protect the body from painful stimuli by causing movement away from it. Autonomic reflexes support homeostasis by maintaining body processes such as blood pressure, heart rate, respiration, and urine formation. We will begin by examining somatic reflexes. Health care providers will often use reflexes in assessing the nervous and muscular systems. The quintessential knee-jerk reflex is an example of a simple reflex used to check the neural pathway between the muscle, spinal cord, and brain. The knee-jerk reflex is often referred to as a deep tendon reflex (DTR). There are many DTRs and they can be understood by examining one in detail.

Reflex Arc

Reflexes are involuntary responses to stimuli that occur unconsciously. The deep tendon reflex consists of a muscle, nerve pathway, and the spinal cord. The muscle contains a sensory receptor that senses changes in the stretch of the muscle. This receptor is called a muscle spindle. The muscle spindle contains motor neurons called gamma motor neurons. These neurons begin in the spinal cord and extend to the muscle spindle. When the tendon of the muscle is tapped by a reflex hammer, the muscle spindle senses the change in length of the muscle and sends a message via a sensory neuron (usually in a spinal nerve) to the spinal cord. There it synapses with a motor neuron (again in a spinal nerve) that sends a message to the muscle to contract (Fig. 10.51). Spinal reflexes are also influenced by the central nervous system. Upper motor neurons extend from the brain to the spinal cord. These neurons have an inhibitory effect on reflexes. Lower motor neurons extend from the spinal cord to the muscle. One reason for eliciting reflexes is to differentiate an upper motor neuron versus lower motor neuron problem. If the nervous system is intact, then the reflex will look normal. This means the brain is providing an inhibitory effect on the reflex. In other words, the brain is inhibiting the reflex so it appears normal. The reflex will look exaggerated with damage to upper motor neurons. This occurs in people who have had a stroke. Diminished or absent reflexes will result from problems with lower motor neurons. In other words, the pathway between the spinal cord and muscle is damaged, so the signal cannot get through. This occurs in peripheral nerve problems such as spinal disc ruptures, spinal stenosis, and demyelinating disorders.

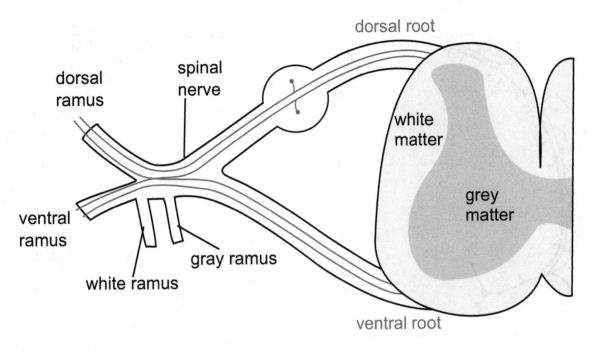

Figure 10.51 Reflex arc.

More Complex Reflexes

Reflexes that respond to painful stimuli are more complex due to the involvement of additional neurons in the spinal cord called interneurons. Examples of these are the withdrawal and crossed-extensor reflexes. The withdrawal reflex incorporates additional interneurons that stimulate the ipsilateral flexor muscles in response to a painful stimulus. For example, touching a hot stove will cause the upper extremity flexors to contract, causing the arm to withdraw from the burner. The crossed-extensor reflex also involves additional interneurons that stimulate the contralateral extensors as well as the ipsilateral flexor muscles. For example, stepping on a nail will cause the lower extremity flexors as well as the contralateral extensor muscles to contract. This allows for further movement away from a painful stimulus.

Functional Areas of the Cerebral Cortex

More complex information processing occurs in the cerebral cortex. We can now revisit the cerebrum to investigate some of the sensory, motor, and association areas. The cerebrum is divided into lobes. The frontal lobe is separated from the parietal lobe by a groove on the lateral aspect of the cerebrum called the central sulcus. The cerebrum also consists of folds called gyri. The first gyrus just anterior to the central sulcus on the frontal lobe is called the precentral gyrus. The gyrus just posterior to the central sulcus is called the post-central gyrus on the parietal lobe. The precentral gyrus is also the primary motor area. The post-central gyrus is called the general sensory area.

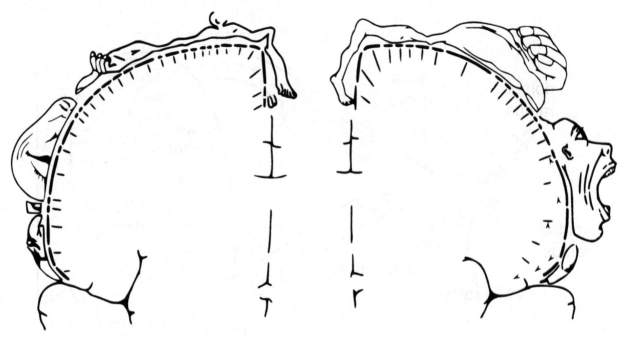

Figure 10.52 Homunculus showing both sensory (blue) and motor (red) areas.

General Sensory Area

The general sensory area on the parietal lobe processes information about pressure, pain, and temperature. This information comes from neurons synapsing in the thalamus, bringing information from spinal tracts to the post-central gyrus. The general sensory area on the post-central gyrus is organized so that the information coming from the feet is processed on the superior aspect of the gyrus, while information from the face is processed on the inferior aspect. The information comes from the contralateral side of the body. A map of the gyrus is called a homunculus (Fig. 10.52).

Other Sensory Areas

The taste area is located at the inferior end of the post-central gyrus and base of the frontal lobe. The sense of smell or olfactory area is located on the inferior aspect of the frontal lobe. The sense of hearing (auditory cortex) is located in the superior aspect of the temporal lobe. The sense of vision (visual cortex) is located in the occipital lobe.

Association Areas

Information about recognition is processed in association areas near the sensory areas. Wernicke's area is located in the lateral parietal and temporal lobes. This area performs speech recognition (Fig. 10.53).

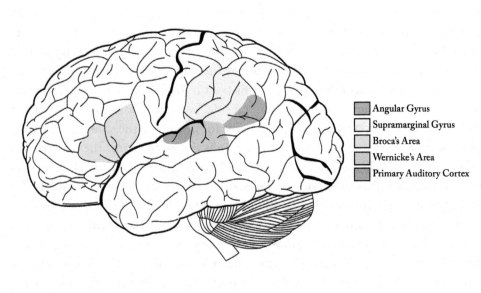

Figure 10.53 Broca's and Wernicke's areas.

Primary Motor Area

The primary motor area is located on the precentral gyrus of the frontal lobe. The neurons are arranged in much the same way as the post-central gyrus, with motor information going to the feet in the superior aspect and motor information going to the face in the inferior aspect. The neurons can be functionally mapped on a homunculus as well. It is important to note the areas on either pre- or post-central gyri are not symmetrical. In other words, there are larger areas corresponding to more complex processing. For example, the motor area for the hands is much larger than the motor area for the knee. This is due to the amount of information processing needed for fine motor movements of the hand versus the knee.

There is also a premotor area located anterior to the primary motor area. This area works with the primary motor area and works to integrate and organize motor information before sending it to the primary motor area. The premotor area also works with decision making occurring in other parts of the frontal lobe. For example, the decision to pick up an object will be made in the frontal lobe and sent to the premotor area where information about various muscle movements is organized. The premotor area then sends the information to the primary motor area in order to stimulate the precise muscles needed to complete the task. The information is then sent, bypassing the thalamus, to the descending motor tracts of the spinal cord, and consequently out to the muscle effectors via spinal nerves.

Broca's area is located in the left cerebral hemisphere. This area helps to coordinate movements of the mouth, larynx, and tongue for speech.

A Detailed Look at a Sensory Motor Pathway

At this point we can take a detailed look at how information flows through the nervous system when spilling a cup of hot tea on my hand and deciding to put it down. I pull up to the drive-through and reach for the cup of hot tea. As I grab the cup, the lid pops off, and hot tea spills on my hand. The sensory information was picked up by temperature and pressure receptors in my hand. The information was converted to electrochemical information called action potentials by the sensory receptors and sent along afferent pathways called nerves to the spinal cord. The spinal nerves transmit the sensory information via the dorsal roots to the posterior portion of the spinal cord so that it can travel via ascending tracts to the brain. The first neuron to carry the information from sensory receptor to the spinal cord is called a primary neuron. This neuron synapses with a secondary neuron that carries the information from the spinal cord to the brain. The temperature information travels via the lateral spinothalamic tract in a secondary neuron. This tract is located in the lateral funiculus of the spinal cord and crosses at cord level to the contralateral side. So if I am holding the cup in my left hand, then the information travels via a cervical spinal nerve (or nerves) to the spinal cord via the dorsal roots to the spinothalamic tract that crosses over to the right side of the cord. The information then ascends to the brainstem (medulla oblongata, pons, midbrain) and synapses with a tertiary neuron in the thalamus. The information then goes to the post-central gyrus of the parietal lobe for processing. The parietal lobe sends the information to other areas of the cortex for interpretation and decision making. The pressure information is picked up by sensory receptors in the skin (Meissner's and Pacinian corpuscles) and sent via a primary neuron to the spinal cord as well. There it synapses with a secondary neuron in the cord and is carried to the thalamus via the dorsal column pathway consisting of the fasciculus gracilis/cuneatus located in the posterior funiculus of the spinal cord. These tracts cross over (decussate) in the medulla oblongata and synapse in the thalamus with a tertiary neuron. This neuron carries the information to the post-central gyrus as well. Now aside from any reflex activity (withdrawal or crossed extensor), I must interpret the sensory information using association areas in the cortex and make a decision using my frontal lobe to set the cup down. The premotor areas will process the decision and work to coordinate the actions of moving the appropriate muscles of my arm to set the cup down. They will send the information to the precentral gyrus of the frontal lobe. The information then reaches the prefrontal gyrus (on the right side) of the frontal lobe and travels toward the brainstem bypassing the thalamus. The primary or upper motor neuron here begins in the precentral gyrus and travels via descending spinal tracts after crossing in the medulla oblongata. The descending tracts include the corticospinal tract. The neurons synapse with lower motor neurons in the anterior horn of the spinal cord. The lower motor neurons carry the information to the skeletal muscles via the ventral root of spinal nerves to spinal nerves to the brachial plexus to terminal nerves to the muscles of my arm and hand. I then can set the cup down.

Processing in the Brainstem

In addition to carrying impulses to and from the brain, the brainstem also processes information. A number of control centers are located in the brainstem. These include centers for controlling heart rate, respiration, digestion, blood gases, and electrolytes. The brainstem also contains the nuclei of all of the cranial nerves, with the exception of the first cranial nerve (olfactory) and eleventh cranial nerve (spinal accessory). One other important function includes modulating sleep and wakefulness. This occurs in the reticular activating system (RAS) (Fig. 10.54). A number of cranial nerves feed information to the RAS including cranial nerves II, V, and VIII. Information from the cerebrum and limbic system also travels to the RAS as well as ascending sensory information. All of these inputs help to arouse consciousness.

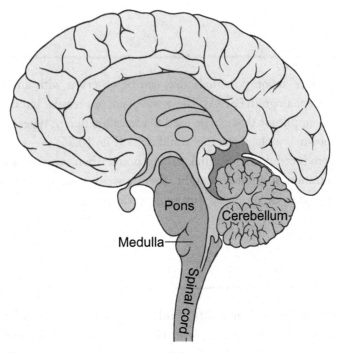

Figure 10.54 The reticular activating system.

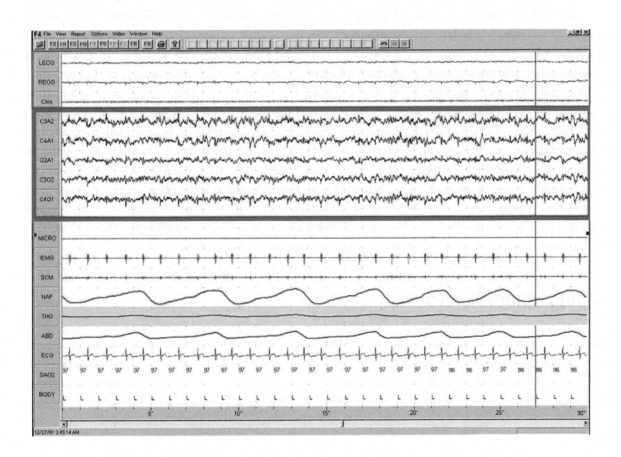

Figure 10.55 An EEG displaying brain waves.

Brain Waves

Clusters of axons produce action potentials that can be observed with an electronic recording device known as an electroencephalogram (EEG). These waves are not regular but do have some distinguishing characteristics, such as amplitude that can be used to identify different types of brain waves. The different types of waves are known as alpha, beta, theta, and delta (Fig. 10.55).

Alpha waves are produced during normal waking hours when subjects are alert. Beta waves are produced during concentration or intense thought. Theta waves are seen in children or in adults who have brain disorders. Delta waves are produced during deep sleep.

Memory

The brain is capable of storing vast amounts of information in its memory. There are two basic types of memory. Short-term memory stores 6–8 pieces of information for brief periods. For example, a telephone number is 7 pieces of information long and can be stored for a short amount of time until the person is asked to remember something else. Long-term memory, as its name states, allows for storage of information for much longer periods of time (as long as a lifetime). Types of long-term memory include declarative and procedural. Declarative is sometimes referred to as explicit and procedural is referred to as implicit. Declarative memory occurs in part of the temporal lobes and the hippocampus and amygdala. The hippocampus is involved in retrieving stored memories and the amygdala stores emotions associated with memories. Declarative memory is also stored in various parts of the cerebrum. Memories are grouped together as well. For example, faces may be stored in a different location than names. Retrieving a memory involves accessing various components and assembling them. Over time, memories decay and can lead to false memories. Procedural memory involves storing skills such as playing an instrument or driving a car. Procedural memories are stored in the premotor area of the cerebrum and cerebellum. Information to be remembered moves from short-term to long-term memory. Neurons in long-term memory actually change in response to storing information. The phenomenon of long-term potentiation occurs when memories are stored. This involves changes in neurotransmitter storage and release as well as protein synthesis. New connections are made and maintained between neurons. This flexible and adaptive characteristic of the brain is known as neural plasticity.

Testing Spinal Nerves

An important part of assessing the nervous system is to assess the function of spinal nerves. There are three primary ways to assess spinal nerves: reflexes, sensory testing, and muscle testing. An overview of assessing a few spinal nerves follows. Spinal nerves are named for the level at which they exit the spine. For example, the C5 spinal nerve exits between vertebral segments C4 and C5.

Testing Dermatomes

Spinal nerves carry sensation from various parts of the body to the spinal cord. These areas are called dermatomes and are named for the spinal nerve associated with that region of the body. Dermatomes are typically assessed by testing pain, light touch, temperature, and discrimination. Pain is tested using sharp and blunt devices touched lightly to the skin. Light touch is tested with a wisp of cotton touched to the skin. Discrimination is often tested with a device that contains two contact points that can be made further apart or closer together.

Reflexes

Deep tendon reflexes associated with a specific spinal nerve are tested with a reflex hammer. Reflexes are graded as:
    0 = absent
    1+ = hypoactive
    2+ = normal
    3+ = hyperactive without clonus
    4+ = hyperactive with clonus

Right and left symmetry is also observed and noted. Remember that upper motor neurons have an influence on spinal reflexes so hyperactive reflexes indicated upper motor neuron problems. Clonus is an involuntary contraction of a muscle when stimulated by reflex testing. Hypoactive reflexes indicate problems with the pathway between the spinal cord and the muscle.

Muscle Testing

Muscle testing (motor strength testing) can be performed by having the subject resist certain movements. The resistance is provided by the examiner. Muscle strength is assessed for symmetry and graded using the following scale:
    0/5 no movement
    1/5 muscle contraction but no joint movement
    2/5 joint movement but not against gravity
    3/5 joint movement against gravity but not against resistance
    4/5 movement against resistance but not normal
    5/5 normal

CRITICAL-THINKING QUESTIONS

1. Studying the embryonic development of the nervous system makes it easier to understand the complexity of the adult nervous system. Give one example of how development in the embryonic nervous system explains a more complex structure in the adult nervous system.
2. Damage to specific regions of the cerebral cortex, such as through a stroke, can result in specific losses of function. What functions would likely be lost by a stroke in the temporal lobe?
3. Why do the anatomical inputs to the cerebellum suggest that it can compare motor commands and sensory feedback?
4. Why can the circle of Willis maintain perfusion of the brain even if there is a blockage in one part of the structure?
5. Meningitis is an inflammation of the meninges, which can have severe effects on neurological function. Why is infection of this structure potentially so dangerous?
6. Why are ganglia and nerves not surrounded by protective structures like the meninges of the CNS are?
7. Testing for neurological function involves a series of tests of functions associated with the cranial nerves. What functions, and therefore which nerves, are being tested by asking a patient to follow the tip of a pen with their eyes?
8. What responses are generated by the nervous system when you run on a treadmill? Include an example of each type of tissue that is under nervous system control.
9. Multiple sclerosis is a demyelinating disease affecting the central nervous system. What type of cell would be the most likely target of this disease? Why?
10. Which type of neuron, based on its shape, is best suited for relaying information directly from one neuron to another? Explain why.
11. What does it mean for an action potential to be an "all or none" event?
12. The conscious perception of pain is often delayed because of the time it takes for the sensations to reach the cerebral cortex. Based on propagation of the axon potential, why would this be the case?
13. Why is the receptor the important element determining the effect a neurotransmitter has on a target cell?
14. A virus causes damage to the sodium channels of the post-synaptic neuron such that they open very slowly in response to a neurotransmitter. How does this modify the generation of an action potential in the post-synaptic neuron?
15. Multiple sclerosis is a disease characterized by degeneration of the myelin sheath of nerve fibers in the brain and spinal cord. Why is the ability to control skeletal muscles affected in this disease?
16. Valium is a drug that triggers the opening of chloride ion channels. Will treatment with Valium depolarize or hyperpolarize a cell? Explain.
17. Predict the effect of a drug that reduces the permeability of the plasma membranes to sodium ions. Would such a drug have any medical applications? Explain.
18. The sciatic nerve of a pithed frog was carefully removed and placed in a nerve chamber with recording electrodes. When both recording electrodes were placed on the surface of the nerve (both were extracellular), a resting membrane potential was not recorded. However, when one electrode was extracellular and the other electrode was intracellular, a resting membrane potential was recorded. Explain these two sets of results.
19. A virus causes damage to the sodium channels of the post-synaptic neuron such that they open very slowly in response to a neurotransmitter. How would this modify the generation of an action potential in the post-synaptic neuron?

Chapter Eleven

Nervous System II: Special Senses

We experience reality through our senses. Senses are the physiological methods of perception, so a sense is a faculty by which outside stimuli are perceived. The senses and their operation, classification, and theory are overlapping topics studied by a variety of fields. Many neurologists disagree about how many senses there actually are due to a broad interpretation of the definition of a sense. Our senses can be split into two different groups. Our exteroceptors detect stimulation from the outsides of our body. For example, smell, taste, and equilibrium. The interoceptors receive stimulation from the inside of our bodies. For instance, blood pressure dropping and changes in glucose and pH levels. Children are generally taught that there are five senses (sight, hearing, touch, smell, taste). However, it is generally agreed that there are at least seven different senses in humans, and a minimum of two more observed in other organisms. Sense can also differ from one person to the next. Take taste for an example: what tastes great to one person will taste awful to someone else. This has to do with how the brain interprets the stimuli that are received. The sensory system relies on specialized structures called sensory receptors. All sensory receptors essentially do the same thing. They collect information in various forms from the environment and convert it to electrochemical impulses (action potentials) for processing by the central nervous system. The environment can be external (outside the body) or internal (inside the body). There are a variety of sensory receptors and they include the following:

- Chemoreceptors that sense changes in chemical concentration.
- Pain receptors (nociceptors) that sense tissue damage.
- Thermoreceptors that sense changes in temperature.
- Mechanoreceptors that sense mechanical deformation of tissue.
- Proprioceptors that sense changes in position of joints.
- Stretch receptors that sense changes in tissue length.
- Photoreceptors that sense changes in light intensity.

Once receptors pick up information and send it to the brain for processing, the brain interprets the information and projects it to the area of stimulation. For example, even though the information regarding pain is processed in the brain, the brain will project the pain to the area of the body that is stimulated. Some receptors can adapt to stimuli. For example, touch receptors in your skin will adapt to the pressure from your clothing so you are not constantly aware of clothing touching every part of your body. Also, the heat felt from entering a hot tub soon diminishes as temperature receptors adapt.

Somatic Sensory System

The somatic senses consist of sensory receptors associated with skin, muscles, joints, and viscera. The somatic senses include touch, pressure, temperature, pain, and stretch. Touch and pressure are sensed by free nerve endings and Merkel's discs located between epithelial cells as well as Meissner's, Pacinian, and Ruffini corpuscles. Meissner's corpuscles are located in hairless portions of skin (lips, fingertips, palms, soles, nipples, external genitals). They are small oval masses of flattened connective tissue that sense primarily light touch (Fig. 11.1). Pacinian corpuscles are located in deeper subcutaneous tissues of hands, feet, genitalia, urethrae, breasts, tendons of muscles, and ligaments of joints. They detect heavy

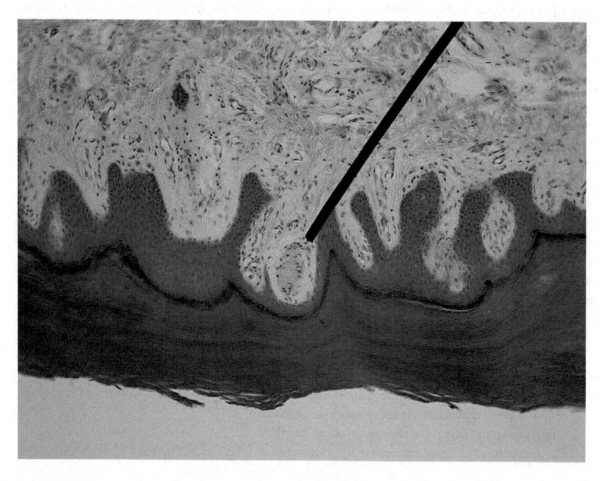

Figure 11.1 Meissner's corpuscle located in the superficial dermis (the superficial layers of the skin are located at the bottom of the slide).

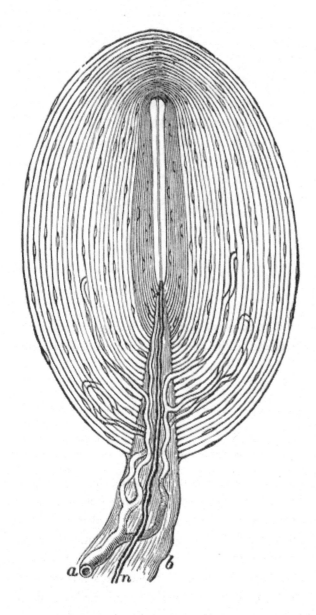

Figure 11.2 Pacinian corpuscles are located in the deeper areas of the dermis.

pressure and vibration (Fig. 11.2). Ruffini corpuscles are also located in the dermis and are sensitive to pressure and skin movement. Merkel's discs sense fine touch and pressure and are located in the stratum germinativum of the epidermis.

There are also warm and cold receptors. Warm receptors are receptive to temperatures greater than 25 degrees Celsius (77°F). Cold receptors are receptive to temperatures between 10 degrees Celsius (50°F) and 20 degrees Celsius (68°F). Pain is experienced if the temperature drops below 10 degrees Celsius. The pain receptors or nociceptors are the free nerve endings. There are no pain receptors in brain. Pain receptors do not adapt to stimuli. Pain from organs or visceral pain can cause the phenomenon of referred pain. In referred pain, the sense of pain is coming from other areas than the location of the viscera. A classic example of referred pain is the pain in the left arm or jaw felt during a heart attack. Referred pain comes from nerve pathways shared by visceral and skin pain receptors. There are two types of pain nerve fiber pathways. A-delta fibers or acute pain fibers are thin myelinated fibers that rapidly conduct pain. The information they carry is interpreted as sharp pain. The sensation of sharp pain tends to not continue after the painful stimulus is removed. Chronic pain fibers or C-fibers are slower than acute fibers and carry information that is interpreted as the sensation of a dull ache. The sensation can be intense and long-lasting, and resists relief.

After reaching the central nervous system, pain follows a specific pathway in spinal cord tracts and is processed in the brain. Pain is processed by the gray matter of the posterior horn of the spinal cord. Pain signals cross over in the spinal cord and travel to the brain via the lateral spinothalamic tracts. The pain signals are then processed in the reticular formation, thalamus, hypothalamus, and cerebral cortex. Areas of gray matter in midbrain, pons, and medulla oblongata also regulate pain impulses from the cord. The impulses travel in the spinal cord via lateral funiculus. The neurons in lateral funiculus can block pain impulses through the secretion of inhibiting neurotransmitters. Some important pain-inhibiting neurotransmitters include enkephalins, endorphins, and serotonin. Enkephalins inhibit acute and chronic pain impulses. Serotonin works by stimulating neurons to release enkephalins. Endorphins are effective in inhibiting chronic pain impulses. Remember, your body is capable of producing the

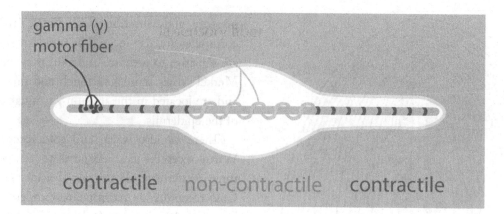

Figure 11.3 Muscle spindle.

pain-inhibiting neurotransmitters. Some pain-controlling therapies are aimed at increasing the number of these neurotransmitters. These include some of the electrical therapies such as interferential current found in physical therapy clinics.

Sensory Receptors in Muscle

We will examine two types of sensory receptors found in muscle tissue. These include the Golgi tendon organs and muscle spindles. Muscle spindles are located near the origin and insertions of muscles (Fig. 11.3). They consist of modified skeletal muscle fibers (intrafusal fibers) enclosed in connective tissue sheath. A nerve fiber wraps around the intrafusal fiber and sends information about muscle tone to the central nervous system. Larger extrafusal fibers surround the intrafusal fibers. Muscle spindles are involved in the stretch reflex. If a muscle is stretched, the spindle is also stretched and sends signals to CNS. The signals oppose muscle lengthening. So if the muscle is stretched or lengthened, signals are sent to CNS telling the muscle to shorten. This produces the muscle jerk in the deep tendon reflex. Golgi tendon organs (GTOs) are located at the muscle–tendon junction. They monitor tension of the muscles generated during muscle contraction. Golgi tendon organs can act as a protective mechanism to overloaded muscles. When muscles become overloaded, the Golgi tendon organs function to inhibit muscle contraction in what is known as "weightlifting failure."

Disorders of Touch

Sensory Processing Disorder (SPD)

In most people, sensory integration occurs naturally without a thought process. But in some people, the sensory integration does not develop properly and becomes distorted. In these people, the brain and central nervous system misinterpret everyday sensory information such as touch, sound, and movement. Research is still being conducted on this disorder but researchers are finding direct links to SPD with other disorders like ADD/ADHD, premature birth, autism, Down's Syndrome, and Fragile X.

Tactile defensiveness

Considered a category of SPD, tactile defensiveness is an overreaction to the sense of touch; it was identified by Dr. Jean Ayers in the 1960s. A person with tactile defensiveness will react with a flight-or-fight reaction to touch stimuli that a normal person would interpret as harmless. Most cases are noticed in infancy or childhood due to the fact that these children do not want to be touched or cuddled as a normal child would. A child with this disorder will probably have these signs or symptoms:

- Does not like to go barefoot or have feet touched
- Does not enjoy baths, haircuts, nail clipping
- Requires tags to be removed from all clothing
- Does not want their face touched
- Hard time eating because of textures, temperatures of the food
- Does not want to touch anything that is messy or has a sticky texture

Congenital insensitivity to pain with anhidrosis (CIPA)

An exceedingly rare disease; there are only about 35 known cases in the United States. CIPA is a severe autosomal recessive condition in which the peripheral nerves demonstrate a loss of unmyelinated and small myelinated fibers. The actual physiopathological mechanism is still unknown and being studied. This is an extremely hard disease to study due to the rarity of cases. Most people with the disease will not live long due to injuries received that go untreated because they are unnoticed and severe.

Insensitivity to pain

Wouldn't it be wonderful if you could no longer feel pain? Is that not something we all would like to have? Or do we have pain for a good reason? Although it is rare there is a disease known as congenital insensitivity to pain. This genetic abnormality causes some people to lack certain components of the sensory system to receive pain. The exact reason for the problem is unknown and varies between people. Sadly people who have the disease often die in childhood. Injuries are very common with people who have congenital insensitivity to pain. They often will lose digits, may suffer from burns, and their knees often have sores from kneeling too long. Clearly pain has a purpose; it is our warning signal when things are awry.

The Special Senses

The special senses include smell, taste, hearing, and vision. The senses of smell and taste work together. Both are sensed as changes in chemical concentration by chemoreceptors. Olfactory organs located on both sides of the nasal septum in the nasal cavity pick up water- and lipid-soluble substances that diffuse in the mucus of the nasal cavity (Fig. 11.4). The information is then converted to action potentials and sent via afferent neurons through the cribriform plate of the ethmoid bone to the olfactory bulbs. The information then leaves the olfactory bulbs and travels via the olfactory tract to the olfactory cortex of the frontal lobe, hypothalamus, and limbic system (Fig. 11.5). The processing of olfactory information by the hypothalamus and limbic system results in a close relationship between the sense

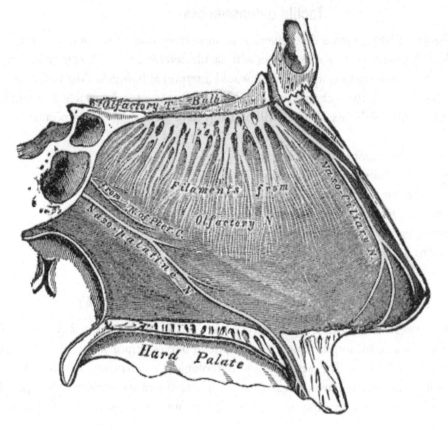

Figure 11.4 Olfactory nerve fibers.

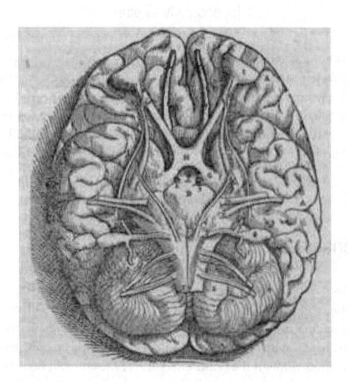

Figure 11.5 Olfactory bulbs (red).

of smell and emotions. We can sense between 2,000 and 4,000 different chemical substances with our olfactory systems. The system is incredibly sensitive and can sense concentrations as small as a few parts per billion.

Disorders of Olfaction

Anosmia

Anosmia is the lack of olfaction, or a loss of the sense of smell. It can be either temporary or permanent. A related term, "hyposmia" refers to a decrease in the ability to smell. Some people may be anosmic for one particular odor. This is called "specific anosmia" and may be genetically based. Anosmia can have a number of detrimental effects. Patients with anosmia may find food less appetizing. Loss of smell can also be dangerous because it hinders the detection of gas leaks, fire, body odor, and spoiled food. The common view of anosmia as trivial can make it more difficult for a patient to receive the same types of medical aid as someone who has lost other senses, such as hearing or sight. A temporary loss of smell can be caused by a stuffy nose or infection. In contrast, a permanent loss of smell may be caused by the death of olfactory receptor neurons in the nose, or by brain injury in which there is damage to the olfactory nerve or damage to brain areas that process smell. The lack of the sense of smell at birth, usually due to genetic factors, is referred to as congenital anosmia. Anosmia may be an early sign of degenerative brain diseases such as Parkinson's disease and Alzheimer's disease. Another specific cause of permanent loss could be from damage to olfactory receptor neurons due to use of nasal sprays. To avoid loss of smell from nasal sprays, use them for only a short amount of time. Nasal sprays that are used to treat allergy-related congestion are the only nasal sprays that are safe to use for extended periods of time.

Phantosmia

Phantosmia is the phenomenon of smelling odors that aren't really present (phantom odors). The most common odors are unpleasant smells such as rotting flesh, vomit, feces, and smoke. Phantosmia often results from damage to the nervous tissue in the olfactory system. The damage can be caused by viral infection, trauma, surgery, and possibly exposure to toxins or drugs. It can also be induced by epilepsy affecting the olfactory cortex. It is also thought the condition can have psychiatric origins.

Taste

Taste is also sensed as changes in chemical concentration by taste receptors. These receptors are located on the surface of tongue (papillae), on the roof of the mouth, linings of cheeks, and walls of the pharynx (Fig. 11.6). The taste receptors form structures called taste buds. The adult human has about 3,000 taste buds. Each taste bud contains different types of cells. Basal cells are the stem cells that mature into gustatory cells that contain microvilli. The microvilli extend into fluid collected in the taste pores, which are small openings in the taste buds (Fig. 11.7). Gustatory cells are replaced frequently and typically last only 10 days. Taste sensation is carried by cranial nerves VII (facial), IX (glossopharyngeal), and X (vagus). The facial nerve receives information from receptors on the anterior two thirds of the tongue. The posterior one third of the tongue is innervated by the

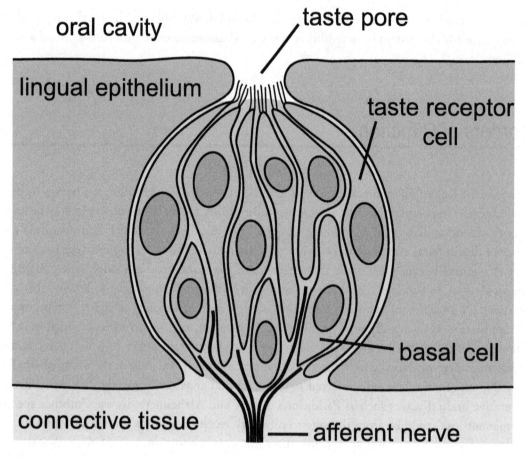

Figure 11.6 Taste bud.

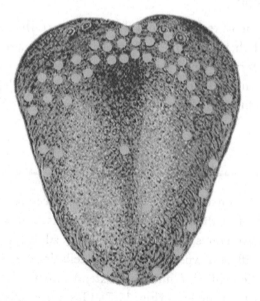

Figure 11.7 Taste buds for bitter.

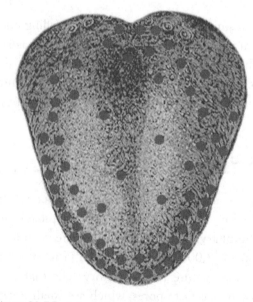

Figure 11.7 Taste buds for salty.

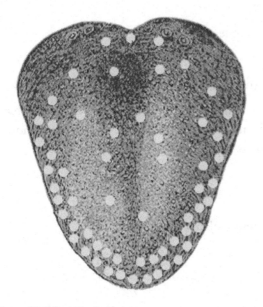
Figure 11.7 Taste buds for sour.

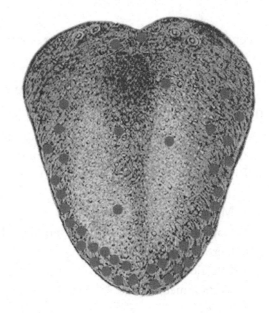
Figure 11.7 Taste buds for sweet.

glossopharyngeal nerve. The vagus nerve innervates the taste buds located on the epiglottis. Taste information is sent to the medulla oblongata and then to the thalamus where it is routed to portions of the primary sensory cortex.

There are 5 primary taste sensations. Tastes are combinations of these 5 primary sensations: 1. Sweet, 2. Sour, 3. Salty, 4. Bitter, 5. Umami. Umami is a hearty, meaty taste that is produced by L-glutamate (think monosodium glutamate). Taste receptors are more sensitive to unpleasant stimuli. For example, we are about a thousand times more sensitive to acids than sweet tastes.

Disorders of the Tongue

Loss of taste

You may lose your sense of taste if the facial nerve is damaged. Then there is also Sjogren's Syndrome where the saliva production is reduced. In most cases the loss of taste is typically a symptom of anosmia, a loss of the sense of smell.

Sore tongue

It is usually caused by some form of trauma, such as biting your tongue, or eating piping-hot or highly acidic food or drink. If your top and bottom teeth don't fit neatly together, tongue trauma is more likely. Some people may experience a sore tongue from grinding their teeth (bruxism). Disorders such as diabetes, anemia, some types of vitamin deficiency, and certain skin diseases can include a sore tongue among the range of symptoms.

Glossodynia

A condition characterized by a burning sensation on the tongue.

Benign migratory glossitis

This condition is characterized by irregular and inflamed patches on the tongue surface that often have white borders. The tongue may be generally swollen, red, and sore. Another name for this condition is geographic tongue. The cause of benign migratory glossitis is unknown.

Risk factors are thought to include:

- Mineral or vitamin deficiencies
- Local irritants, such as strong mouthwashes, cigarettes, or alcohol
- Certain forms of anemia
- Infection
- Certain medications
- Stress

Vision: External and Supportive Structures of the Eye

The eyes are surrounded by a series of supportive structures that protect and move the eye as well as produce secretions. These structures are often referred to as accessory structures. The eyelids (palpebrae) are continuous with the skin. They help to remove debris and allow tears to lubricate the surface of the eye. When closed they protect the eye. The upper and lower eyelids are separated via a gap called the palpebral fissure. The upper and lower lids are connected via the medial and lateral canthus. Sebaceous glands called tarsal glands are located at the inner margins of the eyes. The secretions from these glands help to keep the eyelids from sticking together. The lacrimal caruncle is located at the medial canthus. This structure contains glands that secrete thick mucus. A thin layer of mucus-secreting epithelium covers the inner portion of the eyelids and extends to the outer portion of the eyes. This layer is called the conjunctiva. There are two parts to the conjunctiva. The palpebral conjunctiva is located on the inner surface of the eyelid, while the ocular conjunctiva is located on the eyeball. The ocular conjunctiva extends to the margins of the cornea. An inflammation or infection of the conjunctiva is known as conjunctivitis (pinkeye). This is caused by infection, irritation, or allergies (Fig. 11.8).

Lacrimal Apparatus

The lacrimal apparatus produces tears. It consists of a lacrimal gland, lacrimal canaliculi, and a lacrimal sac (Fig. 11.9). The lacrimal glands secrete tears that function to clean and lubricate the surface of the eye. Tears contain an enzyme called lysozyme that helps to kill bacteria. Tears are spread across the surface of the eye and drained by the lacrimal canals that lead to the nasolacrimal duct.

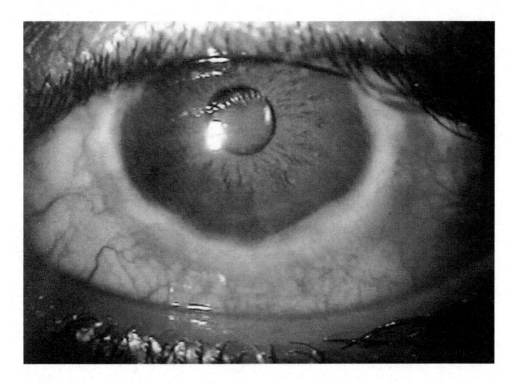

Figure 11.8 Conjunctivitis.

Eye Muscles

There are six extrinsic eye muscles that move the eyeball (Fig. 11.10). There are four rectus (straight fibers) muscles and two obliques. The four rectus muscles are the superior, inferior, medial, and lateral. The two obliques are the superior oblique and inferior oblique. The rectus muscles all originate on the posterior surface of the bony orbit and extend to the surface of the eyeball. The superior oblique attaches to the medial surface of the orbit. Its tendon passes through a fibrocartilaginous pulley called the trochlea. The tendon then attaches to the superolateral surface of the eyeball. The inferior oblique extends from the medial wall of the orbit to its attachment on the inferolateral aspect of the eyeball.

The eye muscles are innervated by cranial nerves III, IV, and VI. Cranial nerve III innervates the superior rectus, inferior rectus, medial rectus, and inferior oblique. Cranial nerve IV innervates the superior oblique, and cranial nerve VI innervates the lateral rectus.

Structures of the Eye

The eye consists of three layers or tunics. These are the outer fibrous tunic, the middle vascular tunic, and the inner tunic (Fig. 11.11). The outer fibrous tunic consists of the cornea and sclera. The sclera (white portion) consists of dense connective tissue containing blood vessels and nerves. The cornea

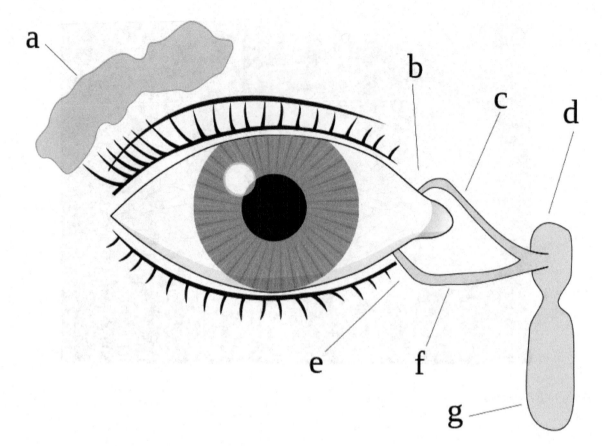

Figure 11.9 Lacrimal apparatus.
a = tear gland/lacrimal gland
b = superior lacrimal punctum
c = superior lacrimal canal
d = tear sac/lacrimal sac
e = inferior lacrimal punctum
f = inferior lacrimal canal
g = nasolacrimal canal

is the transparent portion on the anterior aspect of the eye. The middle vascular tunic consists of the choroid coat, ciliary body, and iris. The middle tunic is also called the uvea because of its similarity to a peeled grape. The choroid coat is a pigmented (dark) layer just deep to the retina. The ciliary body is an extension of the choroid coat. It surrounds the lens and contains smooth muscles that attach to the lens via suspensory ligaments. The ciliary body secretes a watery fluid called aqueous humor. The iris contains smooth muscle that controls the diameter of the pupil. The iris contains chromatophores that contain melanin that gives the iris its color. The inner layer contains the retina and a portion of the optic nerve. Additional structures of the eye that are not part of the tunics help to direct and focus light to the retina. These include aqueous humor, lens, and vitreous body (humor). The aqueous humor is a watery serous fluid secreted by the ciliary body into a space between the iris and lens called the posterior chamber. The fluid flows to the anterior chamber, which lies between the cornea and

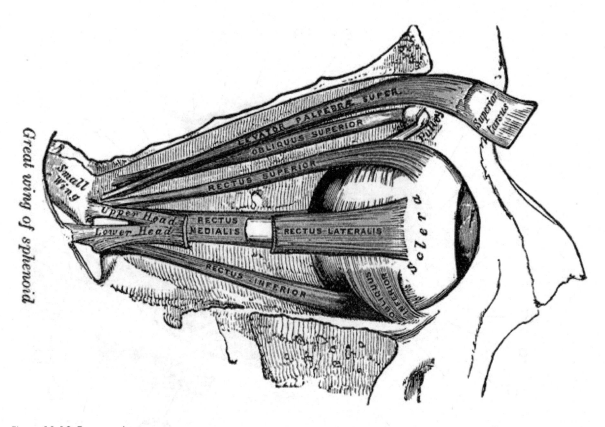

Figure 11.10. Eye muscles.

the iris. The fluid is then absorbed by the Canal of Schlemm (sclera venous sinus). The lens connects to the ciliary body by means of a series of fibers called the suspensory ligament. The suspensory ligament changes the shape of the lens. When the ligament is taut, the lens flattens. Likewise when the ligament relaxes, the lens retains its spheroid shape. The vitreous body is a clear, jellylike fluid that resides in a large hollow area just posterior to the lens. The retina attaches to the eye at only two points: the optic disc and the ora serrata. The optic disc is the area of attachment of the optic nerve. The ora serrata is the anterior margin of the retina. The vitreous body helps to maintain the shape of the retina by pushing against it. Blows to the head can detach the retina. The retina is actually nervous tissue that is consistent with the brain. It develops in utero from the diencephalon. A group of cells that contain a high concentration of photoreceptors is located in the posterior retina. This structure is called the macula lutea and is about 3mm–5mm in diameter. The fovea centralis is located at the center of the macula lutea. The fovea centralis produces the sharpest vision. The optic nerve lies just medial to the fovea centralis. Nerve fibers from the eye converge and exit the eye at the optic disc. Blood vessels also travel through the optic disc. The optic disc contains no photoreceptors and thus produces a blind spot. The brain compensates for the blind spots in each visual field by filling in the field with images similar to the surroundings. The pupil is surrounded by smooth muscles that allow for it to constrict or dilate. The pupillary constrictor muscle narrows the pupil to decrease the amount of light entering the eye. This muscle is innervated by the parasympathetic nervous system. The pupillary dilator does the opposite function and opens the pupil to let in more light in response to stimulation from the sympathetic nervous system.

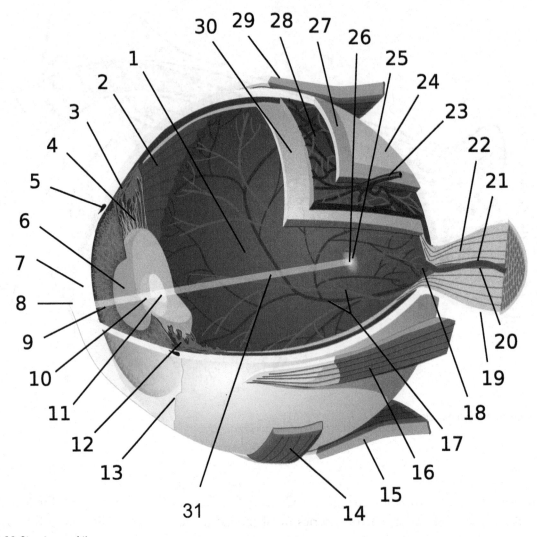

Figure 11.11 Structures of the eye.
1: posterior chamber, 2: ora serrata, 3: ciliary muscle, 4: suspensory ligaments, 5: canal of Schlemm, 6: pupil, 7: anterior chamber, 8: cornea, 9: iris, 10: lens cortex, 11: lens nucleus, 12: ciliary process, 13: conjunctiva, 14: inferior oblique muscle, 15: inferior rectus muscle, 16: medial rectus muscle, 17: retinal arteries and veins, 18: optic disc, 19: dura mater, 20: central retinal artery, 21: central retinal vein, 22: optical nerve, 23: vorticose vein, 24: bulbar sheath, 25: macula, 26: fovea, 27: sclera, 28: choroid, 29: superior rectus muscle, 30: retina

Refraction

Light travels at 300,000 meters per second in a vacuum but will slow down when traveling though other media. As light travels through the eye, it passes though transparent structures that change its speed. The result is a convergence or divergence of light rays. This phenomenon is known as refraction. The pupil will change its diameter to help the lens focus. For example, the pupil will constrict when looking at objects nearby. This helps to reduce the phenomenon of spherical aberration that occurs from an unequal amount of refraction by the lens. Objects are not refracted as well from the margins of the lens as in the middle regions. The result is a blurry image. The pupil constricts to reduce spherical aberration by focusing the light rays on the center of the lens. The lens can change its shape to adjust its curvature. This is called accommodation (Fig. 11.15). For example, contraction of the ciliary muscle occurs when looking at objects close up. This allows the suspensory ligament to relax, which causes the lens to produce a more convex shape. Likewise when you look at objects far away, the ciliary body relaxes, allowing the suspensory ligament to contract. This changes the shape of the lens to be less convex. The more convex the lens, the more it causes light rays to converge. The lens loses its flexibility with age and focusing becomes more difficult. Bifocals are used to correct for the loss of accommodation.

Lenses are used to correct for refraction disorders of the eye (Figs. 11.12, 11.13). Myopia or nearsightedness results from an eyeball that is too long. In this case, images focus in front of the retina. A corrective concave lens is needed to push the images back (Fig. 11.14). Hyperopia or farsightedness results from the eyeball that is too short. This causes images to focus behind the eyeball. A convex lens is needed to pull the images forward. An astigmatism is caused by an irregularly shaped cornea or lens. This can cause blurred vision, eyestrain, or headaches. Astigmatisms are corrected by lenses or refractive surgery.

Photoreceptors of the Retina

Light travels through the transparent structures of the eye (cornea, aqueous humor, lens, vitreous humor) and enters the retina. The retina contains photoreceptors, ganglion cells, and bipolar neurons (Fig. 11.16). Light passes through the retina until it reaches the photoreceptors. Images are sensed by the photoreceptors that transmit impulses to the bipolar neurons that in turn transmit impulses to the

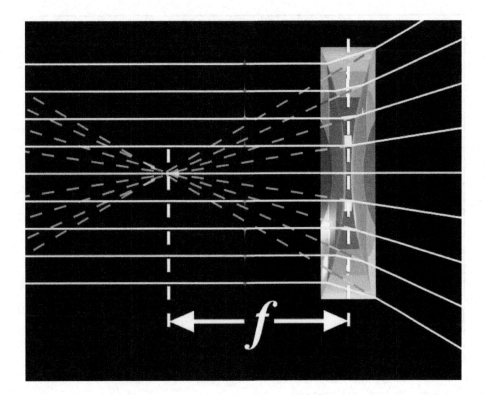

Figure 11.12. A concave lens causes light rays to diverge.

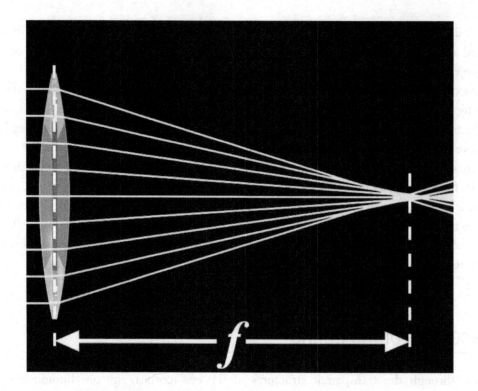

Figure 11.13 A convex lens causes light rays to converge.

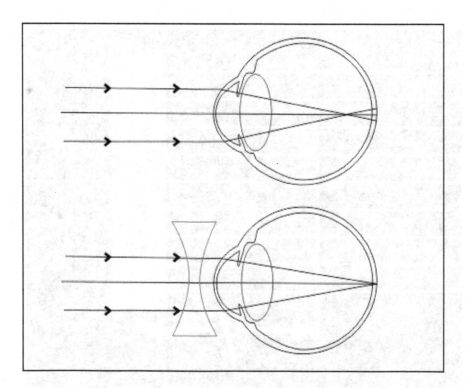

Figure 11.14 Myopia is corrected with a concave lens.

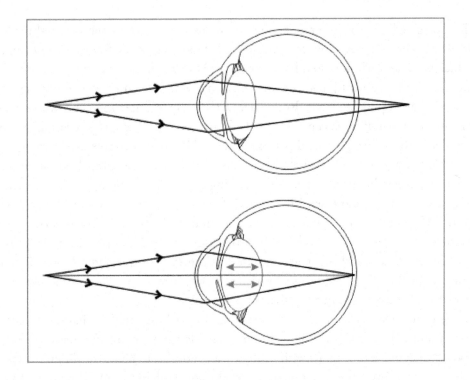

Figure 11.15 Accommodation is the change in shape of the lens.

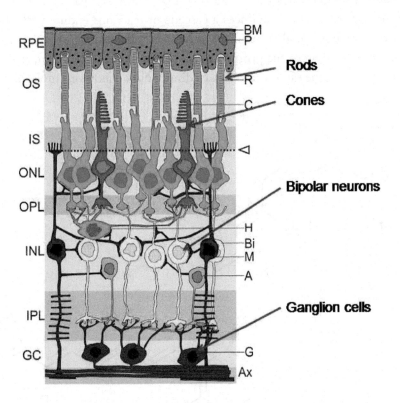

Figure 11.16 Layers of the retina.

ganglion cells. The ganglion cells generate action potentials that travel through cranial nerve II to the occipital lobe. There are two types of photoreceptors, called rods and cones. Rods are more numerous and work to produce black and white vision. Both types of photoreceptors have inner and outer segments joined by cilia. The inner segment joins with the cell body, while the outer segments contain pigments (visual pigments) that respond to light. Rods have a rod-shaped outer segment, while cones have a cone-shaped outer segment. Photoreceptors are able to maintain themselves by continually replenishing their outer segments. Rods are replenished during the day, while cones are replenished at night. Rods and cones have different visual pigments that absorb light at different wavelengths. Rods are more sensitive and respond to gray colors better than cones do, making them better able to produce black and white vision at night. Rods are also more sensitive to peripheral vision. Rods also connect with convergent neural pathways. Many rods will innervate fewer ganglion cells, resulting in loss of sharpness of vision. Rods contain one type of visual pigment. Cones absorb light at a wider range of wavelengths than rods do. They also contain three visual pigments and are less sensitive to dim light. Cones function better in daylight and are better able to produce color vision. Cones also have a one-to-one relationship with ganglion cells. This allows cones to provide sharper color vision.

Visual pigments undergo a chemical reaction in order to produce impulses that send visual information. Retinal is a molecule that absorbs light. Retinal is made from vitamin A and is able to change its shape in response to light. Retinal binds with proteins called opsins to make the different types of visual pigments (Fig. 11.17). Rods contain the pigment rhodopsin (visual purple). Rhodopsin forms in dim light or darkness. Retinal undergoes oxidation and combines with opsin to produce rhodopsin. Retinal begins in one configuration known as the 11-cis retinal form. Once exposed to light, retinal changes to an alternate configuration known as the all-trans configuration. Retinal now separates from opsin. This change in configuration of retinal produces a cascade of reactions ending with the generation of action potentials. Cones require a higher intensity of light in order to trigger action potentials. Cones use different opsins corresponding to the primary colors they absorb (blue, green, red). The reaction of retinal and opsins is essentially the same as in rods. In dim or dark conditions molecules of cGMP bind

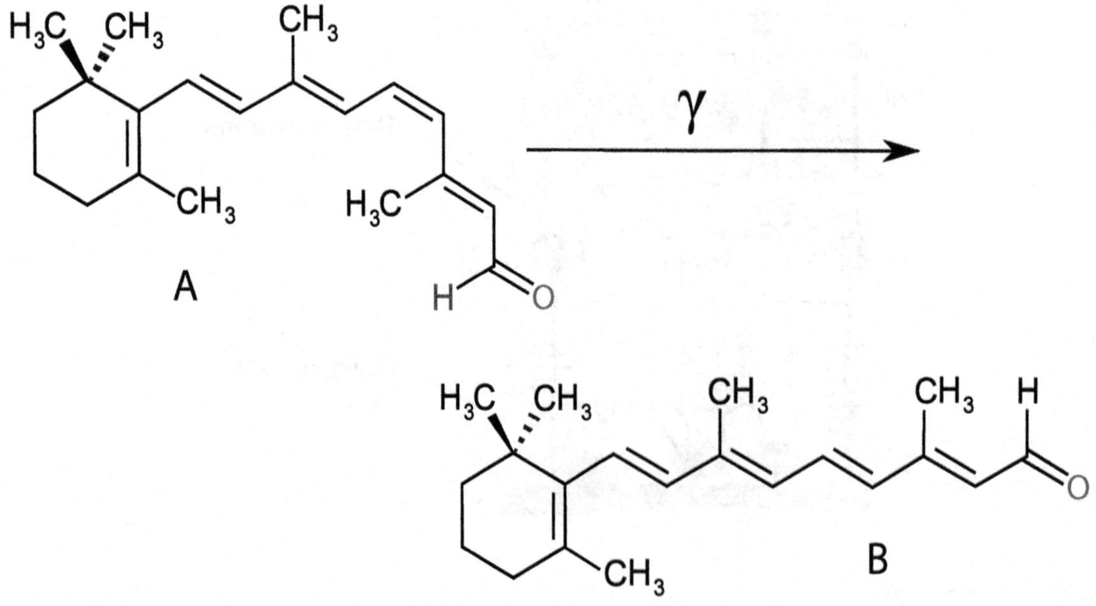

Figure 11.17 Retinal cis (top) and trans (bottom) configurations.

to protein channels in the photoreceptor's cell membrane. The channels open and allow calcium and sodium to continuously enter the cell. This holds the membrane potential to around -40mV and holds the cell in a steady state of depolarization. This keeps the calcium channels at the synaptic terminals open and produces a continuous flow of the neurotransmitter glutamate. The glutamate stimulates receptors on the bipolar neurons.

When light reaches the photoreceptors a series of reactions occur that breaks down the cGMP. The protein channels subsequently close. Potassium channels also located in the membrane remain open and potassium moves out of the cell causing it to be in a hyperpolarized state of about -70mV. This inhibits the release of glutamate. Bipolar cells do not produce action potentials but do produce graded potentials (local currents). These potentials are picked up by the ganglion cells that are able to produce action potentials. The ganglion cells carry the visual information to the optic nerve and on to the optic chiasma and optic tracts. Fibers from the medial side of each eye cross over in the optic chiasma to the contralateral side to the optic tract. Each optic tract contains fibers from the lateral side of the ipsilateral eye and fibers from the medial side of the contralateral eye. Remember that the visual image on the retinal is reversed and upside down. In essence each optic tract ends up carrying the visual information for one half of the visual field. The optic tracts synapse with neurons in the lateral geniculate body of the thalamus. The neurons from the thalamus project back to the primary visual cortex via the optic radiation. Some optic tract fibers are sent to the superior colliculi for control of visual reflexes.

Adaptation

The eyes are capable of adapting to varying intensities of light. You may notice this when walking outdoors on a bright summer day or walking into a dark theater. It takes a few seconds for your eyes to adapt. When we walk from an area of darkness to an area of light, the retina adapts by turning off the rods and decreasing its sensitivity to light. Likewise when we move into areas of darkness, rhodopsin builds up and the rods take over. It is interesting to note that peripheral vision is more acute in dark conditions because of the rods.

Stereoscopic Vision

Both right and left visual fields overlap by about 170 degrees. Each eye has a slightly different perspective of the environment. The neurocortex combines the images and produces depth perception.

Color Blindness

Color blindness results from a lack of one or more types of cones. Since the abnormality is linked to the X chromosome, it is more prevalent in males. The most common type is red-green color blindness in which red and green are seen as the same color. Up to 8–10% of the male population may have some degree of color blindness (Fig. 11.18).

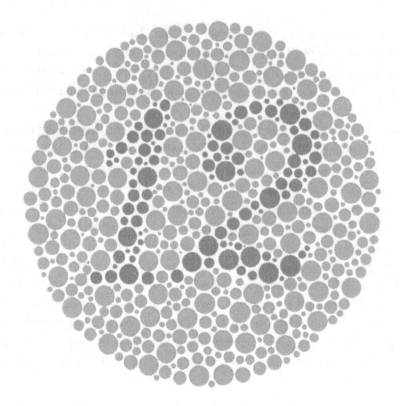

Figure 11.18 Ishihara color blindness test object.

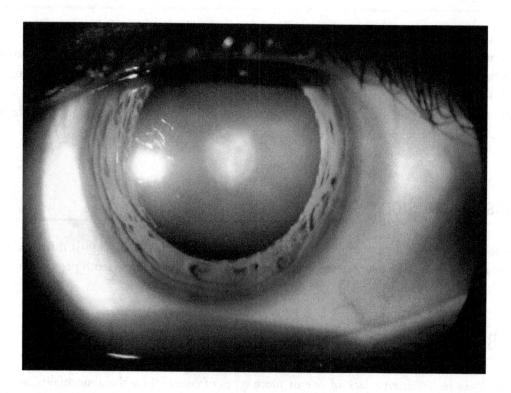

Figure 11.19 Cataract.

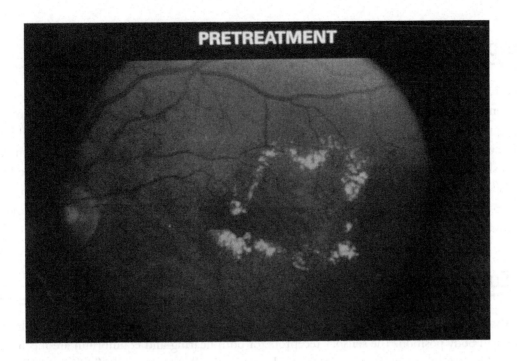

Figure 11.20 Diabetic macular edema.

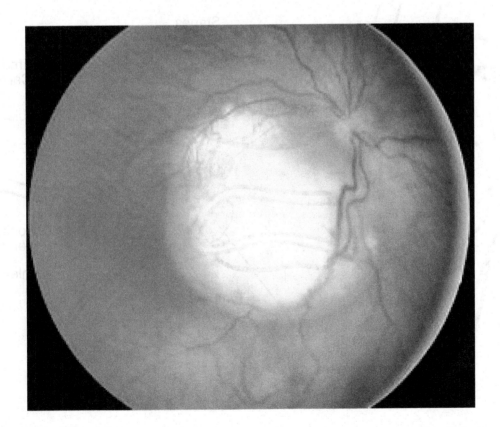

Figure 11.21 Retinoblastoma.

An Experiment

Hold your hand up, about one foot (30 cm) in front of your nose. Keep your head still, and shake your hand from side to side, slowly at first, and then faster and faster. At first you will be able to see your fingers quite clearly. But as the frequency of shaking passes about one hertz, the fingers will become a blur. Now, keep your hand still, and shake your head (up and down or left and right). No matter how fast you shake your head, the image of your fingers remains clear. This demonstrates that the brain can move the eyes opposite to head motion much better than it can follow, or pursue, a hand movement. When your pursuit system fails to keep up with the moving hand, images slip on the retina and you see a blurred hand.

The Ear

The ear is the sense organ that collects and detects sound waves and plays a major role in the sense of balance and body position. The sensory receptors for both hearing and equilibrium are mechanoreceptors found in the inner ear; these receptors are hair cells that have stereocilia (long microvilli) that are extremely sensitive to mechanical stimulations. The ear is divided into three areas: external, middle, and inner ear (Fig. 11.22).

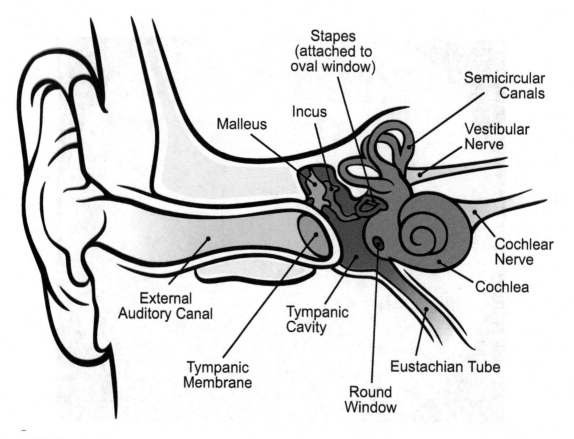

Figure 11.22 Ear anatomy.

External Ear

The external ear consists of the auricle (pinna) and the external auditory meatus (canal). The auricle is the outer portion of the ear consisting of elastic cartilage covered by skin. Its oval rim is called the helix and the earlobe is also known as the lobule. The external auditory meatus is a canal that extends from the outside to the tympanic membrane. It is lined with skin that contains sebaceous glands, hair, and ceruminous glands that secrete a waxy substance called cerumen (ear wax). Cerumen traps foreign particles and helps to protect the canal.

Middle Ear

The tympanic membrane is the boundary between the external and middle ear. It consists of a thin layer of connective tissue. It has a layer of skin on its external surface and a mucous membrane on its internal surface. It is slightly cone-shaped with the apex pointing toward the middle ear. The middle ear resides in a hollow chamber called the tympanic cavity. The cavity is lined with a mucous membrane. The cavity contains a canal called the Eustachian tube (pharyngotympanic tube) that connects with the nasopharynx. The tube is normally closed but opens with chewing, yawning, or swallowing. The tube opens briefly to equalize pressure between the tympanic cavity and the outside. Changes in pressure can disrupt the vibrations of the tympanic membrane and produce muffled sounds. Three small bones called auditory ossicles transmit vibrations from the tympanic membrane to the oval window of the inner ear. These are the malleus, incus, and stapes (hammer, anvil, stirrup). The ossicles also magnify the vibrations from the tympanic membrane by their leverage. Two small muscles, the stapedius and tensor tympani, connect to the ossicles and work to maximize the vibrations carried to the oval window. The stapedius connects to the stapes and the tensor tympani connects to the malleus. These muscles also work to protect the ear from loud noises (tympanic reflex). This works much like putting pressure on the head of a drum while someone is beating it. The effect is to dampen the sound.

Inner Ear

The inner ear resides within a cavity inside of the temporal bone. It consists of the cochlea, vestibule, and semicircular canals. The inner ear is sometimes referred to as the bony labyrinth. On the inside of the bony labyrinth resides a membranous labyrinth that contains fluid (Fig. 11.23). The cochlea is a spiral-shaped structure that connects to the anterior portion of the vestibule. The cochlea winds around a bony structure called the modiolus. The cochlea contains three hollow chambers filled with fluid (Figs. 11.24, 11.25). The innermost chamber is known as the cochlear duct (scala media) and contains the organ of Corti (spiral organ), which senses hearing. The scala vestibule is a chamber that lies superior to the cochlear duct and the scala tympani lies inferior to the cochlear duct. The scala

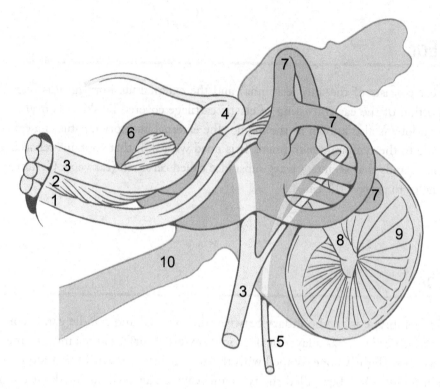

Figure 11.23 Structures of the inner ear.
1: Vestibular portion of CN VIII, 2: Cochlear portion of CN VIII, 3: Intermediate portion of CN VIII, 4: Ganglion geniculi, 5: Chorda tympani, 6: Cochlea, 7: Semicircular canals, 8: Malleus, 9: Tympanic membrane, 10: Eustachian tube

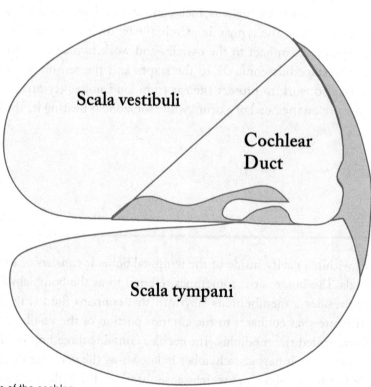

Figure 11.24 Chambers of the cochlea.

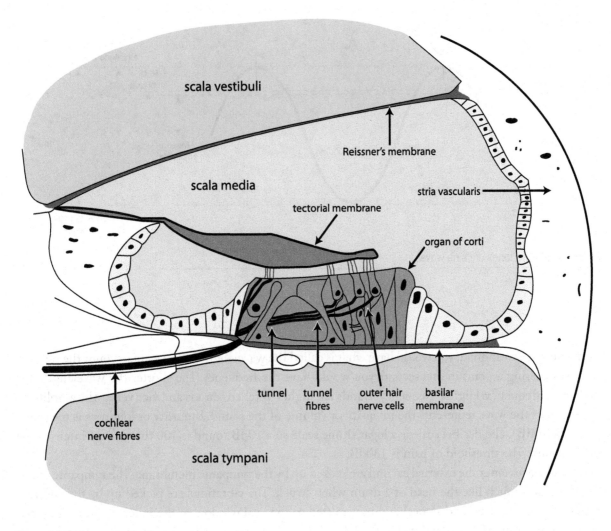

Figure 11.25 Inner ear structures.

tympani ends at the round window. The scala vestibule and scala tympani connect with each other at a point called the helicotrema, which is located at the apex of the cochlear duct. The cochlear duct also contains the vestibular membrane, which is a thin fluid-secreting membrane that produces a fluid called endolymph. The vestibular membrane is located on the superior aspect of the cochlear duct. The inferior aspect contains another membrane called the basilar membrane, which is important in producing hearing.

How the Ear Processes Hearing

We must remember that sound consists of changes in air pressure or vibrations in other media (Figs. 11.26, 11.27). Sound waves consists of areas of high and low pressure that move (propagate) through the air. Sound can be represented as waves, such as sine waves. The peaks of the wave represent areas of high pressure, while the valleys represent low-pressure areas. If you were to measure the distance

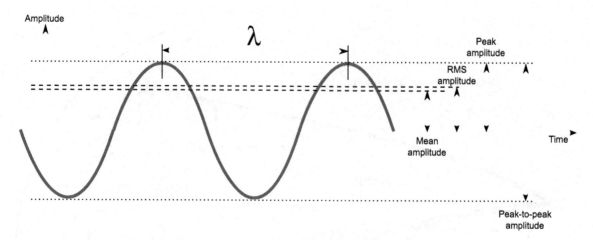

Figure 11.26 Pure tones are sine waves.

from one wave to another, you would have what is called the wavelength. If you were to count the number of waves passing a point in one second, you would have the frequency. The shorter the wavelength, the higher the frequency. Higher-frequency sounds are higher-pitched sounds and vice versa. The amplitude, or height, of the wave represents the intensity or volume of the sound. Intensity or loudness is measured in decibels (dB). The decibel scale is a logarithmic scale so a 10dB sound is 100 times more intense than a 0dB sound. The threshold of pain is 130dB.

Sound waves enter the external ear and are picked up by the tympanic membrane. The tympanic membrane vibrates much like the head of a drum when struck. The vibrations are picked up by the auditory ossicles (malleus, incus, and stapes) and transmitted to the oval window of the inner ear. The vibrations are actually amplified at this point. The fluid-filled chambers of the inner ear pick up the vibrations. The perilymph in the scala vestibule then carries the vibrations toward the helicotrema. Vibrations within the audible range of human hearing (20 frequencies per second to 20,000 frequencies per second) move through the cochlear duct and into the perilymph of the scala tympani. As vibrations move through the

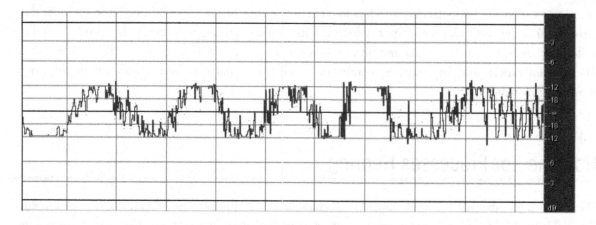

Figure 11.27 Sound is a complex wave consisting of areas of high and low pressure.

cochlear duct, they move the basilar membrane. Different portions of the basilar membrane respond or resonate to different frequencies. For example, areas near the oval window resonate with higher-frequency sounds, while areas near the helicotrema resonate with lower-frequency sounds. The organ of Corti contains specialized hair cells (cochlear hair cells) that are located between the tectorial and basilar membranes. There are three rows of outer hair cells and one row of inner hair cells. The hair cells directly connect with the cochlear portion of the vestibulocochlear nerve (cranial nerve VIII). The hair cells also contain cilia that bend in accordance to vibrations of the basilar membrane. Bending of the cilia of the hair cells in one direction causes depolarization and subsequent release of glutamate. Bending the cilia in the other direction inhibits depolarization. The impulses generated in the cochlea travel to the spiral ganglion and to the superior olivary nucleus, where they synapse with neurons from the lateral lemniscus. The information is then carried to the inferior colliculus and then to the auditory cortex in the temporal lobe. Auditory reflexes are also processed in the medial geniculate body of the thalamus and superior colliculus.

Balance and Equilibrium

The ear also senses changes in position (static equilibrium) and motion (dynamic equilibrium). This processing occurs in the vestibule and semicircular canals, collectively called the vestibular apparatus.

Static Equilibrium

Static equilibrium is sensed by the vestibule. Inside the vestibule are two structures called the utricle and saccule (Fig. 11.28). The utricle and saccule both contain another structure called a macula that contains hair cells much like those of the cochlea. The hairs of the hair cells are connected to the otolithic membrane. The otolithic membrane contains tiny stones called otoliths (otoconia). In the utricle, the macula is in the horizontal plane with the hairs extending vertically when the head is in an upright position. As the head moves in the horizontal plane or tilts, the otoliths pull on the membrane, which in turn bends the hair cells. The bending of the hair cells generates impulses that are transmitted to the vestibular

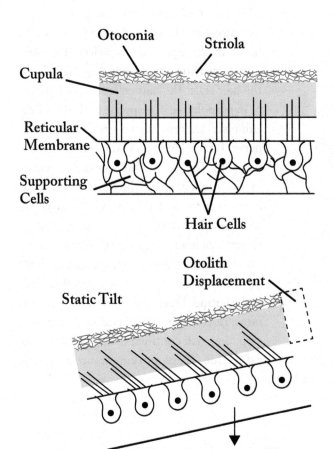

Figure 11.28. Vestibule.

portion of the vestibulocochlear nerve (cranial nerve VIII). The macula in the saccule operates the same way but is oriented in the vertical plane. Thus the saccular macula responds more to vertical motion. This system responds only to changes in motion.

Dynamic Equilibrium

The semicircular canals sense changes in motion. At the base of each semicircular canal is a bulge called an ampulla. Inside the ampulla is a structure called the crista ampullaris. The crista ampullaris also contains hair cells that attach to a gel-like membrane called a cupula (Fig. 11.29). Movement of the head causes fluid (endolymph) inside the semicircular canals to move in the opposite direction. The fluid moves over the cupula, bending the cilia of the hair cells. The hair cells then depolarize or hyperpolarize in response to the bending. Impulses are sent to the vestibulocochlear nerve (cranial nerve VIII), which carries them to the vestibular nuclear complex in the brainstem or the cerebellum. It is important to note that balance and vision are closely related. A reflex movement of the eyes called nystagmus can be created by impulses from the semicircular canals. Let's say that you were sitting on a movable chair and rotating. During the rotation your eyes move in the opposite direction. When you stop, they will continue to look in the same direction for a few moments then quickly look in the opposite direction. These movements are caused by impulses from endolymph movement in the semicircular canals.

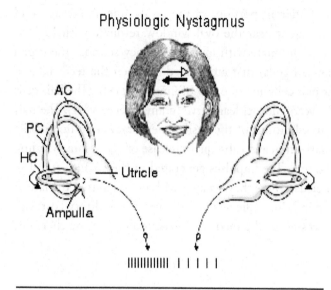

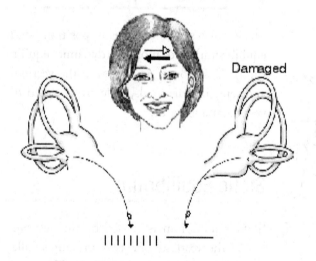

Figure 11.29 Nystagmus is caused by endolymph movement in the semicircular canals.

Ear Disorders

Otitis media is an inflammation of the middle ear (Fig. 11.30). The middle ear can become infected, causing the tympanic membrane to become red and swollen. This condition is typically treated with antibiotics. Chronic conditions may be treated with small tubes inserted in the tympanic membranes to equalize the pressure between the middle ear and outside.

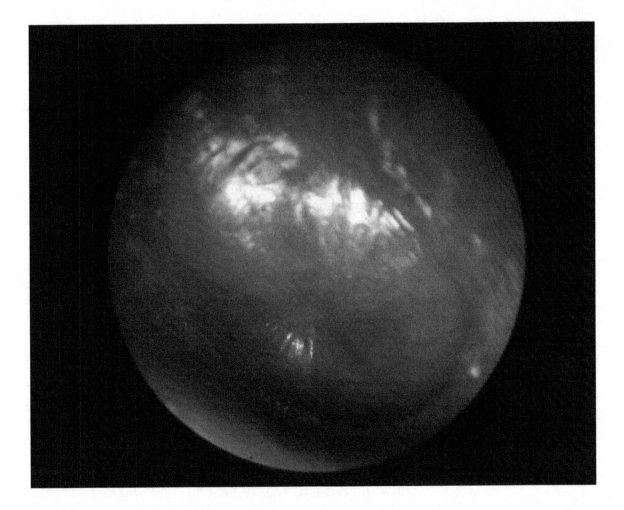

Figure 11.30 Acute otitis media. Note the bulging, red, inflamed eardrum.

Conduction Hearing Loss

Conduction hearing loss results from a blockage somewhere in the ear canal that prevents vibrations from getting to the inner ear. This can result from earwax, ruptured eardrum, or degeneration of the auditory ossicles called otosclerosis.

Sensorineural Hearing Loss

Sensorineural hearing loss results from damage to the neural structures of the inner ear. This could include the cochlear hair cells (from exposure to loud sounds) or portions of the vestibulocochlear nerve. Besides loud sounds, sensorineural loss results from tumors in the nerve, degeneration of the nerve, or congenital problems.

CRITICAL-THINKING QUESTIONS

1. Why does the blind spot from the optic disc in either eye not result in a blind spot in the visual field?
2. A pituitary tumor can cause perceptual losses in the lateral visual field. The pituitary gland is located directly inferior to the hypothalamus. Why would this happen?
3. Images focused on the retina are inverted. How do we see the world in its proper perspective?
4. Why might an individual with a middle ear infection notice a change in their taste sensation?
5. If you spin around on a stool and stop suddenly, the room seems to continue to move. Explain.
6. Why do individuals experience sensitivity to bright light after cataract surgery?
7. Why do wine tasters both smell and taste wine to judge its quality?

Chapter Twelve

Endocrine System

The endocrine system is a control system of ductless glands that secrete hormones within specific organs. Hormones act as "messengers" and are carried by the bloodstream to different cells in the body, which interpret these messages and act on them. The ability to maintain homeostasis and respond to stimuli is due largely to hormones secreted within the body. Without hormones, you could not grow, maintain a constant temperature, produce offspring, or perform the basic actions and functions that are essential for life. The endocrine system provides an electrochemical connection from the hypothalamus of the brain to all the organs that control the body metabolism, growth and development, and reproduction. The endocrine system regulates its hormones through negative feedback, except in very specific cases like childbirth. Increases in hormone activity generally decrease the production of that hormone. The immune system and other factors contribute as control factors also, altogether maintaining constant levels of hormones.

The endocrine system can be considered a "link" between organs and cells. In past sections we saw other similar links between systems. For example, in the nervous system, we examined a number of chemicals called neurotransmitters that were released by neurons that affected other neurons. In the muscular section we saw neurotransmitters affecting muscular contraction. These and other links exist largely to support homeostasis. In homeostasis, changes in the internal or external environment of the body are sensed, invoking some sort of correcting mechanism to keep the system in "balance." This is what the endocrine system does on a regular basis. The endocrine system senses changes in the internal or external environment and responds by secreting hormones. The hormones travel to target cells that contain specific receptors for hormones. The target cells then respond by altering function (Fig. 12.1). Target cells undergo a variety of changes in response to stimulation from hormones; controlling rates of chemical reactions, transporting substances through membranes, regulating fluid, electrolyte balance, etc.

Hormones can be chemically classified into the following groups:

1. **Amino acid–derived**: Hormones that are modified amino acids.
2. **Polypeptide and proteins**: Hormones that are chains of amino acids of fewer than or more than about 100 amino acids, respectively. Some protein hormones are actually glycoproteins, containing glucose or other carbohydrate groups.

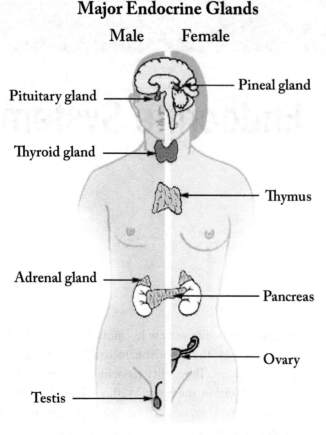

Figure 12.1 Endocrine system.

3. **Steroids**: Hormones that are lipids synthesized from cholesterol. Steroids are characterized by four interlocking carbohydrate rings.

4. **Eicosanoids**: Are lipids synthesized from the fatty acid chains of phospholipids found in plasma membrane. Hormones circulating in the blood diffuse into the interstitial fluids surrounding the cell. Cells with specific receptors for a hormone respond with an action that is appropriate for the cell. Because of the specificity of hormone and target cell, the effects produced by a single hormone may vary among different kinds of target cells. Hormones activate target cells by one of two methods, depending upon the chemical nature of the hormone.

- **Lipid-soluble** hormones (steroid hormones and hormones of the thyroid gland) diffuse through the cell membranes of target cells. The lipid-soluble hormone then binds to a receptor protein that, in turn, activates a DNA segment that turns on specific genes. The proteins produced as result of the transcription of the genes and subsequent translation of mRNA act as enzymes that regulate specific physiological cell activity.

- **Water-soluble** hormones (polypeptide, protein, and most amino acid hormones) bind to a receptor protein on the plasma membrane of the cell. The receptor protein, in turn, stimulates the production of one of the following second messengers: Cyclic AMP (cAMP) is produced when the receptor protein activates another membrane-bound protein called a G-protein. The G-protein activates adenylate cyclase, the enzyme that catalyzes the production of cAMP from ATP. Cyclic AMP then triggers an enzyme that generates specific cellular changes. Inositol triphosphate (IP3) is produced from membrane phospholipids. IP3, in turn, triggers the release of CA2+ from the endoplasmic reticulum, which then activates enzymes that generate cellular changes.

Endocrine glands release hormones in response to one or more of the following stimuli:

1. Hormones from other endocrine glands.
2. Chemical characteristics of the blood (other than hormones).
3. Neural stimulation.

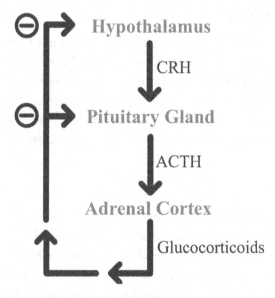

Figure 12.2 A negative feedback pathway.

Most hormone production is managed by a negative feedback system (Fig. 12.2). The nervous system and certain endocrine tissues monitor various internal conditions of the body. If action is required to maintain homeostasis, hormones are released, either directly by an endocrine gland or indirectly through the action of the hypothalamus of the brain, which stimulates other endocrine glands to release hormones. The hormones activate target cells, which initiate physiological changes that adjust the body conditions. When normal conditions have been recovered, the corrective action—the production of hormones—is discontinued.

Thus, in negative feedback, when the original (abnormal) condition has been repaired, or negated, corrective actions decrease or discontinue. For example, the amount of glucose in the blood controls the secretion of insulin and glucagon via negative feedback.

The production of some hormones is controlled by positive feedback. In such a system, hormones cause a condition to intensify, rather than decrease. As the condition intensifies, hormone production increases. Such positive feedback is uncommon, but does occur during childbirth, where hormone levels build with increasingly intense labor contractions. Also in lactation, hormone levels increase in response to nursing, which causes an increase in milk production. The hormone produced by the hypothalamus causing the milk let-down and uterine contraction is oxytocin.

Examples of how hormones affect cells: mechanism-of-action

Prostaglandins:

Prostaglandins are secreted by cells and have a local effect. This means that they are secreted by cells and travel only to nearby cells. This is known as a paracrine secretion. Once the hormone reaches the target cell, it can use what is known as the second messenger system. Prostaglandins help to control smooth muscle contraction and relaxation. Prostaglandins also help to promote inflammation.

Steroid Hormones:

Steroid hormones are transported in the blood. They connect with a special transport protein known as a carrier protein. Once reaching the target cell, the hormone disassociates from the carrier protein. Recall that lipid-soluble substances can diffuse through a cell membrane. Since steroid hormones are considered lipids, they can diffuse through the cell membrane and enter the cell. Once inside the cell, steroid hormones combine with specialized receptors located within the cytoplasm of the cell. Once the hormone combines with the receptor, the receptor-hormone complex moves into the nucleus of the cell. There it invokes changes in DNA transcription that in turn cause changes in the metabolism of the cell characteristic of the hormone.

Non-Steroid Hormones:

Non-steroidal hormones enter the cell differently than steroids. Non-steroidal hormones are not lipid soluble, since they cannot diffuse directly into the cell and must enter via a different process. Non-steroid hormones enter the cell by using what are known as second messengers. Receptors for non-steroidal hormones are located in the cell walls of the target cells. When the hormone connects to the receptor on the outside of the cell membrane, a protein known as a G-protein is activated and moves down the membrane into the cell. The G-protein binds to an enzyme known as adenylate cyclase and activates it. Adenylate cyclase then becomes involved in the reaction:

Adenylate cyclase
ATP cAMP + 2P

cAMP, cyclic adenylate monophosphate, is considered the second messenger in the system. cAMP in turn activates another inactive enzyme called protein kinase. Protein kinase facilitates the phosphorylation of various proteins. Phosphorylation occurs when phosphates are attached to a molecule. The phosphorylated proteins then activate some enzymes and inactivate others inside the cell. This alters the metabolic activity of the cell and the cell responds in accordance with the intended action of the hormone. Results of second messenger activation include altered membrane permeability, activation of enzymes, protein synthesis, modulation of metabolic pathways, promoting movement of cells, and causing secretion of other hormones. cAMP works with a variety of hormones including those from:

- Hypothalamus
- Anterior pituitary
- Posterior pituitary
- Parathyroid
- Adrenals
- Thyroid
- Pancreas
- There are other second messengers besides cAMP. These are:
- Diacyglycerol (DAG)
- Inositol triphosphate (IP3)
- Cyclic guanosine monophosphate (cGMP)

Hormones operating via second messengers have a much greater response, i.e., amplification of response. Many second messengers can be activated by one hormone.

Pituitary Gland

The pituitary gland sits in the sella turcica of the sphenoid bone. It is positioned in close proximity to the hypothalamus and is connected to the hypothalamus by a stalk-like structure called the infundibulum. The pituitary gland is divided into 2 sections. The anterior pituitary (aka adenohypophysis) and the posterior pituitary (aka neurohypophysis) each secrete different hormones. Hormones secreted by the anterior pituitary are influenced by other chemicals secreted by the hypothalamus, known as "releasing factors." Thus the nervous system exhibits some control over secretions of the anterior pituitary. The hypothalamus communicates with the anterior pituitary gland via a capillary network that interconnects the two structures. Blood levels of hormones are monitored by the hypothalamus, causing the secretion of releasing factors that control release of anterior pituitary hormones. The communication between the hypothalamus and posterior pituitary is somewhat different than in the case of the anterior pituitary. The hypothalamus and posterior pituitary are interconnected by a series of specialized nerve cells called neurosecretory cells. Thus the hormones of the posterior pituitary are actually produced in the hypothalamus but are secreted in the posterior pituitary.

Anterior Pituitary Hormones: Growth Hormone (Somatotropin)

Growth hormone is secreted in response to two secretions by the hypothalamus: Growth hormone releasing hormone (GHRH) and somatostatin (SS). Growth hormone affects cellular metabolism by promoting the movement of amino acids into cells for protein synthesis, which affects the overall growth of the organism. Growth hormone releasing hormone secreted by the hypothalamus stimulates the release of growth hormone by the anterior pituitary. Somatostatin inhibits the release of growth hormone. Growth hormone stimulates cells to enlarge and undergo mitosis as well as increasing the rate of protein synthesis and increasing the cellular use of carbohydrates and fats.

Prolactin (PRL)

Prolactin is secreted in response to two secretions by the hypothalamus. Prolactin releasing factor (PRF) stimulates secretion of prolactin by the anterior pituitary. Prolactin inhibiting hormone (PIH) from the hypothalamus inhibits secretion of prolactin by the anterior pituitary. The function of prolactin is to stimulate milk production in females. In males, prolactin decreases the secretion of luteinizing hormone, which facilitates production of the primary male sex hormones or androgens. Too much prolactin secretion in males can cause infertility.

Thyroid Stimulating Hormone (TSH)

Thyroid stimulating hormone is released by the anterior pituitary in response to the release of thyrotropin releasing hormone from the hypothalamus. Thyroid stimulating hormone causes the thyroid gland to release the thyroid hormones triiodothyronine and tetraiodothyronine (T3 and T4). The blood concentration of thyroid hormones provides a negative feedback mechanism to the hypothalamus to help control the release of thyroid stimulating hormone. Secretion of T3 and T4 is also affected by stress. Thyroid stimulating hormone also stimulates growth of the thyroid gland.

Normal TSH levels range from 1 to 4U/ml. The TSH test is important in differentiating primary from secondary hypothyroidism (low thyroid hormone levels). If TSH levels are increased, primary hypothyroidism is indicated. This means that anterior pituitary continues to secrete larger amounts of TSH in response to low levels of the hormones produced by the thyroid gland itself. This indicates that there is a problem with the thyroid gland in producing thyroid hormones. In secondary hypothyroidism, both TSH and thyroid hormone levels are decreased. Causes include pituitary dysfunction and hyperthyroidism.

Adrenocorticotropic Hormone (ACTH)

Adrenocorticotropic hormone is secreted by the anterior pituitary in response to the secretion of corticotropic releasing hormone (CRH) by the hypothalamus. ACTH is picked up by the adrenal cortex and stimulates secretion of hormones by the adrenal cortex. Adrenal cortex hormones then provide feedback to the hypothalamus and anterior pituitary to help regulate secretion of ACTH. Stress also affects secretion of ACTH.

Follicle Stimulating Hormone (FSH)

Follicle stimulating hormone is secreted by the anterior pituitary partly in response to the secretion of a releasing factor known as gonadotropin releasing hormone (GnRH). In females, FSH stimulates growth and development of egg-cell-containing follicles in ovaries and stimulates follicular cells to produce estrogen. In males, FSH stimulates the production of sperm cells in the testes when the male reaches puberty.

Luteinizing Hormone (LH)

Luteinizing hormone secretion is also partly controlled by the release of gonadotropin releasing hormone by the hypothalamus. Luteinizing hormone stimulates the glands of the reproductive system to produce sex hormones.

Posterior Pituitary Hormones

The posterior pituitary contains specialized nerve cells called neurosecretory cells that originate in the hypothalamus. The secretions of these cells function as hormones rather than neurotransmitters in that the target tissues are contained in glands outside of the nervous system.

Antidiuretic hormone (ADH)

Antidiuretic hormone is secreted by the posterior pituitary in response to concentration changes sensed by osmoreceptors located in the hypothalamus. The action of ADH is to cause the kidneys to conserve water. The target tissue of this hormone lies in the kidney, particularly the distal convoluted tubule. ADH acts to make the distal convoluted tubule more permeable to water, which in turn causes conservation of fluids and decreased urine output. Recall, a diuretic increases urine output and consequently decreases overall blood volume. Antidiuretic hormone has the opposite effect, in an effort to conserve fluids and blood volume. If blood solute concentration increases, ADH is released in an attempt to conserve water and produce a more dilute blood. If blood solute concentration decreases, the release of ADH is inhibited.

Oxytocin (OT)

Oxytocin helps to stimulate uterine contractions during labor by causing the smooth muscles in the uterine wall to contract. During pregnancy, the uterus becomes more sensitive to oxytocin. Oxytocin also helps to stimulate release of milk from mammary glands. Although oxytocin is produced in males, its function is not well understood. Oxytocin is sometimes injected into women to stimulate contractions and induce labor. Besides stimulating contractions, oxytocin also causes vasoconstriction of the uterine blood vessels and causes the uterus to shrink. This helps to reduce bleeding.

Thyroid Gland

The thyroid gland is located in the anterior portion of the throat just inferior to the thyroid cartilage (Adam's apple) (Fig. 12.3). It contains distinct regions of tissues known as follicles. The structure of thyroid tissue produces two main cell types: those located within the follicle structure, known as follicular cells; and those not located in the follicle, known as extrafollicular or parafollicular cells. Both cell types secrete hormones. The follicular cells secrete triiodothyronine (T3) and tetraiodothyronine (T4). These hormones are secreted in response to the secretion of thyroid stimulating hormone from the anterior pituitary gland. Both of these hormones affect overall metabolism by increasing the cellular metabolism of carbohydrates, proteins, and lipids. Both T3 and T4 require the presence of iodine in order to be produced. Iodine and thyronine (an amino acid) are joined in the follicular cells. Triiodothyronine has 3 iodines and tetraiodothyronine has 4 iodines. Thyroxine (T4) is the most abundant thyroid hormone. It works to increase metabolism and stimulates the cardiovascular system. It also works to differentiate cells. T3 (triiodothyronine) is secreted in smaller amounts than T4 but has the same effects as T4.

A benign tumor of the thyroid gland called a goiter can develop from the lack of dietary iodine. In this case, since iodine is not present in the diet, TSH continues to be released by the anterior pituitary in an effort to produce T3 and T4. But since there is insufficient iodine to produce T3 and T4, the levels of T3 and T4 decrease. The thyroid gland enlarges or hypertrophies due to the continuous stimulation by the action of TSH. Often, the inclusion of dietary iodine can counteract this phenomenon.

Hyperthyroidism: Typical T4 levels range from 4 to 11 g/dl and T3 levels range from 110 to 230 ng/dl. An increase in T3/T4 indicates hyperthyroidism. In infants this is known as Cretinism. In adults it is known as Grave's disease. Cretinism is characterized by mental retardation, low body temperature, and growth abnormalities. Grave's disease is characterized by exophthalmos (protruding eyes), high metabolic rate, heat sensitivity, restlessness, and weight loss. An increase in thyroid hormone levels can also occur in thyroiditis (an inflammation of the thyroid gland), thyrotoxicosis, and tumors.

Hypothyroidism: Low T3/T4 levels indicate hypothyroidism known as myxedema. Signs of myxedema include a rounded face, and swelling of the hands, feet, and periorbital tissue. If left untreated, myxedema can lead to coma and death. A rare form of hypothyroidism is called Hashimoto's hypothyroidism. This is an autoimmune disorder where the patient's own antibodies bind to receptors on the thyroid and mimic the action of TSH.

Calcitonin

The other thyroid hormone has an effect on blood calcium levels and is called calcitonin. Calcitonin is secreted by the extrafollicular cells (C-cells). Calcitonin decreases blood calcium levels by decreasing osteoclastic activity and increasing osteoblastic activity. Osteoclasts work to release calcium and other minerals from bone into the bloodstream. Osteoblasts work to build up bone by storing these minerals into bone. Calcitonin also affects calcium reabsorption in the kidneys by inhibiting it, thereby causing increased calcium excretion in the urine. Calcitonin works to lower the calcium levels in blood.

Calcitonin is released in response to increases in blood calcium levels. This happens, for example, in pregnancy when an increase in blood calcium is needed for the development of the fetus. Typical calcitonin levels are less than 50 pg/ml. Increased calcitonin levels occur in medullary carcinoma of the thyroid gland; oat cell carcinoma of the lung, breast, and pancreatic cancers; thyroiditis; and pernicious anemia.

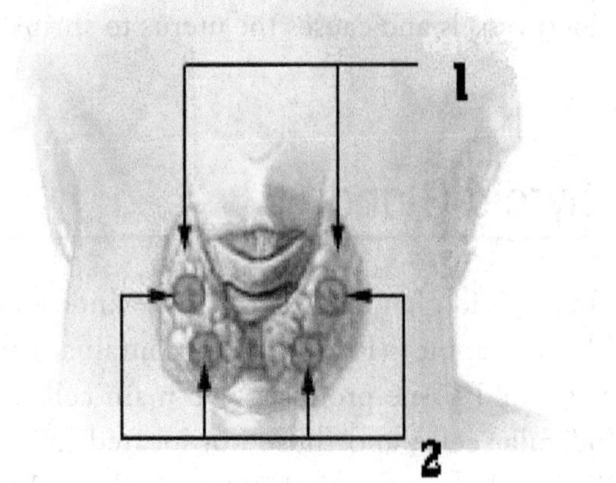

Parathyroid Glands

The parathyroid glands are four small masses of glandular tissue located on the posterior surface of the thyroid gland (Fig. 12.3). These small glands contain secretory cells as well as capillaries. The parathyroid glands secrete one hormone aptly called parathyroid hormone (PTH).

Figure 12.3 Location of the thyroid gland.
1. Thyroid. 2. Parathyroids

Parathyroid hormone works to increase blood calcium levels and decrease blood phosphate levels. PTH accomplishes this by stimulating osteoclastic activity to release calcium and other bone minerals into the bloodstream and inhibiting osteoblastic activity. Recall, osteoblasts work to store minerals in bone. PTH also stimulates the production of vitamin D, which in turn facilitates the absorption of calcium in the intestine. Vitamin D (cholecalciferol) is produced by converting provitamin D stored in the skin to vitamin D. This is done with the help of ultraviolet radiation from the sun. Vitamin D is also stored in tissues after it is converted to a storage form, known as dihydroxycholecalciferol, by the liver. PTH changes dihydroxycholecalciferol to the active form of vitamin D (cholecalciferol), which facilitates calcium absorption in the intestines. PTH also stimulates the release of the phosphate ion in the kidneys. All of these actions work to increase calcium concentration in the blood. Thus calcium levels are controlled by both calcitonin from the thyroid and parathyroid hormone.

Parathyroid Scan: The parathyroid scan is a procedure using radioactive materials (radionucleotides) injected into the patient. The material is absorbed by both the thyroid and parathyroid glands and a subsequent scan can detect the concentration of the radioactive material. Information such as the location, position, and size of the parathyroids can be interpreted from the scan. Normal parathyroid function is indicated if the material is absorbed by the glands. Abnormal function is indicated by an adenoma (a benign tumor).

Adrenal Glands

The adrenal glands are two small pyramid-shaped glands located on top of the kidneys (Fig. 12.4). They consist of 2 functional areas: an outer cortex and an inner medulla. The cortex consists of 3 layers: zona glomerulosa, zona fasciculate, and zona reticularis. The adrenal cortex produces a number of steroids as well as some other hormones. The hormones of the adrenal cortex and medulla are required by the body to sustain life.

Adrenal Cortex Hormones: Aldosterone

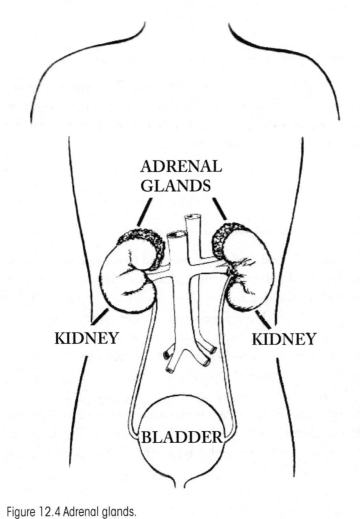

Figure 12.4 Adrenal glands.

Aldosterone is produced by cells of the zona glomerulosa of the adrenal cortex. Aldosterone acts to regulate electrolytes such as magnesium and potassium. These are known as mineral electrolytes, thus aldosterone is known as a mineralocorticoid. Aldosterone causes the kidney to conserve sodium and excrete potassium. The release of aldosterone is more strongly facilitated by the increase in plasma potassium concentration. The decrease in plasma sodium concentration does not affect the secretion of aldosterone as strongly. However, the decrease in sodium concentration can affect the renin-angiotensin system in the kidneys and stimulates the release of aldosterone. Both aldosterone and the renin-angiotensin system work together to conserve blood volume and sodium. Aldosterone works by inhibiting the release of sodium by the kidney and the renin-angiotensin system works by causing vasoconstriction. Aldosterone is also released via stimulation of the adrenal cortex by ACTH.

Cortisol (Cortisone)

Cortisol is also secreted by the adrenal cortex, specifically by the cells of the zona fasciculata. Cortisol has an effect on glucose metabolism, thus it is called a glucocorticoid. Cortisol secretion increases glucose levels in the blood. It does this by stimulating the liver to convert non-carbohydrates into glucose. This process is called gluconeogenesis. It also stimulates the release of fatty acids for use as an energy source. These processes help to regulate the level of blood glucose between meals. Cortisol is released in response to the release of ACTH by the anterior pituitary gland. Remember that ACTH is released in response to release of CRH by the hypothalamus. This system provides a negative feedback mechanism to help control the level of cortisol in the blood. Cortisol is sometimes used to control inflammation. It is injected into the body (cortisone injection). Cortisol works by inhibiting prostaglandin synthesis (prostaglandins work to increase inflammation) and increasing local vasoconstriction of the damaged tissue.

Sex Hormones

The inner layer of the adrenal cortex (zona reticularis) secretes sex hormones. The secretion of sex hormones from the adrenal glands helps contribute to the supply of hormones from the reproductive glands. These hormones are male hormones known as androgens but may be converted to estrogen in the female. These hormones help to develop the primary sex characteristics.

Adrenal Medulla

The adrenal medulla, or inner portion of the adrenal gland, is closely connected to the sympathetic nervous system. The adrenal medulla contains specialized cells called chromaffin cells that secrete chemicals called catecholamines. The catecholamines that are produced are norepinephrine and epinephrine. These chemicals should sound familiar because they were introduced as neurotransmitters in the sympathetic nervous system. Therefore, norepinephrine (NE) and epinephrine (E) have both neurotransmitter and hormonal actions. NE and E are secreted by the adrenal medulla in response to impulses produced by the sympathetic nervous system (SNS). The SNS is connected via nerve fibers to the adrenal medulla. The actions of NE and E from the adrenal medullar are similar to the actions of the (SNS). Thus secretion of NE and E will cause an increase in heart rate, blood pressure, respiration, and a decrease in digestion. The hormonal action of NE and E lasts longer than neurotransmitter action because it takes longer to remove NE and E from the endocrine system. Both the adrenal glands and the SNS work together to provide the sympathetic response.

Pancreas

The pancreas is located in the abdominal cavity at the flexure of the proximal portion of the small intestine called the duodenum (Fig. 12.5). It is connected to the duodenum by ducts. It produces both digestive and hormonal secretions and performs a dual role in these systems. Our focus in the section will be on the hormonal secretions of the pancreas. The internal structure of the pancreas consists of groupings of cells around capillary beds. The groupings of cells are called Islets of Langerhans and consist of 3 distinct types of cells: alpha, beta, and delta cells. Each cell type produces a different secretion. Alpha cells secrete glucagons, beta cells secrete insulin, and delta cells secrete somatostatin. Glucagon (alpha cells) works to increase the level of glucose in the blood. It does this by stimulating

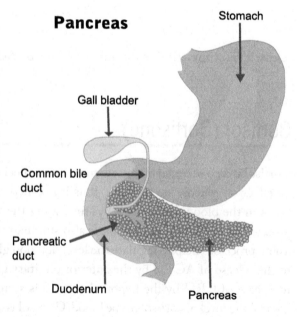

Figure 12.5 Pancreas.

the liver to convert the storage form of glucose (glycogen) into glucose via a process known as glycogenolysis. Glucagon also stimulates the process of gluconeogenesis, which converts non-carbohydrates substances into glucose in the liver and breaks down fats into fatty acids and glycerol. Glucagon is secreted when glucose levels are diminished in the blood. Secretion of glucagon is inhibited by high glucose blood levels. Insulin (beta cells) works to decrease the levels of glucose in the blood. It does this by reversing the processes stimulated by glucagon. Insulin facilitates the storage of glucose in the liver by stimulating the production of glycogen from glucose. Insulin also inhibits the process of gluconeogenesis, stimulates protein synthesis, and increases the storage of lipid in adipose tissue. Insulin also facilitates the release of glucose into body tissues by stimulating facilitative diffusion of glucose carriers in cell membranes. Insulin is secreted when blood glucose levels are high and inhibited when blood glucose levels are low. Somatostatin (secreted by the delta cells) inhibits both glucagon and insulin secretion. Thus it also works to control glucose levels in the blood.

Diabetes is a disease that affects insulin production in the pancreas. There are 2 types. Type I diabetes is caused by an autoimmune disorder in which the body's own immune cells attack the cells in the Islets of Langerhans in the pancreas (the beta cells that produce insulin). Type I diabetes is also known as juvenile onset diabetes because it manifests before the early 20s. Symptoms of Type I diabetes include weight loss, glucose in the urine (glucosuria), and poor wound healing or ulcers. If untreated, a buildup of ketone bodies occurs as a result of excess fat metabolism. This can lower pH, causing metabolic acidosis. The lowered pH can adversely affect neurons causing coma and death. Treatment includes daily insulin injections. Type II diabetes, also known as adult onset diabetes, develops in middle-aged adults from loss of insulin receptors in the cell membranes. Risk factors include weight gain and sedentary lifestyle. Treatment includes weight loss and exercise.

Pineal Gland

The pineal gland is a small pine cone–shaped gland located between the cerebral hemispheres (Fig. 12.6). It attaches to the posterior portion of the thalamus. The pineal gland secretes melatonin. Melatonin is synthesized from the neurotransmitter serotonin and is involved in the regulation of sleep–wake cycles known as circadian rhythms. Melatonin secretion increases with a decrease in light. Melatonin also helps to regulate the menstrual cycle.

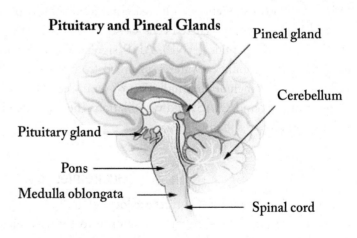

Figure 12.6 Pineal gland.

Thymus Gland

The thymus gland is located posterior to the sternum. It is larger at birth and shrinks throughout adulthood (Fig. 12.7). The thymus gland secretes thymosins. Thymosins function in facilitating the production of a type of white blood cell known as a T-lymphocyte, which functions in immunity.

The Reproductive Glands

The ovaries and placenta in the female, as well as the testes in the male, secrete hormones that have a role in the endocrine system. The ovaries and placenta secrete estrogen and progesterone. The placenta also secretes a gonadotropin. The testes secrete testosterone. These hormones and glands will be discussed in more detail in the reproductive system section.

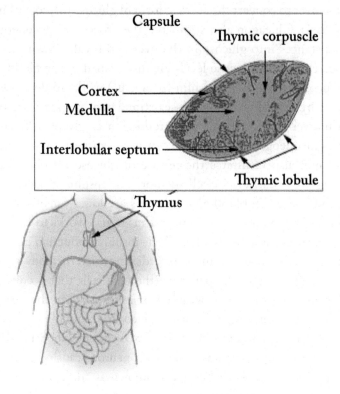

Figure 12.7 Thymus gland.

Organs with Secondary Endocrine Functions

We have already encountered a few of the many organs of the body that have secondary endocrine functions. Here, you will learn about the hormone-producing activities of the heart, gastrointestinal tract, kidneys, skeleton, adipose tissue, skin, thymus, and liver.

Heart

When the body experiences an increase in blood volume or pressure, the cells of the heart's atrial wall stretch. In response, specialized cells in the wall of the atria produce and secrete the peptide hormone atrial natriuretic peptide (ANP). ANP signals the kidneys to reduce sodium reabsorption, thereby decreasing the amount of water reabsorbed from the urine filtrate and reducing blood volume. Other actions of ANP include the inhibition of renin secretion and the initiation of the renin-angiotensin-aldosterone system (RAAS) and vasodilation. Therefore, ANP aids in decreasing blood pressure, blood volume, and blood sodium levels.

Gastrointestinal Tract

The endocrine cells of the GI tract are located in the mucosa of the stomach and small intestine. Some of these hormones are secreted in response to eating a meal and aid in digestion. An example of a hormone secreted by the stomach cells is gastrin, a peptide hormone secreted in response to stomach distention that stimulates the release of hydrochloric acid. Secretin is a peptide hormone secreted by the small intestine as acidic chyme (partially digested food and fluid) moves from the stomach. It stimulates the release of bicarbonate from the pancreas, which buffers the acidic chyme, and inhibits the further secretion of hydrochloric acid by the stomach. Cholecystokinin (CCK) is another peptide hormone released from the small intestine. It promotes the secretion of pancreatic enzymes and the release of bile from the gallbladder, both of which facilitate digestion. Other hormones produced by the intestinal cells aid in glucose metabolism, such as by stimulating the pancreatic beta cells to secrete insulin, reducing glucagon secretion from the alpha cells, or enhancing cellular sensitivity to insulin.

Kidneys

The kidneys participate in several complex endocrine pathways and produce certain hormones. A decline in blood flow to the kidneys stimulates them to release the enzyme renin, triggering the renin-angiotensin-aldosterone (RAAS) system, and stimulating the reabsorption of sodium and water. The reabsorption increases blood flow and blood pressure. The kidneys also play a role in regulating blood calcium levels through the production of calcitriol from vitamin D3, which is released in response to the secretion of parathyroid hormone (PTH). In addition, the kidneys produce the hormone erythropoietin (EPO) in response to low oxygen levels. EPO stimulates the production of red blood cells (erythrocytes) in the bone marrow, thereby increasing oxygen delivery to tissues. You may have heard of EPO as a performance-enhancing drug (in a synthetic form).

Skeleton

Although bone has long been recognized as a target for hormones, only recently have researchers recognized that the skeleton itself produces at least two hormones. Fibroblast growth factor 23 (FGF23) is produced by bone cells in response to increased blood levels of vitamin D3 or phosphate. It triggers the kidneys to inhibit the formation of calcitriol from vitamin D3 and to increase phosphorus excretion. Osteocalcin, produced by osteoblasts, stimulates the pancreatic beta cells to increase insulin production. It also acts on peripheral tissues to increase their sensitivity to insulin and their utilization of glucose.

Adipose Tissue

Adipose tissue produces and secretes several hormones involved in lipid metabolism and storage. One important example is leptin, a protein manufactured by adipose cells that circulates in amounts directly proportional to levels of body fat. Leptin is released in response to food consumption and acts by binding to brain neurons involved in energy intake and expenditure.

Binding of leptin produces a feeling of satiety after a meal, thereby reducing appetite. It also appears that the binding of leptin to brain receptors triggers the sympathetic nervous system to regulate bone metabolism, increasing deposition of cortical bone. Adiponectin—another hormone synthesized by adipose cells—appears to reduce cellular insulin resistance and to protect blood vessels from inflammation and atherosclerosis. Its levels are lower in people who are obese, and rise following weight loss.

Skin

The skin functions as an endocrine organ in the production of the inactive form of vitamin D3, cholecalciferol. When cholesterol present in the epidermis is exposed to ultraviolet radiation, it is converted to cholecalciferol, which then enters the blood. In the liver, cholecalciferol is converted to an intermediate that travels to the kidneys and is further converted to calcitriol, the active form of vitamin D3. Vitamin D is important in a variety of physiological processes, including intestinal calcium absorption and immune system function. In some studies, low levels of vitamin D have been associated with increased risks of cancer, severe asthma, and multiple sclerosis. Vitamin D deficiency in children causes rickets, and in adults, osteomalacia—both of which are characterized by bone deterioration.

Thymus

The thymus is an organ of the immune system that is larger and more active during infancy and early childhood, and begins to atrophy as we age. Its endocrine function is the production of a group of hormones called thymosins that contribute to the development and differentiation of T-lymphocytes, which are immune cells. Although the role of thymosin is not yet well understood, it is clear that they contribute to the immune response. Thymosins have been found in tissues other than the thymus and have a wide variety of functions, so the thymosins cannot be strictly categorized as thymic hormones.

Liver

The liver is responsible for secreting at least four important hormones or hormone precursors: insulin-like growth factor (somatomedin), angiotensinogen, thrombopoietin, and hepcidin. Insulin-like growth factor-1 is the immediate stimulus for growth in the body, especially of the bones. Angiotensinogen is the precursor to angiotensin, mentioned earlier, which increases blood pressure. Thrombopoietin stimulates the production of the blood's platelets. Hepcidins block the release of iron from cells in the body, helping to regulate iron homeostasis in our body fluids.

Development and Aging of the Endocrine System

The endocrine system arises from all three embryonic germ layers. The endocrine glands that produce the steroid hormones, such as the gonads and adrenal cortex, arise from the mesoderm. In contrast, endocrine glands that arise from the endoderm and ectoderm produce the amine, peptide, and protein hormones. The pituitary gland arises from two distinct areas of the ectoderm: the anterior pituitary gland arises from the oral ectoderm, whereas the posterior pituitary gland arises from the neural ectoderm at the base of the hypothalamus. The pineal gland also arises from the ectoderm. The two structures of the adrenal glands arise from two different germ layers: the adrenal cortex from the mesoderm and the adrenal medulla from ectoderm neural cells. The endoderm gives rise to the thyroid and parathyroid glands, as well as the pancreas and the thymus.

As the body ages, changes occur that affect the endocrine system, sometimes altering the production, secretion, and catabolism of hormones. For example, the structure of the anterior pituitary gland changes as vascularization decreases and the connective tissue content increases with increasing age. This restructuring affects the gland's hormone production. For example, the amount of human growth hormone that is produced declines with age, resulting in the reduced muscle mass commonly observed in the elderly. The adrenal glands also undergo changes as the body ages; as fibrous tissue increases, the production of cortisol and aldosterone decreases. Interestingly, the production and secretion of epinephrine and norepinephrine remain normal throughout the aging process. A well-known example of the aging process affecting an endocrine gland is menopause and the decline of ovarian function. With increasing age, the ovaries decrease in both size and weight and become progressively less sensitive to gonadotropins. This gradually causes a decrease in estrogen and progesterone levels, leading to menopause and the inability to reproduce. Low levels of estrogens and progesterone are also associated with some disease states, such as osteoporosis, atherosclerosis, and hyperlipidemia, or abnormal blood lipid levels. Testosterone levels also decline with age, a condition called andropause (or viropause); however, this decline is much less dramatic than the decline of estrogens in women, and much more gradual, rarely affecting sperm production until very old age. Although this means that males maintain their ability to father children for decades longer than females, the quantity, quality, and motility of their sperm are often reduced. As the body ages, the thyroid gland produces less of the thyroid hormones, causing a gradual decrease in the basal metabolic rate. The lower metabolic rate reduces the production of body heat and increases levels of body fat. Parathyroid hormones, on the other hand, increase with age. This may be because of reduced dietary calcium levels, causing a compensatory increase in parathyroid hormone. However, increased parathyroid hormone levels combined with decreased levels of calcitonin (and estrogens in women) can lead to osteoporosis, as PTH stimulates demineralization of bones to increase blood calcium levels. Notice that osteoporosis is common in both elderly males and females. Increasing age also affects glucose metabolism, as blood glucose levels spike more rapidly and take longer to return to normal in the elderly. In addition, increasing glucose intolerance may occur because of a gradual decline in cellular insulin sensitivity.

CRITICAL-THINKING QUESTIONS

1. Describe several main differences in the communication methods used by the endocrine system and the nervous system.
2. Compare and contrast endocrine and exocrine glands.
3. True or false: Neurotransmitters are a special class of paracrines. Explain your answer.
4. Describe the mechanism of hormone response resulting from the binding of a hormone with an intracellular receptor.
5. Compare and contrast the anatomical relationship of the anterior and posterior lobes of the pituitary gland to the hypothalamus.
6. Explain why maternal iodine deficiency might lead to neurological impairment in the fetus.
7. Define hyperthyroidism and explain why one of its symptoms is weight loss.
8. Describe the role of negative feedback in the function of the parathyroid gland.
9. Explain why someone with a parathyroid gland tumor might develop kidney stones.
10. What are the three regions of the adrenal cortex and what hormones do they produce?
11. If innervation to the adrenal medulla were disrupted, what would be the physiological outcome?
12. Seasonal affective disorder (SAD) is a mood disorder characterized by, among other symptoms, increased appetite, sluggishness, and increased sleepiness. It occurs most commonly during the winter months, especially in regions with long winter nights. Propose a role for melatonin in SAD and a possible non-drug therapy.
13. Retinitis pigmentosa (RP) is a disease that causes deterioration of the retinas of the eyes. Describe the impact RP would have on melatonin levels.
14. What would be the physiological consequence of a disease that destroys the beta cells of the pancreas?

15. In Hashimoto's disease, thyroid gland function is depressed.
    A. Would the levels of T3 and T4 be higher or lower than normal? Explain.
    B. Would the level of TSH be higher or lower than normal? Explain.
16. Suppose that a person's immune system made antibodies that destroyed receptors for testosterone in the target cells of this hormone. What effect(s) would this have on the functioning of the body?
17. John had his posterior pituitary removed by accident during surgery. Immediately after surgery, John's ADH levels declined dramatically. About 10 days after surgery, John's ADH levels began to climb again. While his ADH levels never quite reached pre-surgical levels, they were certainly adequate. How do you explain these results?

Section Four

Circulation and Body Defense

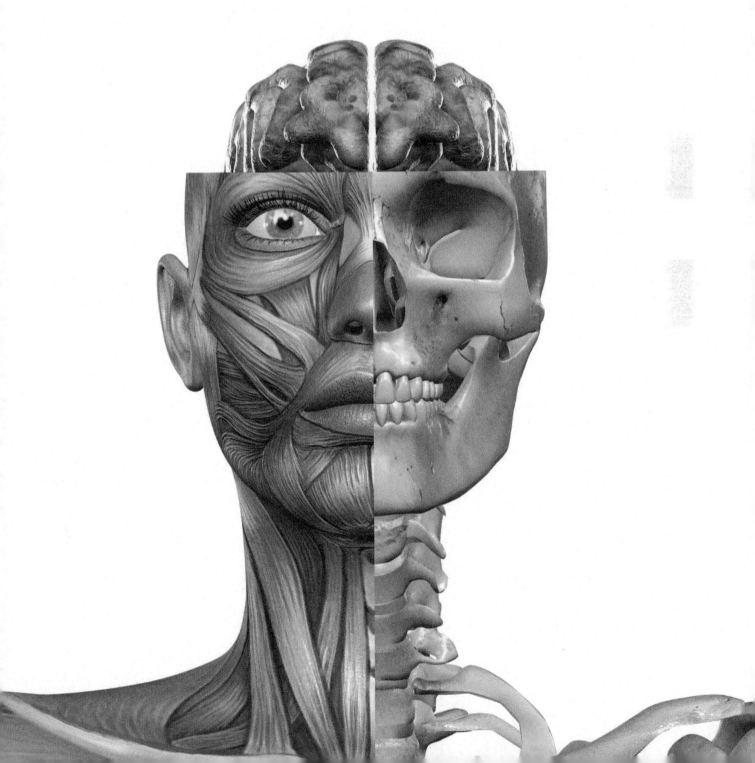

Chapter Thirteen

Blood

The primary function of blood is to supply oxygen and nutrients to tissues and to remove waste products. Blood also enables hormones and other substances to be transported between tissues and organs. Problems with blood composition or circulation can lead to downstream tissue malfunction. Blood is also involved in maintaining homeostasis by acting as a medium for transferring heat to the skin and by acting as a buffer system for bodily pH. The blood is circulated through the lungs and body by the pumping action of the heart. The right ventricle pressurizes the blood to send it through the capillaries of the lungs, while the left ventricle re-pressurizes the blood to send it throughout the body. Pressure is in essence lost in the capillaries, hence gravity and especially the actions of skeletal muscles are needed to return the blood to the heart. Blood is considered a connective tissue. Its primary constituents include specialized cells such as red blood cells (RBCs), white blood cells (WBCs), cell fragments known as platelets, and straw-colored liquid called plasma. The cells are known as the "formed elements" of blood and each has specialized functions.

Cells and Plasma

When blood is separated—the solid from the liquid portion—it is about 45% cells and 55% plasma by volume. This separation is done in clinical laboratories and is called a hematocrit. The fluid portion of blood is known as plasma. Plasma contains a variety of substances including water, proteins, carbohydrates, lipids, amino acids, vitamins, hormones, electrolytes, and waste products. The solid portion consists of cells. All blood cells come from the same cell of origin known as a hemocytoblast (stem cell) (Fig. 13.1). The hemocytoblast can differentiate into any of the mature blood cells by responding to factors called colony stimulating factors.

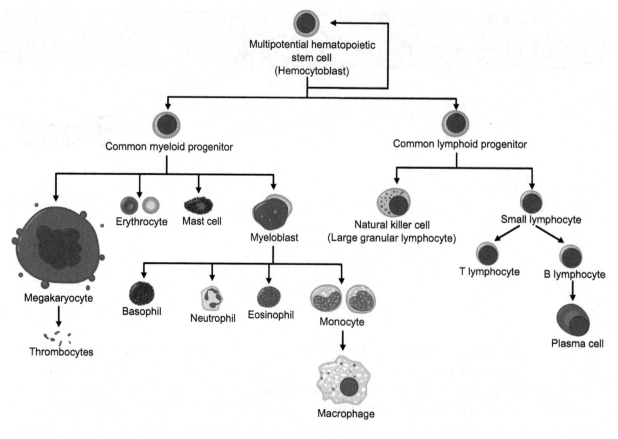

Figure 13.1 Hematopoiesis.

Red Blood Cells

Red blood cells (erythrocytes) make up the largest percentage of the formed elements of blood. The red blood cell's primary function is to transport oxygen and carbon dioxide. They have a unique biconcave disc shape that functions to increase the surface area of the cell in order to allow a greater amount of oxygen binding (Fig. 13.2). The red color of the cell is due to the presence of a protein molecule called hemoglobin. Hemoglobin makes up about one third of the cell's volume. Hemoglobin is the transport molecule for oxygen and carbon dioxide. Oxyhemoglobin is formed when oxygen combines with hemoglobin. Oxyhemoglobin is a bright red color. Deoxyhemoglobin is formed when oxygen is released from hemoglobin. Deoxyhemoglobin is a dull red color. Deoxygenated blood looks bluish under the skin because the skin filters out some of the light. Hemoglobin also transports carbon dioxide (Fig. 13.3). Carbaminohemoglobin is formed when carbon dioxide combines with hemoglobin. A small amount of carbon dioxide is transported this way. Since the primary function of red blood cells is to transport gases, there is no need for a nucleus or mitochondria.

Red Blood Cell Life Cycle

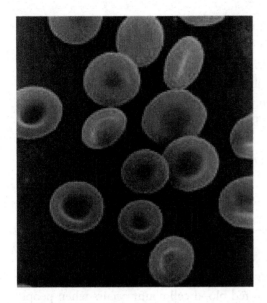

Figure 13.2 Red blood cells have a unique biconcave disc shape.

Red blood cells develop in the bone marrow. The process of blood cell formation is called hematopoiesis. The process of red blood cell formation is known as erythropoiesis. In erythropoiesis, immature red blood cells called erythroblasts differentiate from hemocytoblasts in the presence of the hormone erythropoietin. Erythropoietin is secreted by the kidneys and liver in response to low oxygen concentration in the blood. Erythroblasts still contain a nucleus and mitochondria and produce the hemoglobin. When they mature into erythrocytes, they shed the nucleus and mitochondria. The red blood cells then circulate in the bloodstream for approximately 120 days. About 2.5 million red blood cells are produced each second to keep up with the number of cells being recycled. During their lifetime they slowly wear out from passing through small capillary membranes. Each cell can travel through the body as many as 75,000 times. After about 120 days they become worn enough to pass through capillary membranes in the spleen and liver. There they are broken down and phagocytized by macrophages (white blood cells) and the hemoglobin is recycled. Hemoglobin is broken down in the

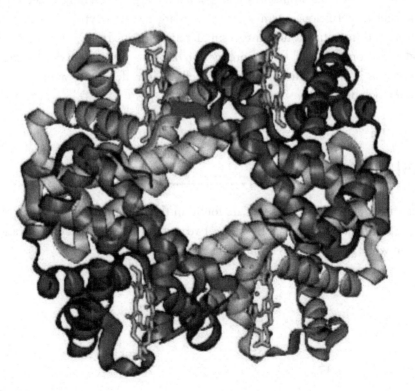

Figure 13.3 Hemoglobin molecule.

liver and spleen into four globin molecules and one heme molecule. The heme molecule breaks down into iron and biliverdin. Some of the iron combines with a molecule of transferrin and is recycled in the bone marrow. The remaining iron is combined with ferrin and stored in the liver. The biliverdin is converted to bilirubin (an orange pigment) and both bilirubin and biliverdin are secreted in the bile and end up in the digestive tract. An indicator of the amount of red blood cells in the bloodstream is known as the red blood cell count. The red blood cell count is defined as the number of red blood cells in a cubic millimeter of blood.

Red blood cell count = number of RBCs in cubic millimeter of blood

- 4,600,000–6,200,000 in males
- 4,200,000–5,400,000 in females
- 4,500,000–5,100,000 in children

An increase in the red blood cell count is known as erythrocytosis or erythema. This is caused by lung disease, poisoning, and dehydration. A temporary increase in red blood cell count occurs when people living at sea level visit high-altitude environments. A decrease in the red blood cell count is known as anemia. Anemia can be caused by a number of factors such as vitamin B12 deficiency, blood loss, and iron deficiency. Certain substances are needed in order to make red blood cells. These include vitamin B12, folic acid, and iron. Intrinsic factor is needed in order for vitamin B12 absorption to occur. Intrinsic factor is secreted by the stomach. Damage to the stomach lining can decrease the secretion of intrinsic factor and produce a vitamin B12 deficiency. Vitamin B12 and folic acid are needed in order to allow the hemocytoblast to fully develop into a mature red blood cell. If there is a deficiency of vitamin B12 from either a dietary problem or stomach lining problem, a type of anemia called pernicious anemia can develop. The cells in pernicious anemia are usually larger (macrocytic). This is because the larger hemocytoblast has not differentiated fully into a smaller erythrocyte. Iron is also needed for the production of red blood cells. Iron is needed for the production of hemoglobin. Iron deficiencies can also result in an anemia known as iron deficiency anemia. The cells in iron deficiency anemia are usually smaller and contain less hemoglobin.

White Blood Cells

White blood cells are called leukocytes. They are found in the blood but many also work outside of the circulatory system in organs and tissue. There are two categories of leukocytes. Granulocytes have a granular cytoplasm while agranulocytes do not.

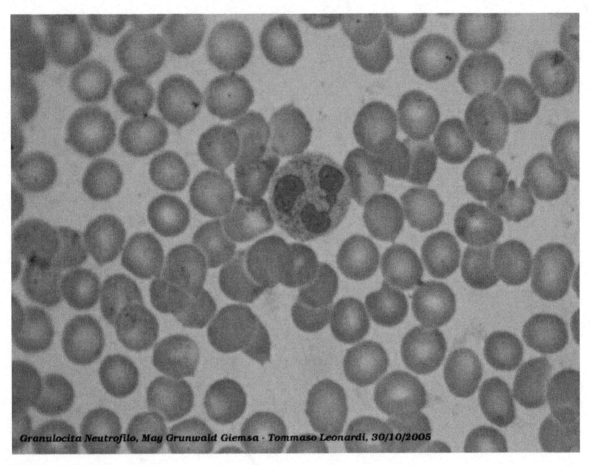

Figure 13.4 Neutrophil.

Granulocytes

Granulocytes tend to be significantly larger than red blood cells. There are three main types of granulocytes. These are the neutrophils, basophils, and eosinophils. Granulocytes also have a significantly shorter lifespan (measured in hours) than red blood cells (120 days). Neutrophils are larger than red blood cells and contain a segmented nucleus (Fig. 13.4). They are the majority of leukocytes. Their function is primarily phagocytosis of bacteria and viruses. They are the first cells to arrive at an infection (Fig. 13.5).

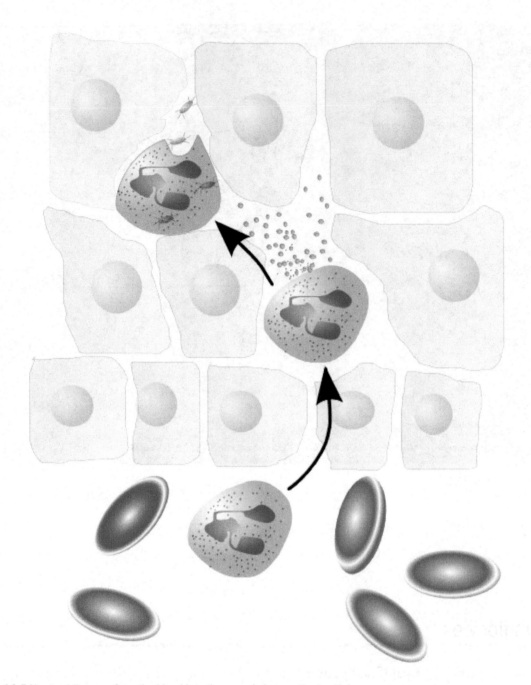

Figure 13.5 Neutrophils move from the blood into tissue and phagocytize bacteria.

Figure 13.6 Eosinophil.

Eosinophils

Eosinophils have a bi-lobed nucleus and granular cytoplasm (Fig. 13.6). They are relatively rare cells that only make up 1% to 3% of the leukocyte population. Their function is to moderate allergic reactions and defend against parasites.

Basophils

Basophils have the same size and shape of nuclei as eosinophils (Fig. 13.7). They have fewer and much larger granules. The granules release histamine (a vasodilator) and heparin (an anticoagulant). Basophils function in inflammation. Think of the cardinal signs of inflammation: heat, redness, pain, and swelling. The basophils release heparin and histamine that function to bring more blood to the area. The additional blood produces the signs of inflammation. Basophils are also relatively rare cells and represent only 1% of the leukocyte population.

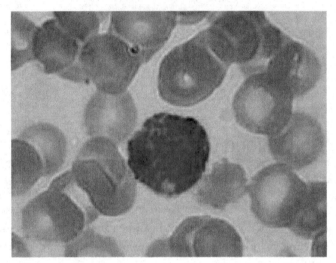

Figure 13.7 Basophil.

Agranular Leukocytes

There are two agranular leukocytes. These include monocytes and lymphocytes. Monocytes are the largest blood cells (Fig. 13.8). Their nuclei can be spherical, kidney-shaped, oval, or lobed. Monocytes have a function very similar to neutrophils. Monocytes work to clean up debris and phagocytize bacteria. They constitute 3% to 9% of the leukocyte population. Lymphocytes are about the same size as red blood cells (Fig. 13.9). They constitute 25% to 30% of the leukocyte population. There are two main types of lymphocytes. These include T-lymphocytes and B-lymphocytes. Both function in immunity. T-lymphocytes attack pathogens and help activate B-lymphocytes. B-lymphocytes produce antibodies when activated that attack pathogens.

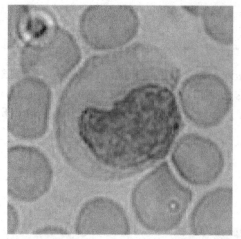

Figure 13.8 Monocyte.

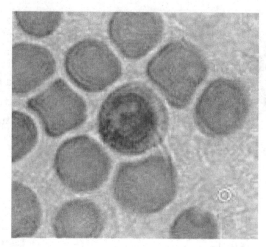

Figure 13.9 Lymphocyte.

White Blood Cell Count

The typical WBC count is about 5,000 to 10,000 cells per cubic millimeter. A high count is called leukocytosis and can be caused by infection, exercise, loss of body fluids, and emotional stress. A low count is called leucopenia and can be caused by viruses such as influenza, measles, mumps, and chicken pox, and toxins such as lead poisoning. A test that breaks out the relative percentages of WBCs is called a differential.

Platelets

Platelets are cell fragments called thrombocytes (Fig. 13.10). Hemocytoblasts (stem cells) differentiate into megakaryocytes that fragment into platelets. Platelets are about one half the size of red blood cells. There are about 130,000 to 360,000 platelets per cubic millimeter of blood. Platelets help to stop bleeding by sticking together to form plugs and secreting the hormone serotonin, which acts to vasoconstrict the vessels.

Blood Plasma

Plasma is the fluid portion of blood that carries the cells and platelets. Plasma is straw-colored and clear and contains water with a variety of substances. Plasma contains various proteins including fibrinogen, globulins, and albumin. Plasma also contains dissolved gases such as carbon dioxide and oxygen, and nutrients such as carbohydrates, amino acids, and lipids. Lipids are packaged in lipoproteins, such as very low density lipoproteins (VLDL), low density lipoproteins (LDL), high density lipoproteins (HDL), and chylomicrons. Other constituents of plasma include electrolytes such as sodium, potassium, calcium,

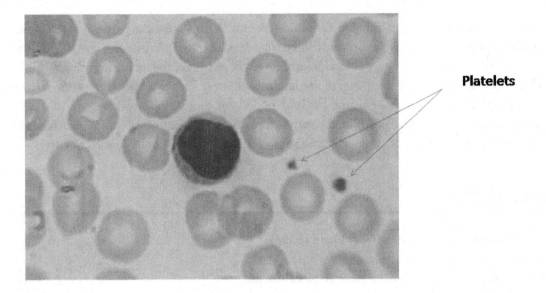

Figure 13.10 Platelets.

magnesium, chloride, sulfates, phosphates, and bicarbonate ions, and nitrogenous substances such as uric acid, urea, creatine, and creatinine.

Hemostasis

The blood system has some self-protection mechanisms built into it. These come into play during bleeding. The stopping of bleeding is called hemostasis. There are three basic mechanisms of hemostasis: blood vessel spasm, platelet plug formation, and clotting. Blood vessel spasm occurs in response to a damaged vessel. Blood vessels have a smooth muscle layer that constricts when the vessel is damaged. Platelets also release serotonin that facilitates constriction of the vessel. This action helps to stop the bleeding. Platelets become sticky when they contact damaged blood vessels. They can stick together to form a plug. The platelet plugs help to plug small holes in vessels. Clotting is the third mechanism of hemostasis. There are two pathways consisting of a cascade of reactions involving molecules called clotting factors (designated by Roman numerals). The two pathways are called the intrinsic and extrinsic pathways. Both pathways converge at a common point to form a fibrin clot. The extrinsic pathway is triggered when blood leaves the damaged blood vessel. The clotting factor tissue thromboplastin (Factor III) is triggered to begin the cascade of reactions. Tissue thromboplastin activates factor VII, which in turn activates factor X. Factor X combines with factor V and calcium ions to activate prothrombin activator. Prothrombin activator allows for the conversion of prothrombin (factor II) to thrombin. Thrombin in turn allows for the conversion of fibrinogen (factor I) to fibrin. Factor XIII stabilizes the fibrin clot. Fibrinogen is a soluble plasma protein, while fibrin is insoluble. The intrinsic pathway is triggered when blood contacts damaged blood vessel walls but does not leave the vessel. The first clotting factor to become activated in intrinsic clotting is called the Hageman factor (factor XII). This factor activates factor XI, which activates factor IX. Factor IX combines with factor VIII to activate factor X. Factor X combines with factor V and calcium to activate prothrombin activator. At this point the pathway is the same as that for extrinsic clotting. The intrinsic pathway can be triggered by atherosclerosis and stasis. Atherosclerotic plaquing can cause turbulent flow of blood in arteries that can trigger the intrinsic pathway. Also, lack of blood movement in circulation can trigger the intrinsic pathway. This is why it is recommended to walk around when taking long trips or sitting for long periods of time. Clot formation and dissolution are continuously occurring in the body. Blood contains a plasma protein called plasminogen that is inactive. A number of substances can activate plasminogen to form plasmin. Plasmin dissolves clots.

Clotting terms

- Hematoma—clot resulting from blood leakage
- Thrombus—clot forming in vessel
- Embolus—thrombus broken loose in bloodstream.
- Embolism—clot lodged in blood vessel, cutting off circulation
- Infarction—clot forming in vessel to organ (heart, lung, brain)
- Atherosclerosis—accumulation of fatty deposits in arterial linings; causes thrombosis

Blood Groups

Blood can be categorized or typed according to a set of antigens present on the surface of red blood cells (Fig. 13.11). Blood typing can be used to determine compatibility in case a transfusion is needed. There are three antigens that can be used to determine compatibility. These are A, B, and Rh.

Type A blood contains antigen A on the surface of the red blood cells. Type B contains antigen B and type AB contains both antigens A and B. Type O blood contains neither antigen. Type A blood also contains antibodies that are not compatible with antigen B. These are known as antibody anti-B. Likewise type B contains antibodies that are incompatible with type A. Mixing type A and B produces a reaction known as agglutination in which the blood coagulates. Type AB blood contains no antibodies that will react with the type A or B antigens. Type O blood contains both antibody anti-A and antibody anti-B. By just using the ABO system we can determine compatibility. Type A is compatible with itself and type O in case of an emergency. Even though type O contains antibodies that can cause agglutination, the risk is less because of the decreased concentration of these. Type B is compatible with itself and type O in case of an emergency. Type AB is compatible with all of the blood types and type O is compatible only with itself. The third antigen that can be used for typing is the Rh antigen. If the Rh antigen is present on the surface of the red blood cells, the blood is Rh positive. If it is not, the blood is Rh negative. Anti-Rh antibodies typically do not appear in the blood. However, they can develop in an Rh negative person who has been exposed to Rh positive blood. Rh positive blood is compatible with either Rh positive or Rh negative blood. However, Rh negative blood is compatible only with Rh negative.

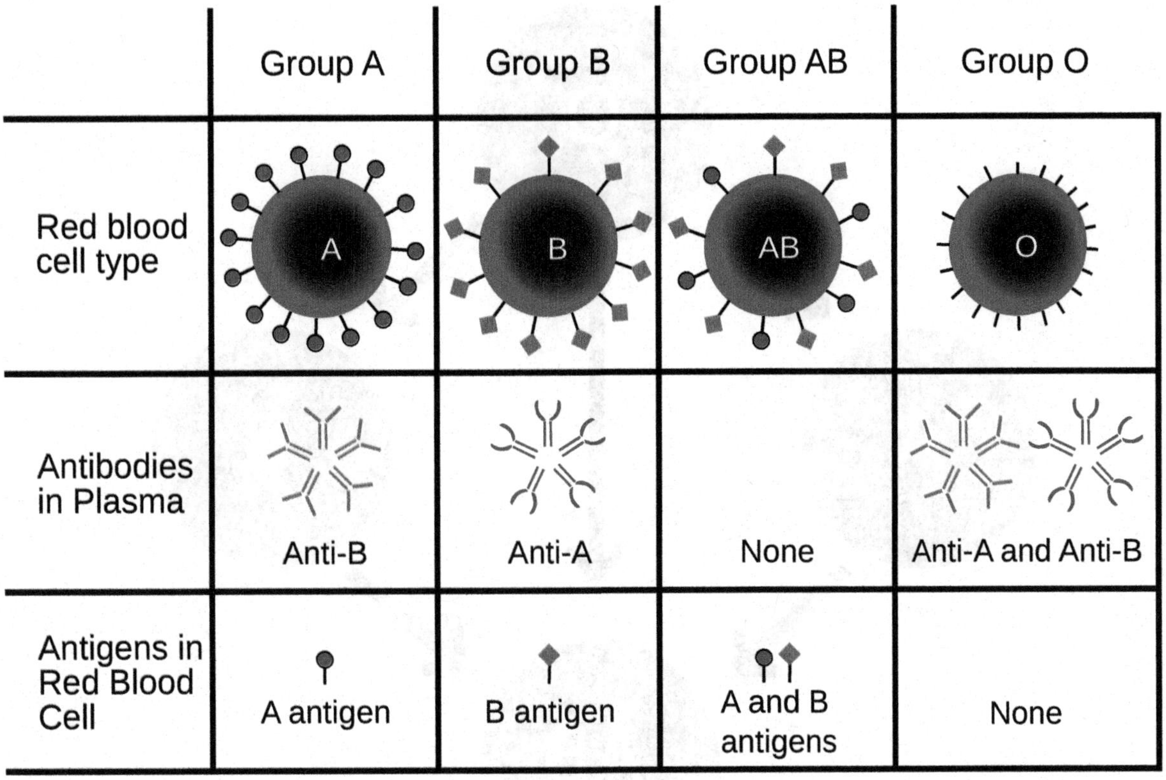

	Group A	Group B	Group AB	Group O
Red blood cell type	A	B	AB	O
Antibodies in Plasma	Anti-B	Anti-A	None	Anti-A and Anti-B
Antigens in Red Blood Cell	A antigen	B antigen	A and B antigens	None

Figure 13.11 Blood types.

Blood Typing Compatibility

Type	A+	A−	B+	B−	AB+	AB−	O+	O−
A+	+	+	−	−	−	−	+	+
A−	−	+	−	−	−	−	−	+
B+	−	−	+	+	−	−	+	+
B−	−	−	−	+	−	−	−	+
AB+	+	+	+	+	+	+	+	+
AB−	−	+	−	+	−	+	−	+
O+	−	−	−	−	−	−	+	+
O−	−	−	−	−	−	−	−	+

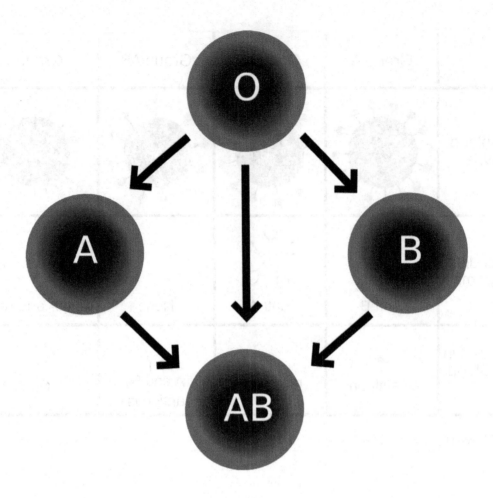

Inheritance

Blood types are inherited and represent contributions from both parents. The ABO blood type is controlled by a single gene with three alleles: i, IA, and IB. The gene encodes an enzyme that modifies the carbohydrate content of the red blood cell antigens. IA gives type A, IB gives type B, i gives type O.

Blood group inheritance:
Mother/Father O A B AB
O O O, A O, B A, B
A O, A O, A O, A, B, AB A, B, AB
B O, B O, A, B, AB O, B A, B, AB
AB A, B A, B, AB A, B, AB A, B, AB

IA and IB are dominant over i, so ii people have type O, IAIA or IAi have A, and IBIB or IBi have type B. IAIB people have both phenotypes because A and B are co-dominant, which means that type A and B parents can have an AB child. Thus, it is extremely unlikely for a type AB parent to have a type O child (it is not, however, direct proof of illegitimacy): the cis-AB phenotype has a single enzyme that creates both A and B antigens. The resulting red blood cells do not usually express A or B antigen at the same level that would be expected on common group A or B red blood cells, which can help solve the problem of an apparently genetically impossible blood group.

Rh Factor

Many people have the Rh factor on the red blood cell. Rh carriers do not have the antibodies for the Rh factor, but can make them if exposed to Rh. Most commonly, Rh is seen when anti-Rh antibodies cross from the mother's placenta into the child before birth. The Rh factor enters the child destroying the child's red blood cells. This is called hemolytic disease.

Compatibility in Blood/Plasma Transfusions

Blood transfusions between donor and recipient of incompatible blood types can cause severe acute immunological reactions, hemolysis (RBCT destruction), renal failure, shock, and sometimes death. Antibodies can be highly active and can attack RBCs and bind components of the complement system to cause massive hemolysis of the transfused blood. A patient should ideally receive their own blood or type-specific blood products to minimize the chance of a transfusion reaction. If time allows, the risk will further be reduced by cross-matching blood, in addition to blood-typing both recipient and donor. Cross-matching involves mixing a sample of the recipient's blood with a sample of the donor's blood and checking to see if the mixture agglutinates, or forms clumps. Blood bank technicians usually check for agglutination with a microscope, and if it occurs, that particular donor's blood cannot be transfused to that particular recipient. Blood transfusion is a potentially risky medical procedure and it is vital that all blood specimens be correctly identified, so in cross-matching, labeling is standardized using a barcode system known as ISBT 128.

When considering a plasma transfusion, keep in mind that plasma carries antibodies and no antigens. For example, you can't give type O plasma to a type A, B, or AB, because a person with type O blood has A and B antibodies and the recipient would have an immune response. On the other hand, an AB donor could give plasma to anyone, since they have no antibodies.

Hemolytic Disease of the Newborn

Often a pregnant woman carries a fetus whose blood type is different from hers, and sometimes the mother forms antibodies against the red blood cells of the fetus, leading to low fetal blood counts, a condition known as hemolytic disease of the newborn. Hemolytic disease of the newborn (also known as HDN) is an alloimmune condition that develops in a fetus when the IgG antibodies produced by the mother and passing through the placenta include ones that attack the red blood cells in the fetal circulation. The red cells are broken down and the fetus can develop reticulocytosis and anemia. The fetal disease ranges from mild to very severe and fetal death from heart failure—hydrops fetalis—can occur. When the disease is moderate or severe, many erythroblasts are present in the fetal blood and so these forms of the disease can be called erythroblastosis fetalis. Before birth, options for treatment include intrauterine transfusion or early induction of labor when pulmonary maturity has been attained, fetal distress is present, or 35 to 37 weeks of gestation have passed. The mother may also undergo plasma exchange to reduce the circulating levels of antibody by as much as 75%. After birth, treatment depends on the severity of the condition, but could include temperature stabilization and monitoring, phototherapy, transfusion with compatible packed red blood, exchange transfusion with a blood type compatible with both the infant and the mother, sodium bicarbonate for correction of acidosis, and/or assisted ventilation. Rh negative mothers who have had a pregnancy with or are pregnant with an Rh positive infant are given Rh immune globulin (RhIG), also known as Rhogam, during pregnancy and after delivery to prevent sensitization to the D antigen. It works by binding any fetal red cells with the D antigen before the mother is able to produce an immune response and form anti-D IgG. A drawback to pre-partum administration of RhIG is that it causes a positive antibody screen when the mother is tested, which is indistinguishable from immune reasons for antibody production.

Diseases of the Blood

Von Willebrand Disease

The most common inherited bleeding disorder, von Willebrand disease affects men and women equally. Von Willebrand disease is similar to hemophilia in that it involves a deficiency in the ability of blood to clot properly. Those affected by von Willebrand disease will have low levels of von Willebrand factor (a protein that helps the blood to clot), a malfunctioning von Willebrand factor, or both. While it is mostly an inherited disease (with factors contributed by both parents), von Willebrand disease may be an acquired syndrome in rare cases. There are three types of von Willebrand disease: Type 1, which is the mildest and most common form of the disease; Type 2, which has four subtypes (2A, 2B, 2M, and 2N) and ranges from mild to moderate in severity; and finally, Type 3, which is very rare and is the most severe form. In type 1 von Willebrand disease, there is a low level of von Willebrand factor. The level of factor VIII may also be lower than normal. About 3 out of 4 people diagnosed with von Willebrand disease have type 1. In type 2 von Willebrand disease, a defect in von Willebrand factor causes it to not work properly. Type 2 is divided into 2A, 2B, 2M, and 2N. Each is treated differently, so knowing the exact type is important. People with type 1 and type 2 von Willebrand disease may have the following mild to moderate bleeding symptoms: easy bruising, nosebleeds, bleeding from the gums after a dental procedure, heavy menstrual bleeding in women, blood in their stools or urine (from bleeding in the intestines, stomach, kidneys or bladder), excessive bleeding after a cut or other accident or surgery. People with type 3 von Willebrand disease usually have no von Willebrand factor and very low factor VIII. Symptoms of type 3 von Willebrand disease might include any of the symptoms of types 1 and 2, and also include severe bleeding episodes for no reason, which can be life threatening if not treated immediately. Bleeding into soft tissues or joints (hemarthrosis), causing severe pain and swelling, is another symptom. Many people with von Willebrand disease do not require treatment to manage the disease. However, if treatment is necessary, it may include a range of different interventions depending on the severity. These involve medicine to increase the level of von Willebrand factor in the blood (DDAVP), medicine to prevent the breakdown of clots (called antifibrinolytic drugs), medicine to control heavy menstrual bleeding in women (often birth control pills), or injection of clotting factor concentrates (containing von Willebrand factor and factor VIII).

Disseminated Intravascular Coagulation

Disseminated intravascular coagulation (DIC), also called consumptive coagulopathy, is a pathological process in the body where the blood starts to coagulate throughout the whole body. This depletes the body of its platelets and coagulation factors, and there is a paradoxically increased risk of hemorrhage. It occurs in critically ill patients, especially those with gram-negative sepsis (particularly meningococcal sepsis) and acute promyelocytic leukemia.

Hemophilia

Hemophilia is a disease where there is low or no blood protein, causing an inability to produce blood clots. There are two types of hemophilia: Type A, which is a deficiency in factor VIII and Type B (Christmas disease), a deficiency on factor IX. Because people with hemophilia do not have the ability to make blood clots, even a little cut may kill them, or the smallest bump or jar to the body could cause severe bruising that doesn't get better for months. Hemophilia is passed down from mothers to their sons. Hemophilia is sometimes known as the "Royal Disease." This is because Queen Victoria, Queen of England (1837–1901), was a carrier of hemophilia. The hemophilia disease was passed down to her son Leopold who ended up dying at age 31. Queen Victoria also had two daughters who were carriers. These daughters passed hemophilia into the Spanish, German, and Russian royal families. One of the most famous stories is that of the Russian royal family. Alexandra, granddaughter to Queen Victoria, married Nicholas (Tsar of Russia in the 1900s). Alexandra was a carrier of the disease and passed the disease to their first son, Tsarevich Alexei, who was heir to the throne of Russia. The family tried to keep their son's secret from the people, but Alexei suffered with serious bruises and extreme pain. The family found help from a monk named Rasputin. He kept their secret and gained a great deal of power over the family, making them think he was their only hope. During this time of great turmoil in Russia, Nicholas and Alexandra spent most of their attentions on their son, and not on the people. It wasn't long before the Bolshevik Revolution of 1917 began.

Factor V Leiden

The opposite of hemophilia, Factor V Leiden is the name given to a variant of human factor V that causes a hypercoagulability disorder. In this disorder, the Leiden variant of factor V, cannot be inactivated by activated protein C. Factor V Leiden is the most common hereditary hypercoagulability disorder among Eurasians. It is named after the city Leiden (The Netherlands), where it was first identified in 1994 by Prof R. Bertina et al. Those who have it are at a slightly higher risk of developing blood clots than those without. Those who test positive for factor V should avoid oral contraceptives, obesity, smoking, and high blood pressure.

Anemia

Anemia, from the Greek (Anaimía) meaning "without blood," refers to a deficiency of red blood cells (RBCs) and/or hemoglobin. This results in a reduced ability of blood to transfer oxygen to the tissues, causing hypoxia. Since all human cells depend on oxygen for survival, varying degrees of anemia can have a wide range of clinical consequences. Hemoglobin (the oxygen-carrying protein in the red blood cells) has to be present to ensure adequate oxygenation of all body tissues and organs. The three main classes of anemia include excessive blood loss (acutely, such as a hemorrhage, or chronically, through low-volume loss), excessive blood cell destruction (hemolysis), or deficient red blood cell production (ineffective hematopoiesis). In menstruating women, dietary iron deficiency is a common cause of deficient red blood cell production.

Sickle cell

Sickle cell disease is a general term for a group of genetic disorders caused by sickled hemoglobin (Hgb S or Hb S). In many forms of the disease, the red blood cells change shape upon deoxygenation because of polymerization of the abnormal sickle hemoglobin. This process damages the red blood cell membrane, and can cause the cells to become stuck in blood vessels. This deprives the downstream tissues of oxygen and causes ischemia and infarction. The disease is chronic and lifelong. Individuals are most often well, but their lives are punctuated by periodic painful attacks. In addition to periodic pain, there may be damage of internal organs, and/or stroke. Lifespan is often shortened, with sufferers living an average of 40 years. It is common in people from parts of the world where malaria is or was common, especially in sub-Saharan Africa or in descendants of those peoples. Sickle cell disease is inherited in the autosomal recessive pattern, depicted above. The allele responsible for sickle cell anemia is autosomal recessive. A person who receives the defective gene from both father and mother develops the disease; a person who receives one defective and one healthy allele remains healthy, but can pass on the disease and is known as a carrier. If two parents who are carriers have a child, there is a 1-in-4 chance of their child developing the illness and a 1-in-2 chance of their child just being a carrier.

Polycythemia

Polycythemia is a condition in which there is a net increase in the total circulating erythrocyte (red blood cell) mass of the body. There are several types of polycythemia.

Primary Polycythemia

In primary polycythemia, there may be 8 to 9 million and occasionally 11 million erythrocytes per cubic millimeter of blood (a normal range for adults is 4–5 million), and the hematocrit may be as high as 70 to 80%. In addition, the total blood volume can increase to as much as twice as normal. The entire vascular system can become markedly engorged with blood, and circulation times for blood throughout the body can increase up to twice the normal value. The increased numbers of erythrocytes can increase of the viscosity of the blood to as much as five times normal. Capillaries can become plugged by the very viscous blood, and the flow of blood through the vessels tends to be extremely sluggish. As a consequence, people with untreated Polycythemia are at a risk of various thrombotic events (deep venous thrombosis, pulmonary embolism), heart attack and stroke, and have a substantial risk of Budd-Chiari syndrome (hepatic vein thrombosis). The condition is considered chronic; no cure exists. Symptomatic treatment (see below) can normalize the blood count and most patients can live a normal life for years.

Secondary Polycythemia

Secondary polycythemia is caused by either appropriate or inappropriate increases in the production of erythropoietin that result in an increased production of erythrocytes. In secondary polycythemia, there may be 6 to 8 million and occasionally 9 million erythrocytes per cubic millimeter of blood. A type of secondary polycythemia in which the production of erythropoietin increases appropriately is called physiologic polycythemia. Physiologic polycythemia occurs in individuals living at high altitudes (4,275 to 5,200 meters), where oxygen availability is less than at sea level. Many athletes train at higher altitudes to take advantage of this effect—a legal form of blood doping. Actual polycythemia sufferers have been known to use their condition as an athletic advantage for greater stamina. Other causes of secondary polycythemia include smoking, renal or liver tumors, or heart or lung diseases that result in hypoxia. Endocrine abnormalities, prominently including pheochromocytoma and adrenal adenoma with Cushing's Syndrome, are also secondary causes. Athletes and bodybuilders who abuse anabolic steroids or erythropoietin may develop secondary polycythemia.

Relative Polycythemia

Relative polycythemia is an apparent rise of the erythrocyte level in the blood; however, the underlying cause is reduced blood plasma. Relative polycythemia is often caused by fluid loss, i.e., burns, dehydration, and stress polycythemia.

Leukemia

Leukemia is a cancer of the blood or bone marrow characterized by an abnormal proliferation of blood cells, usually white blood cells (leukocytes). It is part of the broad group of diseases called hematological neoplasms. Damage to the bone marrow, by way of displacing the normal marrow cells with increasing numbers of malignant cells, results in a lack of blood platelets, which are important in the blood clotting process. This means people with leukemia may become bruised, bleed excessively, or develop pin-prick bleeds (petechiae). White blood cells, which are involved in fighting pathogens, may be suppressed or dysfunctional, putting the patient at the risk of developing infections. The red blood cell deficiency leads to anemia, which may cause dyspnea. All symptoms may also be attributable to other diseases; for diagnosis, blood tests and a bone marrow biopsy are required.

CRITICAL-THINKING QUESTIONS

1. Why would it be incorrect to refer to the formed elements as cells?
2. Myelofibrosis is a disorder in which inflammation and scar tissue formation in the bone marrow impair hemopoiesis. One sign is an enlarged spleen. Why?
3. Would you expect a patient with a form of cancer called acute myelogenous leukemia to experience impaired production of erythrocytes, or impaired production of lymphocytes? Explain your choice.
4. A young woman has been experiencing unusually heavy menstrual bleeding for several years. She follows a strict vegan diet (no animal foods). She is at risk for what disorder, and why?
5. A patient has thalassemia, a genetic disorder characterized by abnormal synthesis of globin proteins and excessive destruction of erythrocytes. This patient is jaundiced and is found to have an excessive level of bilirubin in his blood. Explain the connection.
6. A patient was admitted to the burn unit the previous evening suffering from a severe burn involving his left upper extremity and shoulder. A blood test reveals that he is experiencing leukocytosis. Why is this an expected finding?
7. A lab technician collects a blood sample in a glass tube. After about an hour, she harvests serum to continue her blood analysis. Explain what has happened during the hour that the sample was in the glass tube.

8. Following a motor vehicle accident, a patient is rushed to the emergency department with multiple traumatic injuries, causing severe bleeding. The patient's condition is critical, and there is no time for determining his blood type. What type of blood is transfused, and why?
9. In preparation for a scheduled surgery, a patient visits the hospital lab for a blood draw. The technician collects a blood sample and performs a test to determine its type. She places a sample of the patient's blood in two wells. To the first well she adds anti-A antibody. To the second she adds anti-B antibody. Both samples visibly agglutinate. Has the technician made an error, or is this a normal response? If normal, what blood type does this indicate?
10. Explain why patients with impaired renal function might be anemic.
11. A patient has been on antibiotics for two weeks for a urinary tract infection. The patient develops complete bone marrow suppression as a result of this therapy. Speculate how each of the following lab values might be affected.
    A. red cell count
    B. white cell count
    C. hemoglobin and hematocrit
    D. blood type
    E. clotting time

Chapter Fourteen

Cardiovascular System

The primary function of the heart is to pump blood through the arteries, capillaries, and veins. There are an estimated 60,000 miles of vessels throughout an adult body. Blood transports oxygen, nutrients, disease-causing viruses, bacteria, and hormones, and has other important functions as well.

The heart is located in the thoracic cavity in an area known as the mediastinum (Fig. 14.1). The mediastinum contains the heart, esophagus, trachea, vessels, nerves, and membranes surrounding the heart. Dissecting the membrane known as the parietal pericardium reveals the heart. The heart is shaped like a blunt cone and is approximately the size of a fist. The point of the cone is called the apex and the other end is called the base. The apex points downward (Fig. 14.2). The heart is positioned centrally with the apex pointing to the left. More of the heart resides left of the midline of the thoracic cavity than on the right.

Membranes of the Heart

The heart is surrounded by a double-layered sac consisting of two membranes. The outer membrane consists of fibrous connective tissue and is known as the fibrous or parietal pericardium. The inner membrane is thinner and consists of simple squamous epithelium. It is known as the visceral or serous pericardium. The visceral pericardium is consistent with the great vessels of the heart and the diaphragm. Serous fluid known as pericardial fluid exists between the membranes. The fluid helps to reduce friction when the heart beats. The visceral pericardium can become inflamed and produce extra fluid in a condition known as pericarditis. This can result from infection or diseases of the connective tissues. Pericarditis can also result from damage caused by radiation therapy. Pericarditis can cause severe sharp pains in the chest and back.

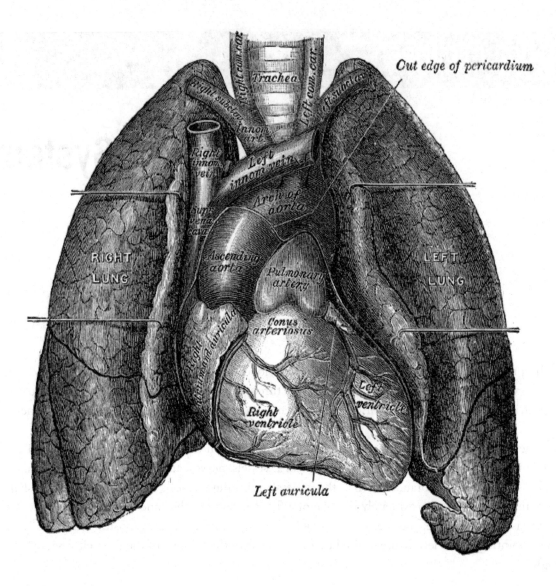

Figure 14.1 The mediastinum.

Layers of the Heart

The heart consists of three layer of tissue. These are the endocardium, myocardium, and epicardium. The endocardium is the internal layer consisting of simple squamous epithelium and connective tissue. This layer is consistent with the valves of the heart. The middle myocardium is a thick layer of cardiac muscle. The outer epicardium is the visceral pericardium and consists of a thin serous membrane.

Heart Structures

The heart consists of four chambers. Two of these chambers receive blood and are called atria. The other two chambers are larger for pumping blood outside of the heart and are called ventricles. Each side of the heart has an atrium and ventricle. The atria are separated by a mass of tissue called the

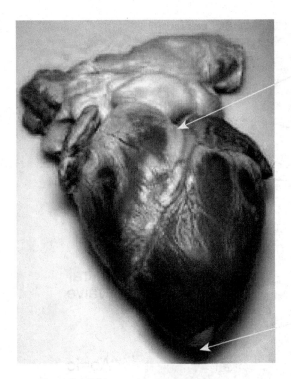

Figure 14.2 Base and apex of heart.

interatrial septum. The interatrial septum contains a small indentation that is a remnant of fetal circulation known as the fossa ovalis. In utero the fossa ovalis is the foramen ovale that serves as a passageway for blood to bypass the lungs. The foramen ovale closes at birth. On the outer surface of the heart between the atria and ventricles are the auricles, which are extensions of the atria. The ventricles are also separated by a thick mass of muscles known as the interventricular septum. On the surface of the heart a sulcus, known as the coronary sulcus, separates the atria and ventricles. The anterior interventricular sulcus divides the right and left ventricles anteriorly. The posterior interventricular sulcus divides right and left ventricles posteriorly.

Blood Flow

One way to learn the heart structures is to follow a drop of blood through the heart (Fig. 14.3). Deoxygenated blood enters the right atrium via two large veins called the superior and inferior venae cavae. Blood then flows from the right atrium to the right ventricle past a one-way valve known as the tricuspid valve. The tricuspid valve has three cusps with each connected to the internal wall of the right ventricle via connective tissue structures called chordae tendineae. The chordae tendineae connect to finger-like projections of muscle called papillary muscles. The tricuspid valve is driven by pressure and allows blood to flow in only one direction (from atrium to ventricle). Contraction of the right ventricle pushes blood to the pulmonary trunk and past the pulmonary semilunar valve on its way to the lungs. The pulmonary trunk is a thick artery that splits into right and left pulmonary arteries that serve the right and left lungs. The pulmonary semilunar valve contains three cusps that allow blood to flow in only one direction. When pressure builds in the ventricle, the cusps open, allowing blood to move into the pulmonary trunk. When pressure causes movement of blood back toward the heart, the valves close. Blood moves from the pulmonary arteries to the lungs for oxygenation. Oxygenated blood is carried by four pulmonary veins to the left atrium. Blood then moves from the left atrium to the left ventricle past the bicuspid valve (mitral valve). The bicuspid valve is a one-way valve with two cusps that attach to the ventricle wall via chordae tendineae and papillary muscles. Contraction of the left ventricle causes blood to flow into the aorta past the aortic semilunar valve. The aortic semilunar valve has three cusps and allows blood to flow only away from the heart. Oxygenated blood now flows through the aorta to the body.

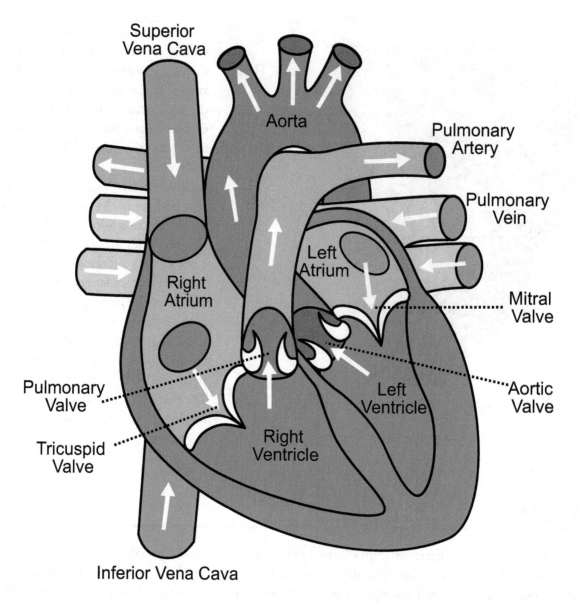

Figure 14.3. Blood flow through the heart.

Coronary Arteries

The coronary arteries supply the heart muscle with oxygenated blood. The right coronary artery branches off the aorta and resides in the coronary sulcus. It divides into the right marginal and posterior interventricular arteries. The right marginal supplies the right atrium and ventricle, while the posterior interventricular supplies the posterior sides of both ventricles. The left coronary also branches from the aorta and divides to form the anterior interventricular artery (left anterior descending artery), the left marginal artery, and circumflex artery. The anterior interventricular artery supplies the anterior side of the ventricles. The left marginal artery supplies the lateral wall of the left ventricle and the circumflex artery supplies the posterior wall of the heart. The left side of the heart is drained by the great cardiac vein and the right side is drained by the small cardiac vein. Both veins empty into the coronary sinus, which empties into the right atrium.

The Cardiac Cycle

During one heartbeat the heart must receive blood in the atria and then move this blood to the ventricles and out to the lungs or the body. The cardiac cycle is the sequence of events that makes this possible (Figs. 14.4, 14.5). There are three phases in the cardiac cycle. In phase one (called the rest phase) the heart is relaxed or in diastole. Blood passively flows into the atria and subsequently into the ventricles. Actually, about 70% of blood flows into the ventricles without any contraction of the atria. As was stated earlier, the valves operate by changes in pressure. During phase one, the pressure is greater in the atria than in the ventricles. This causes both atrioventricular valves (bicuspid and tricuspid) to open. The relaxation of the ventricles (ventricular diastole) creates a lower pressure in the ventricles than in the pulmonary trunk and aorta. This causes the semilunar valves (pulmonary and aortic) to close. The next phase in the cycle is characterized by systole of the atria. The ventricles are still in diastole in this phase.

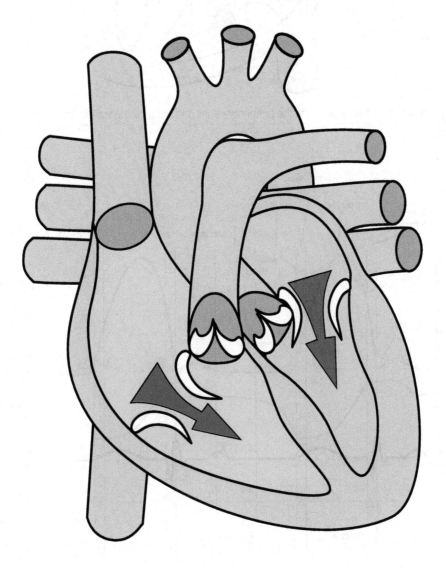

Figure 14.4 Blood passively flows into the heart during the rest phase, and then is pushed into the ventricles during atrial systole.

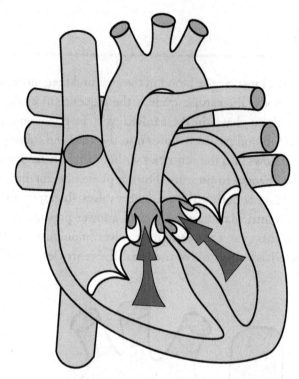

Figure 14.5 Blood moves from the ventricles to the pulmonary trunk and aorta during ventricular systole.

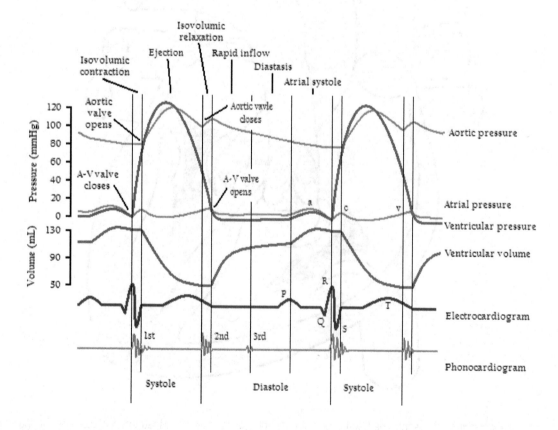

Figure 14.6. Events of the cardiac cycle.

The contraction of the atria pushes the remaining 30% of the blood into the ventricles. The atrioventricular valves remain open and the semilunar valves remain closed. The final phase is characterized by atrial diastole and ventricular systole. Contraction of the ventricles causes increased pressure in the ventricles. This increase in pressure causes the atrioventricular valves to close and the semilunar valves to open. The heart again enters the rest phase and the cycle repeats (Fig. 14.6).

Auscultation

A lot of information can be obtained from listening to the heart with a stethoscope. This procedure is called auscultation. The heart sounds are produced by changes in blood flow during different parts of the cardiac cycle. There are four heart sounds that repeat continuously but only two are usually heard in healthy adults. The first sound or S1 (lubb) is produced by turbulent flow of blood resulting from closure of the atrioventricular valves (bicuspid, tricuspid) in ventricular systole. It is louder and longer than the second sound (S2). The S2 sound (dupp) results from closure of the semilunar valves (pulmonary, aortic) during ventricular diastole. A third sound (S3) can sometimes be heard between S2 and S1. This sound is the result of ventricular filling and can be heard in children and athletes. It also arises in congestive heart failure. The fourth sound (S4) occurs just before S1 and is not often heard. It is considered abnormal in adults. The sound is produced by forceful contraction of the atria forcefully pushing blood against a failing ventricle. The period between S1 and S2 represents ventricular systole. Likewise, the period between S2 and S1 represents ventricular diastole.

Blood Vessels

Blood vessels carry blood from the heart to the lungs and tissues of the body and back to the heart. The system of blood vessels is called the vascular system and is considered a closed system. Oxygenated blood is carried to tissues where substances are exchanged. Substances needed for cell maintenance and growth move out of the blood, while waste products and substances needed for regulation of the body move in.

The Arterial System

The arterial system consists of arteries, arterioles, and capillaries. The largest arteries consist of three layers (Fig. 14.7). The outer layer, tunica externa, consists of elastic and collagen fibers. The larger vessels also contain minute blood vessels that carry nutrients to the tissue. Nerve fibers also innervate arteries. The middle layer, or tunica media, is thicker in arteries than in veins. It consists primarily of smooth muscle with some elastic fibers.

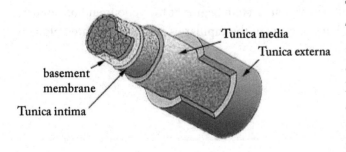

Figure 14.7 Artery.

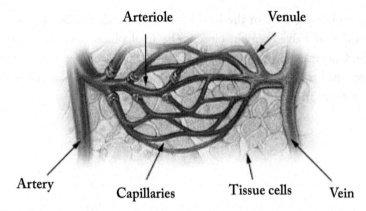

Figure 14.8 Capillary bed.

The smooth muscle in the tunica media allows for constriction (vasoconstriction) and dilation (vasodilation) of the arteries. The nervous system has some control over the diameter of arteries in order to control blood pressure. Also, blood vessels constrict when damaged to reduce the loss of blood. The inner layer, or tunica interna, consists of an inner thin layer of simple squamous epithelium called the endothelium anchored to another layer by a basement membrane. The basement membrane anchors the endothelium to a layer called the internal elastic lamina, consisting of elastic fibers. Arteries branch to form smaller structures called arterioles. Arterioles help to control blood flow to various parts of the body by way of vasoconstriction and vasodilation. The end of the arteriole that connects with the capillaries narrows and becomes a metarteriole that contains a round smooth muscle called a precapillary sphincter. The precapillary sphincters help to control the flow of blood to the capillary beds (Fig. 14.8). One metarteriole may supply up to 100 capillaries, forming what is known as a capillary bed. Capillaries are the smallest blood vessels in the body. They carry blood to the venous system and allow for the exchange of substances between the blood and the tissues. Capillaries form complex networks and it is estimated that there are about one billion capillaries in the human body. Blood flow to capillaries is controlled by small smooth muscles called precapillary sphincters. Capillaries are extensions of the endothelium of arteries. They consist of simple squamous epithelium and a basement membrane that allows a good degree of permeability for substance exchange. Capillaries are more numerous in areas with high metabolic activity such as muscle and nerve tissue. Permeability also varies according to metabolic demand. For example, capillaries in the liver and spleen are more permeable than those in smooth or skeletal muscle. Most substances are exchanged by diffusion but other transport mechanisms include filtration and osmosis.

The Venous System

Capillaries contain an arterial and a venous end. The venous system begins at the venous end of capillaries (Fig. 14.9). Oxygen and carbon dioxide exchange occurs in the capillaries and deoxygenated blood is now carried into the venous vessels.

Venules

Venules begin at the ends of capillaries and carry blood to the veins. Venules are very small and similar in structure to capillaries. They allow for substance exchange and merge with larger-diameter veins.

Veins

Veins like arteries contain three layers. However, the middle layer, or tunic media, is not as thick as in arteries. Veins have larger lumina than arteries have and many veins contain valves that allow blood to flow only to the heart. Veins can vasoconstrict and do so in situations of blood loss in order to conserve blood. When significant blood is lost, the sympathetic nervous system stimulates veins to constrict in an effort to return blood to the heart. This allows for nearly normal blood flow when up to 25% of blood is lost.

Blood Flow in Veins

Pressure in the arterial system is greatest at its source and decreases throughout the system. For example, blood pressure is greater in the aorta than in arterioles and even less in capillaries. In fact, we would not want a high blood pressure in capillaries, as their thin walls would burst (a condition that happens in hypertension). Thus there is minimal blood pressure in capillaries. The question is then, if blood pressure is so low in capillaries, how does blood get back to the heart? The answer lies in the structure of veins. Veins contain one-way valves that allow blood flow only to the heart. Many veins lie next to muscles. Muscle contraction produces an external pumping force on veins that helps to move blood to the heart. The movement of the diaphragm also contributes to venous flow.

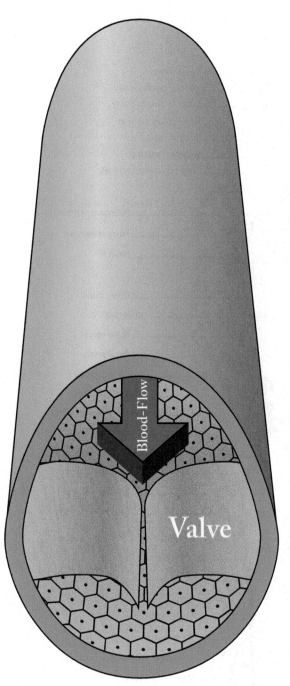

Figure 14.9 Vein.

Major Arteries and Veins

This section will cover some of the major arteries and veins (Fig. 14.10). Remember the heart has two jobs. It must pump deoxygenated blood to the lungs for oxygenation and then pump the oxygenated blood out to the tissues of the body. In order to complete both jobs, there are two pathways or circuits in which the blood flows. The pulmonary circuit begins at the right side of the heart with the pulmonary trunk and ends at the left atrium. The systemic circuit begins at the left side of the heart and ends at the right side.

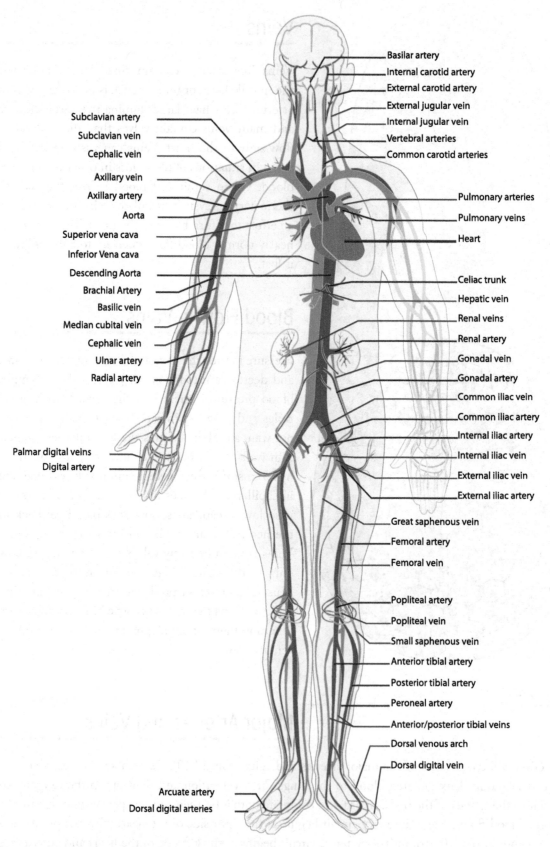

Figure 14.10 Circulatory system.

The Pulmonary Circuit

Deoxygenated blood enters the right side of the heart at the right atrium. The blood moves to the right ventricle and then exits via the pulmonary trunk. The pulmonary circuit begins at the pulmonary trunk. The pulmonary trunk then divides into right and left pulmonary arteries. The pulmonary arteries enter the lungs and form smaller and smaller branches. The smallest branches consist of the pulmonary arterioles that bring oxygenated blood to the capillaries that feed the alveoli. The alveoli are minute structures in the lung that exchange oxygen and carbon dioxide.

Once the blood becomes oxygenated, it exits the alveoli and enters venules that branch to larger vessels called pulmonary veins. There are two pulmonary veins for each lung that carry blood to the left atrium. It is important to note that in the pulmonary circuit, deoxygenated blood is carried by arteries and oxygenated blood is carried by veins. The opposite is true of the systemic circuit.

The Systemic Circuit

The systemic circuit begins at the left ventricle with the aorta and ends at the right atrium at the superior and inferior venae cavae.

Arteries

One way to think of the circulatory system is to compare it with a freeway system. There are major freeways linking to smaller highways that link to even smaller roads. The circulatory system has a similar structure with main routes linking to smaller ones and so on.

Aorta

As the aorta exits, it is known as the ascending aorta. Near the aortic valve is an enlargement known as the aortic sinus. The aortic sinus contains the aortic bodies. The aortic bodies are chemoreceptors that sense changes in chemical concentration and feed this information back to the nervous system. The aorta then curves, forming the arch of the aorta, and extends inferiorly to become the thoracic aorta. It then passes below the diaphragm to become the abdominal aorta.

Branches of the Aorta

The right and left coronary arteries arise from the aorta shortly after it emerges from the aortic valve. Along the arch of the aorta are three branches. From left to right, these are the brachiocephalic trunk, left common carotid, and left subclavian arteries. The brachiocephalic trunk then divides into the right common carotid and right subclavian arteries (Fig. 14.11). The thoracic aorta contains both visceral branches to organs and parietal branches to structures of the body wall. The visceral branches are the pericardial, bronchial, esophageal, and mediastinal. The parietal branches are the posterior intercostals, subcostal, and superior phrenic arteries.

The thoracic aorta then moves through the diaphragm and becomes the abdominal aorta. The abdominal aorta ends with a bifurcation producing the right and left common iliac and middle sacral arteries. The visceral branches of the abdominal aorta are the celiac trunk, right and left suprarenal, renal

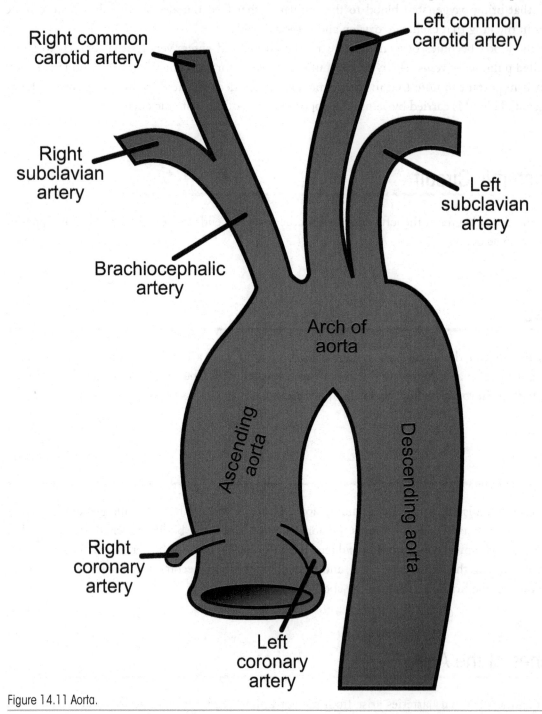

Figure 14.11 Aorta.

and gonadal arteries, and the superior and inferior mesenteric arteries. The parietal branches are the inferior phrenic, lumbar, and median sacral. The celiac trunk divides into the splenic, left gastric, and common hepatic arteries. The splenic artery supplies the spleen and some of the arteries that attach to the stomach. The left gastric artery supplies the stomach and part of the esophagus. The common hepatic artery supplies arteries for the liver, stomach, gallbladder, and small intestine. The right and left common iliac arteries divide and become the internal and external iliac arteries at the level of the lumbosacral junction. The internal iliac arteries supply the urinary bladder, genitalia, walls of the pelvis, and medial thigh. The external iliac arteries continue to the lower extremities. The external iliacs emerge from under the inguinal ligament as the femoral arteries. The deep femoral artery branches from the femoral artery and gives rise to the femoral circumflex artery. The femoral artery continues distally and gives rise to a branch known as the descending genicular artery that supplies the area around the knee. The femoral artery then pierces the adductor longus muscle and emerges as the popliteal artery, which branches to become anterior and posterior tibial arteries. The fibular artery (peroneal artery) branches from the posterior tibial artery. The anterior tibial artery becomes the dorsalis pedis artery at the ankle, which branches and supplies the foot. The posterior tibial divides into the medial and lateral plantar arteries. These arteries supply the plantar area of the foot. The smaller divisions of the plantar arteries connect with the dorsalis pedis artery to form the dorsal and plantar arches of the foot.

Arteries of the Head and Upper Extremity

As stated earlier, among the first branches of the aorta is the brachiocephalic trunk, left common carotid artery, and left subclavian artery. The brachiocephalic trunk divides to form the right common carotid and right subclavian arteries. The common carotid moves superiorly and divides into the internal and external carotid arteries. The carotid sinus is located at the junction of the internal and common carotid arteries. The external carotid continues on the outer part of the skull and gives rise to arteries that supply the esophagus, neck, pharynx, larynx, mandibular region, and face. The internal carotid enters the skull through the carotid canal and divides into three branches. These are the ophthalmic, anterior cerebral, and middle cerebral arteries. The vertebral arteries branch from the subclavian arteries and extend upward through the transverse foramen of the cervical vertebra and enter the skull at the foramen magnum. Both vertebral arteries merge to form the basilar artery. Both the vertebral arteries and basilar arteries give rise to branches that supply various parts of the brain before dividing to form the posterior cerebral arteries, which then branch to form the posterior communicating arteries. The anterior portion of the cerebrum is supplied by the internal carotid arteries and the remaining portion of the brain is supplied by the vertebral arteries. The internal carotid arteries connect with the basilar artery via two posterior communicating arteries. Since the resulting vascular structure forms a ring, it is called the Circle of Willis (cerebral arterial circle) (Fig. 14.12). This structure allows for some redundancy in supply to the brain, as it can receive blood from either the vertebral arteries or internal carotid arteries. The subclavian artery continues under the clavicle and gives rise to the internal thoracic, vertebral, and thyrocervical trunk. The subclavian artery emerges from under the clavicle to form the axillary artery, which produces the humeral circumflex artery. The axillary artery continues along the arm to become the brachial artery, which gives rise to the deep brachial artery and the ulnar collateral arteries. At the elbow the brachial artery divides to form the radial and ulnar arteries. At the wrist the radial and ulnar arteries reconnect to form the superficial and deep palmar arches, which in turn supply the digital arteries of the fingers.

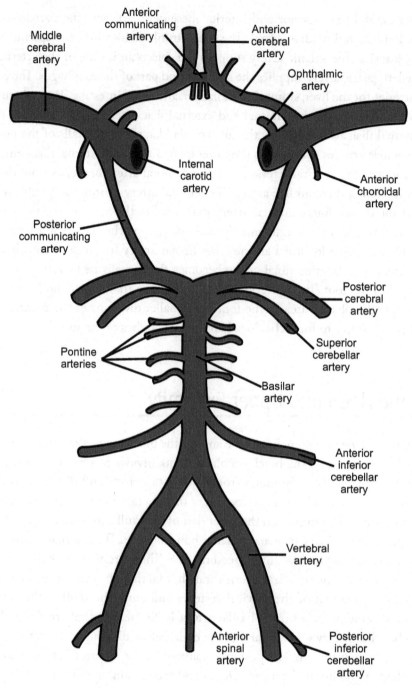

Figure 14.12 Circle of Willis.

Veins of the Systemic Circuit

The veins of the systemic circuit drain into the superior and inferior venae cavae that connect with the right atrium of the heart. The superior vena cava is formed by the union of the right and left brachiocephalic veins. Each brachiocephalic vein is formed by two branches, including the subclavian and jugular veins. The subclavian veins drain the veins of the head and neck. The subclavian drains the shoulder and upper extremity.

Veins of the Upper Extremity

Following the subclavian vein laterally, we see that it makes connections with two superficial veins and one deep vein in the upper extremity. The superficial veins are the cephalic vein laterally and the basilic vein medially. Both of these veins extend to the distal upper extremity. The deep vein is the axillary vein, which is a continuation of the subclavian vein as it emerges from the inferior aspect of the clavicle. The axillary vein then becomes the brachial vein, which divides into radial and ulnar veins. Like the arteries of the upper extremities, the radial and ulnar veins also connect at the deep palmar arch. The median cubital vein resides in the anterior portion of the elbow and connects with the cephalic and basilic veins. The cephalic and basilic veins both connect at the superficial palmar arch. The median antebrachial vein is located in the forearm area and connects the radial and ulnar veins.

Veins of the Head and Neck

The head and neck are drained by the internal and external jugular veins, and the vertebral veins. The drainage begins with the dural sinuses. The dural sinuses are located in the dura mater of the brain. The superior and inferior dural sinuses are located in the falx cerebri. The inferior sagittal sinus, located deep between the two hemispheres, drains into the straight sinus. The superior sagittal and straight sinuses in turn drain into the transverse sinuses. These drain into the sigmoid sinuses, which are continuous with the internal jugular vein. The cavernous sinuses drain the ophthalmic veins and also connect with the internal jugular veins. The internal jugular veins exit the brain at the jugular foramen. As they proceed inferiorly to the subclavian, they receive blood from the superficial temporal and facial veins. The external jugular vein is a superficial vein of the head that drains the superficial structures of the face and head. They drain into the subclavian veins. The vertebral veins drain the area around the cervical vertebrae. This is unlike the vertebral arteries that supply a good deal of the brain with oxygenated blood.

Veins of the Thorax

The azygos vein originates from the right ascending lumbar vein and right posterior intercostals vein. The azygos vein is not a paired vein and runs along the right side of the thoracic vertebrae. It empties into the superior vena cava. The hemiazygos vein lies on the left side of the vertebral column and originates from the posterior intercostals and left ascending lumbar veins. A continuation of the hemiazygos vein is the accessory hemiazygos vein, which extends superiorly.

Veins of the Abdomen and Lower Extremity

The veins of the abdomen and lower extremity drain into the inferior vena cava. Deoxygenated but nutrient-rich blood from the digestive system empties into the hepatic portal system. This blood flows to the liver, which extracts the nutrients for use in metabolism. The veins that drain into the hepatic portal vein have the same names as the arteries in the digestive system. These include the superior mesenteric, inferior mesenteric, and splenic veins. A series of lumbar veins are located in the posterior abdominal wall and drain into the inferior vena cava and the ascending lumbar veins. The gonadal veins drain the testes and ovaries. The right gonadal vein drains into the inferior vena cava, while the left gonadal vein empties into the left renal vein. The renal veins drain the kidneys. The hepatic veins drain blood from the liver to the inferior vena cava. The cystic veins drain the gallbladder and connect with the hepatic veins, and the inferior phrenic veins drain the diaphragm.

A number of veins of the lower extremity originate from the medial and lateral plantar veins, which drain into the posterior tibial vein. The fibular or peroneal vein drains into the posterior tibial vein. The dorsalis pedis vein at the ankle becomes the anterior tibial vein, which becomes the popliteal vein at the posterior knee. The popliteal vein ascends into the thigh region and becomes the femoral vein. The femoral vein becomes the external iliac vein that combines with the internal iliac vein to become the common iliac vein that unites with the inferior vena cava. The superficial dorsal venous arch of the foot becomes the great saphenous vein, which is a long superficial vein located on the medial side of the leg that connects with the external iliac vein. The small saphenous vein also originates from the dorsal venous arch and courses more laterally before connecting with the great saphenous vein.

Cardiovascular System Physiology

The cardiovascular system maintains the flow of oxygenated blood to the tissues by adjusting pressure throughout the system. In this section, we will investigate how the cardiovascular system produces and maintains blood pressure in a variety of circumstances. We covered some characteristics of cardiac muscle in the tissue chapter. You may recall that cardiac muscle is very similar to skeletal muscle. It consists of long red cells containing densely packed actin and myosin protein filaments, which give it a striated appearance. Cardiac muscle cells contain only one nucleus, while their muscle counterparts are multinucleated. Cardiac muscle also contains a specialized cell junction called an intercalated disc. Intercalated discs help to transmit action potentials from cell to cell in order to produce a more ordered contraction of large areas of muscle tissue. Cardiac muscle contraction physiology is also very similar to skeletal muscle. Depolarization of cardiac muscle cells causes the release of calcium. Calcium in turn binds to troponin surrounding actin, causing it to move and expose myosin-binding sites. Myosin and actin connect and slide past each other, powered by ATP.

Cardiac muscle has a resting membrane potential of about -90mV. The threshold for a typical ventricular muscle cell is about -75mV. An action potential that reaches the threshold causes rapid depolarization and movement of sodium inside the cell. This changes the membrane potential to +30mV, at which point the sodium channels close. The sodium channels are known as fast channels because of their quick reaction to stimuli. Once the membrane reaches +30mV, slow calcium channels open in order to maintain the transmembrane potential at about 0mV. The slow calcium channels react slowly to stimuli and remain open for longer periods of time (about 175 milliseconds (ms)). At the end of their cycle, the calcium channels close and slow potassium channels open, allowing the diffusion of potassium ions out of the cell. The cells then repolarize back to the resting membrane potential (Fig. 14.13). Cardiac muscle cells also exhibit relative and absolute refractory periods much like skeletal muscle cells do. During the absolute refractory period, the membrane cannot respond to stimuli. This is due to the sodium channels being open. The absolute refractory period in ventricular muscle cells is about 200 ms. This is followed by a relative refractory period in which a strong stimulus can produce an action potential. The relative

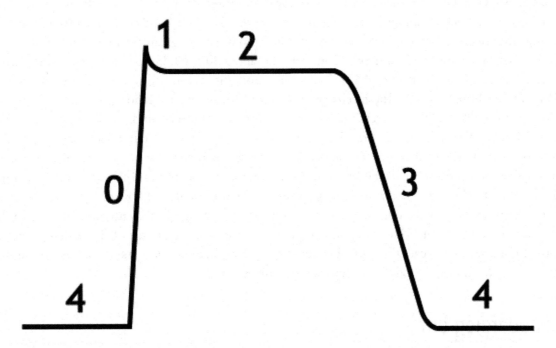

Figure 14.13 Action potential in cardiac muscle. 0. Rapid depolarization produced by opening of fast sodium channels; 1. Maximum depolarization to +30mV; 2. Plateau produced by slow calcium channels; 3. Repolarization produced by slow potassium channels; 4. Resting membrane potential of -90mV.

refractive period is characterized by closed sodium channels that can open. The relative refractive period lasts for about 30 ms. Cardiac muscle contraction, like skeletal muscle, relies on the influx of calcium. In cardiac muscle there are two sources of calcium. These include the influx of calcium from slow calcium channels, as mentioned, and the calcium stored in the sarcoplasmic reticulum. The long action potential in cardiac muscle also allows it to continue contraction until relaxation occurs. There is no summation in cardiac muscle. This prevents cardiac muscle cells from undergoing tetanic contractions.

The Cardiac Conducting System

Cardiac muscle tissue is capable of contracting on its own without stimulation from the nervous or endocrine systems. This phenomenon is known as automaticity. Automaticity occurs because of specialized cells that produce action potentials. In a normal heartbeat, conduction begins with an area of special cells called pacemaker cells in the posterior wall of the right atrium. This area is known as the sinoatrial (SA) node. It is sometimes referred to as the pacemaker node. The pacemaker cells cannot maintain a normal resting membrane potential but cycle from depolarization to repolarization. The SA node can generate action potentials automatically at a rate of 60–100 beats per minute. The impulse from the SA node is transferred by an intermodal pathway consisting of conducting cells to the atrioventricular (AV) node located in the floor of the right atrium. The impulse is delayed about 100 ms as it passes through the AV node. This allows for the completion of atrial contraction before the beginning of ventricular contraction. The AV node is also capable of producing action potential on its own at a rate of 40–60 bpm. If for some reason the SA node becomes damaged, the AV node will cause the heart to contract at 40–60 bpm. The AV node can conduct impulses at a maximum rate of 230 bpm. The heart begins to decrease its pumping efficiency at about 180 bpm. The heart cannot produce rates greater than 230 bpm unless it is damaged. The maximal rate of ventricular contraction is about 300–400 bpm. However, contractions at these rates are very inefficient (Fig. 14.14). The impulse from the AV node travels to the atrioventricular (AV) bundle or Bundle of His. These cells are also capable of producing action potentials at a rate of 20–40 bpm. The AV bundle connects the atria and ventricles. The AV bundle sends impulses to the right and left bundle branches. The branches extend to the apex of the heart and distribute impulses to the ventricles via Purkinje fibers and to the papillary muscles via moderator bands. This distribution of impulses allows for contraction of the papillary muscles before the ventricles. This allows for tensioning of the chordae tendinae of the atrioventricular valves to help prevent backflow of blood to the atria. Purkinje fibers are fast-conducting cells and allow for even emptying of the ventricles. Damage to the heart can manifest in what is known as an ectopic pacemaker. This is an area of tissue that generates abnormal impulses that bypass the normal conducting system. Ectopic pacemakers can disrupt normal ventricular contraction and produce dangerous arrhythmias.

Electrocardiogram

The heart generates significant electrical impulses that can be measured. Devices that measure the heart's electrical impulses produce a recording called an electrocardiogram or ECG (sometimes called an EKG) (Fig. 14.15). The information from an ECG can be used to determine problems with conduction, nodes,

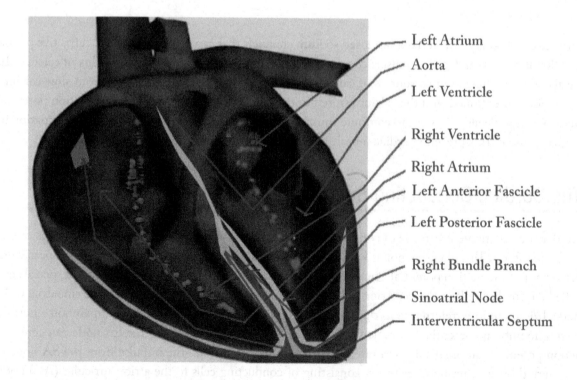

Figure 14.14 Cardiac conduction system.

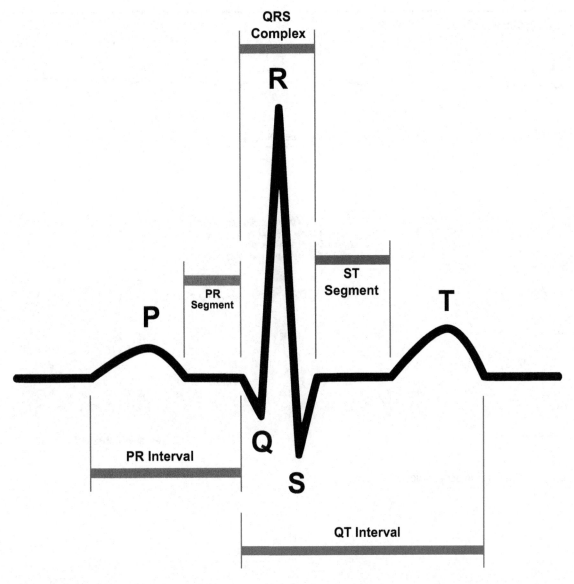

Figure 14.15 ECG.

or contraction of the heart. There are a variety of locations of electrodes that produce different views of the impulses. We will examine a standard ECG view. The electrical impulses in an ECG produce waves, which are a summation of electrical impulses in a given time frame. The P wave is the first wave seen in an ECG and represents atrial depolarization. Atrial depolarization occurs just before atrial contraction (atria contract about 25 ms after the beginning of the P wave). The P wave is followed by the QRS complex. The QRS complex represents ventricular depolarization. Atrial repolarization is also occurring during this time but is overshadowed by the powerful ventricular signal. The T wave follows the QRS complex and results from ventricular repolarization. Some common measurements include the P–R interval and the Q–T interval. The P–R interval extends from the beginning of the P wave to the beginning of the QRS complex. A prolonged P–R interval can indicate a conduction problem. The Q–T interval extends from the end of the P–R interval to the end of the T wave. The Q–T interval represents ventricular systole. A prolonged Q–T interval can indicate heart damage or electrolyte problems (Figs. 14.16–14.19).

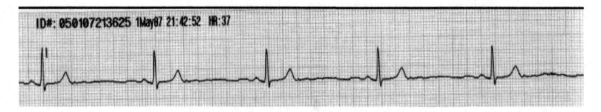

Figure 14.16 Sinus bradycardia. (Note the long interval between beats.)

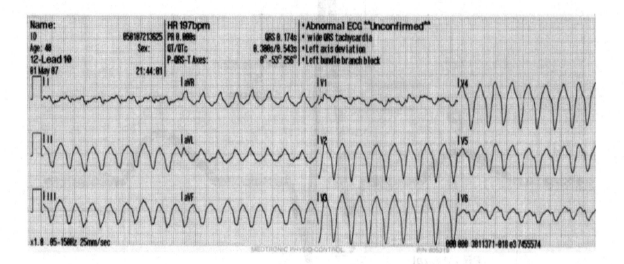

Figure 14.17 Ventricular tachycardia.

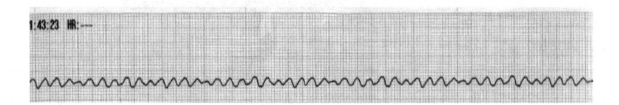

Figure 14.18 Ventricular fibrillation. Note: there is no organized contraction with this arrhythmia.

Figure 14.19 Asystole represents no contraction.

Cardiac Output

The primary goal of the cardiovascular system is to maintain the flow of oxygenated blood to the tissues. In order to accomplish this task, the heart must maintain what is known as a good cardiac output. Cardiac output is the amount of blood pumped by each ventricle in one minute. It is a measure of ventricular efficiency. Cardiac output can be described by this equation:

CO = SV x HR
CO = Cardiac output
SV = Stroke volume
HR = Heart rate

Stroke volume is the amount of blood ejected by a ventricle in one contraction. Heart rate is in beats per minute. Thus cardiac output is the amount of blood ejected by a ventricle in one minute. Generally SV is about 70–80ml. For example, if HR is 70 bpm and SV is 80 ml, then cardiac output is 5600 ml per minute or 5.6 L per minute. Factors that influence SV and HR then will affect cardiac output. SV is affected by end diastolic volume (EDV) or the amount of blood in the ventricle just before contraction, as well as end systolic volume (ESV), which is the amount of blood remaining in the ventricle after contraction (systole). SV can be calculated from EDV and ESV by the following: SV = EDV − ESV

The following is an example to illustrate how cardiac output works to maintain blood flow. Let's say you have two patients. One is a highly trained endurance athlete; the other suffers from congestive heart failure. The athlete enters your office and you begin your cardiac assessment by taking her pulse. You record the resting pulse as 55 bpm. Your next patient enters the room and you also take his pulse and record it at 100 bpm. How is cardiac output responsible for the difference in the two pulses?

In both circumstances the heart is working to provide adequate amounts of oxygenated blood to the tissues. In other words the heart is working to maintain a good cardiac output. The athlete's stroke volume is high because of her athletic conditioning. Therefore, the heart rate will be low in order to maintain good cardiac output. The CHF patient's stroke volume is much lower than the athlete's. Heart rate must then be higher in order to maintain cardiac output. Starling's law of the heart relates stretch of the ventricular walls to stroke volume. The degree of stretch of the ventricular walls is called preload. An increase in preload results in an increase in stroke volume, which, in turn, increases cardiac output. Stroke volume then is affected by venous return, which can vary from 2 L/min to about 24 L/min. Other factors that affect cardiac output include neural and hormonal control mechanisms. We will explore these next.

Nervous System Influences

We have examined how the heart beats on its own accord by generating impulses in the nodes. However, heart rate must sometimes vary in order to meet the varying demands of the body. This control of heart rate comes from the nervous and endocrine systems. In fact, the autonomic nervous system is constantly adjusting heart rate in order to maintain good blood pressure and flow of blood to the tissues. The autonomic nervous system connects to the heart by means of the cardiac plexus. The cardiac plexus sends postganglionic sympathetic neurons to the SA and AV nodes and the atrial muscle cells. The postganglionic neurons are referred to as cardiac accelerator nerves and they originate in the cervical and upper thoracic paravertebral ganglia. The parasympathetic nervous system sends postganglionic neurons to the cardiac plexus via the vagus nerve (CN X). The autonomic nervous system impulses originate in the cardiac control centers in the medulla oblongata. There is a cardioaccelleratory center and a cardioinhibitory center. The cardioaccelleratory center controls the sympathetic pathway and increases heart rate, while the cardioinhibitory center controls the parasympathetic pathway and decreases heart rate. The centers have input from higher cortical regions of the brain as well as the hypothalamus.

Sensory information about the cardiovascular system originates in baroreceptors and chemoreceptors. These receptors are innervated by the glossopharyngeal nerve (CN IX). The baroreceptors monitor changes in pressure, while chemoreceptors monitor changes in blood levels of oxygen, carbon dioxide, and pH. An increase in carbon dioxide or decrease in blood pH will stimulate the sympathetic nervous system, which will result in an increase in heart rate and force of contraction. Chemoreceptors in the carotid sinus and aortic body sense changes in oxygen concentration. A drop in oxygen levels causes vasoconstriction and a decrease in heart rate. This allows for movement of blood without an increase in oxygen use of the heart. For example, when a subject rises from a supine to a sitting position, there is a temporary drop in blood pressure in the head. This is sensed by baroreceptors in the carotid sinus. The sensory information travels via the glossopharyngeal nerve to the cardiac control centers to produce a subsequent increase in heart rate via the sympathetic pathway.

The sympathetic neurotransmitter released by the postganglionic neurons is norepinephrine (NE). NE binds to beta adrenergic receptors, causing sodium and calcium channels to open. The resulting influx of sodium decreases the period of depolarization, causing the threshold to be reached more quickly. This results in an increase in heart rate. Likewise the parasympathetic postganglionic neurons secrete acetylcholine (Ach) that causes potassium gates to open, resulting in a longer time period for depolarization. This produces a slower heart rate. The Bainbridge reflex (atrial reflex) occurs with an increase in stretch of the atrial walls. The reflex results in an increase in sympathetic activity and subsequent increase in heart rate. NE and ACh have both neurotransmitter and hormonal action. NE and epinephrine are both secreted by the adrenal medulla. These hormones are secreted in response to stress and exercise. Thyroid hormone also has a similar action to NE and increases heart rate.

Ischemic Response

The ischemic response, or central nervous system ischemic response, occurs with a decrease in blood flow to the medulla oblongata. The ischemic response occurs when blood pressure decreases to about 50 mm Hg. It produces systemic vasoconstriction in an effort to support blood flow. If ischemia continues then the vasomotor center will cease to function, resulting in vasodilation and death.

Blood Pressure Control

Blood pressure is a measure of the force on blood vessel walls. There are two numbers associated with blood pressure. The systolic pressure is the higher number and results from ventricular contraction. The diastolic pressure represents the pressure in the system during ventricular diastole. Blood pressure can be measured with a stethoscope and a device called a sphygmomanometer. Typically the pressure in the brachial artery is measured. The examiner listens to (auscultates) the artery at the elbow while the cuff is squeezed above the elbow until the brachial artery collapses. The examiner then slowly releases the cuff and listens for Korotkoff sounds, which are produced by turbulent blood flow. The first sound heard represents the systolic pressure. The cuff is loosened until turbulent flow ceases. The pressure at which the sounds disappear represents the diastolic blood pressure. The normal systolic pressure is around 120 mm of mercury. The normal diastolic pressure is around 80 mm Hg. The difference between the systolic and diastolic pressures is called the pulse pressure. With a normal blood pressure of 120/80 the pulse pressure is 40 mm Hg. Stroke volume and vascular compliance both affect pulse pressure. When stroke volume decreases then so does pulse pressure. Likewise when vascular compliance decreases then pulse pressure increases. This occurs with aging and atherosclerotic plaquing. Mean arterial pressure is a measure of pressure in the arteries and is somewhere between the average systolic and diastolic pressures. Mean arterial pressure (MAP) can be determined by the following:

MAP = SV x HR X PR
SV = stroke volume
HR = heart rate
PR = peripheral resistance (resistance in vascular system)

This means that anything affecting stroke volume, peripheral resistance, or heart rate will also affect blood pressure. The mechanisms that control these variables will also work to control blood pressure. Blood pressure is directly related to cardiac output. Therefore, the previously discussed factors that affect cardiac output also affect blood pressure. For example, we know that an increase in cardiac output results in an increase in blood pressure. We also know that an increase in heart rate increases cardiac output, given no changes in stroke volume. If sympathetic activity increases, say, due to periods of stress, then we know that heart rate will increase because of stimulation from the medulla's cardiac control center activating the sympathetic pathway to the SA node. The increase in heart rate increases cardiac output, which in turn increases blood pressure. In essence long periods of stress can cause an increase in blood pressure. The same process occurs with pain. Pain will activate the sympathetic pathway and increase heart rate and cardiac output. Blood pressure will also increase with pain.

Fluid Volume and Blood Pressure

Overall, fluid volume is directly related to blood volume. An increase or decrease in overall fluid volume subsequently increases or decreases blood volume. Likewise blood volume is directly related to blood pressure. Mechanisms that control fluid volume also have an effect on blood pressure. We will examine three mechanisms in this section: the renin-angiotensin system, atrial natriuretic hormone, and antidiuretic hormone. The renin-angiotensin system (renin-angiotensin-aldosterone system) begins with the secretion of renin by the kidneys in response to a decrease in blood pressure (we will explore this system in more detail in the urinary system chapter). Renin activates a plasma protein called angiotensinogen by causing it to cleave a portion known as angiotensin I (one). Angiotensin I travels through the bloodstream to the lungs where it encounters an enzyme known as angiotensin converting enzyme (ACE). The angiotensin converting enzyme again cleaves angiotensin I, producing angiotensin II. Angiotensin II causes systemic vasoconstriction as well as stimulates the release of aldosterone, an adrenal cortex hormone. Aldosterone targets the kidneys to conserve sodium and secrete potassium. The conservation of sodium causes an increase in fluid volume by way of osmosis. The increase in fluid volume causes a subsequent increase in blood volume and blood pressure (Fig. 14.20). Atrial natriuretic hormone (sometimes called a peptide) is secreted by the walls of the atria in response to atrial stretch. If blood volume increases, so does venous return, causing increased atrial stretch. The subsequent release of ANH targets the kidneys to eliminate sodium. Water follows sodium by osmosis, causing a decrease in fluid volume,

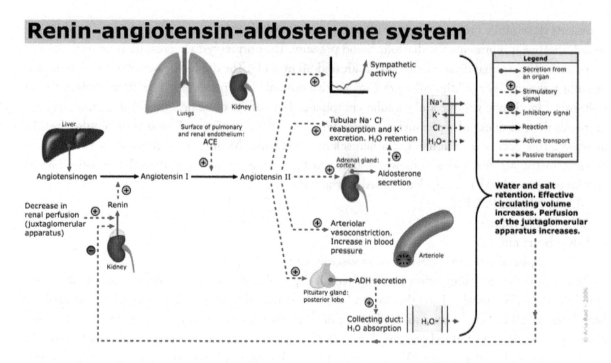

Figure 14.20 The renin-angiotensin-aldosterone system.

blood volume, and blood pressure. Antidiuretic Hormone (ADH)/(vasopressin) is a hormone secreted by the posterior portion of the pituitary gland in response to increases in blood solute concentration. The hypothalamus contains neurons that sense changes in blood solute concentration. Like angiotensin II, ADH causes vasoconstriction, although it is not as powerful as angiotensin II. ADH targets the kidneys to conserve fluid. Less urine is produced when ADH is secreted, as fluid volume is conserved.

Peripheral Resistance

Fluid moves by virtue of a pressure gradient. In other words, fluid moves from areas of higher to lower pressure. The left ventricle must produce a pressure that is greater than the fluid pressure in the arterial side of the vascular system in order to move blood through the system. The pressure that is resident in the vascular system that the heart must overcome in order to move blood is called afterload. The resistance to blood flow in the vascular system is also known as peripheral resistance. Peripheral resistance is directly proportional to blood pressure. Pressure in the vascular system is greatest in the aorta and decreases as blood moves from the arteries to arterioles, capillaries, and the venous system. The pressure can be as low as 0 mm Hg at the right atrium. The pressure in each part of the arterial system is directly proportional to the resistance to blood flow. The larger structures such as the larger arteries have little resistance, while the smaller structures such as the capillaries have a much larger resistance to blood flow. Pressure in the arterioles is about 85 mm Hg and in the capillaries about 30 mm Hg. Pressure is controlled in part in the vascular system by changing the diameter of the vessels. Arteries and arterioles have a larger smooth muscle layer than veins have and are capable of constricting or dilating according to the body's needs for oxygenated blood. The vessels receive input from the sympathetic nervous system. An increase in sympathetic stimulation will cause vasoconstriction, while a decrease causes vasodilation. There is continuous stimulation from the sympathetic nervous system that produces a continuous partial vasoconstriction in order to maintain pressure. This is known as vasomotor tone. When the arteries are fully stimulated by the sympathetic nervous system, they are about one half of their normal diameter. Increases in vasoconstriction produce subsequent increases in peripheral resistance and blood pressure. Likewise a decrease in sympathetic stimulation results in vasodilation that in turn decreases peripheral resistance and blood pressure. Peripheral resistance also increases with the disease process known as atherosclerosis (Fig. 14.21). Atherosclerosis is a thickening of the tunic media of arteries along with damage to the endothelium. This disease has been linked to high levels of small particle low density lipoproteins. These lipids deposit on blood vessel walls and undergo phagocytosis by white blood vessels. The result is the deposition of what is known as plaque. Plaque narrows the lumen of the arteries and increases peripheral resistance and blood pressure.

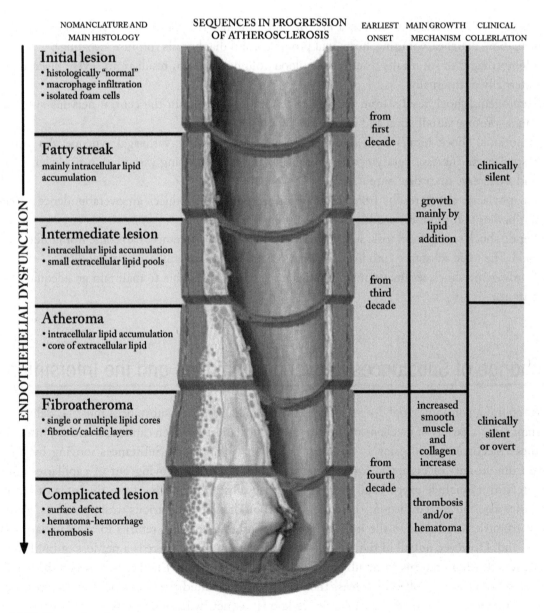

Figure 14.21 Atherosclerosis.

Shock

Shock results when there is an inadequate supply of oxygenated blood to the tissues. Circulatory shock can be described in three stages: compensated, progressive, and irreversible. Compensated shock is characterized by a moderate decrease in blood pressure. Compensated shock stimulates all of the mechanisms for maintaining blood pressure. Blood pressure is gradually restored to normal levels in compensated shock. If blood pressure control mechanisms are not adequate to restore blood pressure, then progressive shock results. Blood flow decreases to levels that produce ischemia in heart tissue, resulting in damage if not restored quickly. Without successful intervention, the body progresses into irreversible shock. Irreversible shock is fatal and does not respond to medical treatment.

Types of shock include the following:

1. Hypovolemic shock results from a loss of fluid volume. This can result from severe dehydration, urination, diarrhea, vomiting, hormonal problems, and diaphoresis (profuse sweating).
2. Hemorrhagic shock results from loss of blood volume. This can result from severe bleeding either externally or internally.
3. Neurogenic shock results from damage to vasomotor centers in the central nervous system. This causes profuse vasodilation and decrease in blood pressure.
4. Emotional shock results from fainting. This is also known as vasovagal syncope. An emotional response can increase parasympathetic input to the heart, causing vasodilation in skeletal muscles and a decrease in cardiac output.
5. Anaphylactic shock results from an allergic reaction that produces an overabundance of certain antibodies that produce vasodilation and capillary permeability.
6. Septic shock results from toxic substances in the blood. The substances can come from infections and food. The toxic substances produce vasodilation and increased capillary permeability.
7. Cardiogenic shock results from heart damage. The heart is unable to maintain an adequate cardiac output.

Exchange of Substances Between Capillaries and the Interstitium

Cells require a constant supply of substances from capillaries. Thus substances must be delivered at a constant rate. A variety of mechanisms help to move substances between capillaries and the interstitium. Of these mechanisms, diffusion is the most prevalent (Fig. 14.22). Substances moving by diffusion move from an area of higher to lower concentration. Substances moving out of capillaries and into the interstitium include oxygen and glucose. Likewise substances moving from the interstitium to the capillaries also move by diffusion. These include carbon dioxide and various waste products. Fluid also moves from the capillaries to the interstitium at the arterial end and returns to the capillaries at the venous end. However, not all of the fluid is returned to circulation. There is a net loss of fluid from the capillaries. This fluid returns to circulation via the lymphatic system. Fluid flow between the capillaries and interstitium is controlled by forces. These forces include fluid pressure and osmotic pressure. The pressure moving fluid out of the capillaries is known as net hydrostatic pressure. This pressure must overcome another pressure that works to pull fluid back into capillaries. This pressure is known as net osmotic pressure.

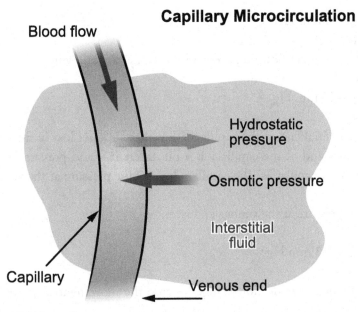

Figure 14.22 Movement of fluid in capillary.

We can calculate the total pressure by subtracting net osmotic pressure from net hydrostatic pressure.

Net Filtration Pressure = Net Hydrostatic Pressure − Net Osmotic Pressure

Our calculation gives us Net Filtration Pressure, which accounts for all of the pressures moving fluids.

Net hydrostatic pressure consists of two pressures: capillary hydrostatic pressure (CHP), which is the blood pressure at the arterial end of the capillary, and interstitial fluid pressure (IFP). Interstitial fluid pressure is the pressure in the interstitium that opposes movement of fluid from capillary to interstitium. This pressure is a negative pressure because of the action of lymphatic vessels. The lymphatic vessels create a suction effect that pulls fluid from the capillaries. For example, CHP is usually around 30 mm Hg and IFP is usually around -3 mm Hg. We can calculate net hydrostatic pressure (NHP) as follows:

NHP = CHP − IFP
NHP = 30 mm Hg − (-3 mm Hg)
NHP = 33 mm Hg

This represents the pressure pushing fluid out of the capillaries. The pressure pulling fluid back into the capillaries is the net osmotic pressure (NOP). NOP represents the difference in osmotic pressures between the capillaries and the interstitium. There are two pressures that make up NOP. These include blood colloid osmotic pressure and interstitial osmotic pressure. Blood colloid osmotic pressure (BCOP) represents the pulling force resulting from the presence of plasma proteins in the blood. Fluid moves toward an area of higher concentration of solute due to osmosis. The plasma proteins act like a solute. BCOP then works to move fluid from the interstitium to the capillaries. Interstitial colloid osmotic pressure (ICOP) results from the presence of proteins in the interstitium. The BCOP is much larger than ICOP because proteins are large molecules that do not pass through capillary walls and thus stay in the blood. We can calculate the net osmotic pressure by subtracting the ICOP from the BCOP. For example, the BCOP is usually around 28 mm Hg and the ICOP is usually around 8 mm Hg.

Net osmotic pressure (NOP) = Blood colloid osmotic pressure (BCOP) − interstitial colloid osmotic pressure (ICOP).

NOP = 28 mm Hg − 8 mm Hg
NOP = 20 mm Hg
Now we can calculate the net filtration pressure (NFP) by the following.
NFP = NHP − NOP
NFP = 33 mm Hg − 20 mm Hg
NFP = 13 mm Hg

This represents the arterial end of the capillary. Thus there is a net loss of fluid from this end of the capillary bed. The venous end of the capillary is a bit different. Fluid pressure decreases between the arterial and venous ends of capillaries. The capillary hydrostatic pressure at the venous end decreases to about 10 mm Hg.

The net hydrostatic pressure at the venous end is:
NHP = CHP − IFP
NHP = 10 mm Hg − (-3 mm Hg)
NHP = 13 mm Hg

The colloid osmotic pressures do not change because the movement of proteins from capillary to interstitium does not change.

To calculate net filtration pressure at the venous end:
NFP = NHP − NOP
NFP = 13 mm Hg − 20 mm Hg
NFP = -7 mm Hg

The negative pressure at the venous end of capillaries causes fluid to move into the capillaries. The various movements of fluid between the capillaries and interstitium maintain a balance. Disrupting this balance can result in edema.

Cardiovascular Disease

Cardiovascular disease refers to the class of diseases that involve the heart and/or blood vessels. While the term technically refers to any disease that affects the cardiovascular system, it is usually used to refer to those related to atherosclerosis (arterial disease). These conditions have similar causes, mechanisms, and treatments. Over 50 million Americans have cardiovascular problems, and most other Western countries face high and increasing rates of cardiovascular disease. It is the number-one cause of death and disability in the United States and most European countries. By the time that heart problems are detected, the underlying cause (atherosclerosis) is usually quite advanced, having progressed for decades.

Hypertension

Hypertension, or high blood pressure, is a medical condition wherein the blood pressure is chronically elevated. Hypertension is defined by some authors as systolic pressure over 130 and diastolic over 85 mmHg. Hypertension often has an insidious or un-noticed onset and is sometimes called the silent killer because stretching of the arteries causes microscopic tears in the arterial wall and accelerates degenerative changes. Persistent hypertension is one of the risk factors for strokes, heart attacks, heart failure, and arterial aneurysm, and is a leading cause of chronic renal failure.

Atherosclerosis

Atherosclerosis is a disease affecting the arterial blood vessel. It is commonly referred to as a "hardening" or "furring" of the arteries. It is caused by the formation of multiple plaques within the arteries. Arteriosclerosis ("hardening of the artery") results from a deposition of tough, rigid collagen inside the vessel wall and around the atheroma. This increases the stiffness and decreases the elasticity of the artery wall. Atherosclerosis typically begins in early adolescence, is usually found in most major arteries, and yet is asymptomatic and not detected by most diagnostic methods during life. It most commonly becomes seriously symptomatic when interfering with the coronary circulation supplying the heart or cerebral circulation supplying the brain, and is considered the most important underlying cause of strokes, heart attacks, various heart diseases including congestive heart failure, and most cardiovascular diseases in general.

Plaque

Plaque atheroma, commonly known as plaque, is an abnormal inflammatory accumulation of macrophage white blood cells within the walls of arteries.

Circulatory Shock

Circulatory shock is a severe condition that results from reduced blood circulation.

Thrombus

A thrombus, or blood clot, is the final product of the blood coagulation step in hemostasis. It is achieved via the aggregation of platelets that form a platelet plug, and the activation of the humoral coagulation system (i.e., clotting factors). A thrombus is physiologic in cases of injury, but pathologic in case of thrombosis. Preventing blood clots reduces the risk of stroke, heart attack, and pulmonary embolism. Heparin and warfarin are often used to inhibit the formation and growth of existing blood clots, thereby allowing the body to shrink and dissolve the blood clots through normal methods.

Embolism

An embolism occurs when an object (the embolus) migrates from one part of the body (through circulation) and causes a blockage (occlusion) of a blood vessel in another part of the body. Blood clots form the most common embolic material by far: other possible embolic materials include fat globules (a fat embolism), air bubbles (an air embolism), septic emboli (containing pus and bacteria), and amniotic fluid.

Stroke

A stroke, also known as cerebrovascular accident (CVA), is an acute neurological injury whereby the blood supply to a part of the brain is interrupted. Strokes can be classified into two major categories: ischemic and hemorrhagic. Approximately 80% of strokes are due to ischemia.

Ischemic Stroke: In ischemic stroke, which occurs in approximately 85–90% of strokes, a blood vessel becomes occluded and the blood supply to part of the brain is totally or partially blocked. Ischemic stroke is commonly divided into thrombotic stroke, embolic stroke, systemic hypoperfusion (Watershed or Border Zone stroke), or venous thrombosis.

Hemorrhagic Stroke: A hemorrhagic stroke, or cerebral hemorrhage, is a form of stroke that occurs when a blood vessel in the brain ruptures or bleeds. Like ischemic strokes, hemorrhagic strokes interrupt the brain's blood supply because the bleeding vessel can no longer carry the blood to its target tissue. In addition, blood irritates brain tissue, disrupting the delicate chemical balance, and, if the bleeding continues, it can cause increased intracranial pressure, which physically impinges on brain tissue and restricts blood flow into the brain. In this respect, hemorrhagic strokes are more dangerous than their more common counterpart, ischemic strokes. There are two types of hemorrhagic stroke: intracerebral hemorrhage and subarachnoid hemorrhage. The term "brain attack" is starting to come into use in the United States for stroke, just as the term "heart attack" is used for myocardial infarction, where a cutoff of blood causes necrosis to the tissue of the heart. Many hospitals have brain attack teams within their neurology departments specifically for swift treatment of stroke. If symptoms of stroke are detected at early onset, special "clot busting" drugs may be administered. These clot busters will dissolve clots before they can cause tissue death and restore normal circulation. One of the initial drugs used to dissolve clots was streptokinase, although its use creates a possibility of clot destruction throughout the entire body, leading to serious hemorrhage. There are newer, third-generation thrombolytics that are safer.

Heart Attack

Acute myocardial infarction (AMI or MI), commonly known as a heart attack. A heart attack occurs when the supply of blood and oxygen to an area of heart muscle is blocked, usually by a clot in a coronary artery. Often, this blockage leads to arrhythmias (irregular heartbeat or rhythm) that cause a severe decrease in the pumping function of the heart and may bring about sudden death. If the blockage is not treated within a few hours, the affected heart muscle will die and be replaced by scar tissue. It is the leading cause of death for both men and women all over the world

Angina Pectoris

Angina pectoris is chest pain due to ischemia (a lack of blood and hence oxygen supply) of the heart muscle, generally due to obstruction or spasm of the coronary arteries (the heart's blood vessels).

Congestive Heart Failure

Congestive heart failure (CHF), also called congestive cardiac failure (CCF) or just heart failure, is a condition that can result from any structural or functional cardiac disorder that impairs the ability of the heart to fill with or pump a sufficient amount of blood throughout the body. It is not to be confused with "cessation of heartbeat," which is known as asystole, or with cardiac arrest, which is the cessation of normal cardiac function in the face of heart disease. Because not all patients have volume overload at the time of initial or subsequent evaluation, the term "heart failure" is preferred over the older term "congestive heart failure." Congestive heart failure is often undiagnosed due to a lack of a universally agreed definition and difficulties in diagnosis, particularly when the condition is considered "mild."

Aneurysm

An aneurysm (or aneurism) is a localized dilation or ballooning, much like a bulge on an overinflated inner tube, of a blood vessel by more than 50% of the diameter of the vessel and can lead to instant death at any time. The larger an aneurysm becomes, the more likely it is to burst. Aneurysms most commonly occur in arteries at the base of the brain (the circle of Willis) and in the aorta (the main artery coming out of the heart)—this is an aortic aneurysm. Aneurysms are also described according to their shape: Saccular or fusiform. A saccular aneurysm resembles a small sack; a fusiform aneurysm is shaped like a spindle.

Varicose Veins

Varicose veins are veins on the leg that are large, twisted, and ropelike, and can cause pain, swelling, or itching. They are an extreme form of telangiectasia, or spider veins. Varicose veins result due to insufficiency of the valves in the communicating veins. These are veins that link the superficial and deep veins of the lower limb. Normally, blood flows from the superficial to the deep veins, facilitating return of blood to the heart. However, when the valve becomes defective, blood is forced into the superficial veins by the action of the muscle pump (which normally aids return of blood to the heart by compressing the deep veins). People who have varicose veins are more at risk of getting a deep vein thrombosis (DVT) and pulmonary embolisms.

Phlebitis

Phlebitis is an inflammation of a vein, usually in the legs. This is usually the most serious if found in a deep vein. However, most people with the condition, perhaps 80 to 90 percent, are women. The disease may also have a genetic component, as it is known to run in families.

Congenital Heart Defects

Heart defects present at birth are called congenital heart defects. Slightly less than 1% of all newborns have congenital heart disease. Eight defects are more common than all others and make up 80% of all congenital heart diseases, whereas the remaining 20% consist of many independently infrequent conditions or combinations of several defects.

Acyanotic Defects

Acyanotic heart defects are those in which there is a normal amount of oxygen in the bloodstream. The most common congenital heart defect is a ventral septal defect (VSD), which occurs in about 20% of all children with congenital heart disease. In VSD, blood from the left ventricle is shunted to the right ventricle, resulting in oxygenated blood returning into pulmonic circulation. One of the potential problems of VSD is pulmonary hypertension.

Cyanotic Defects

Cyanotic heart defects refer to defects that result in decreased amounts of oxygen in the blood. In cyanotic heart defects, deoxygenated blood from the right ventricle flows into the systemic circulation.

Interesting Facts

- Heart disease is the number-one killer in American women.
- 16.7 million deaths are the result of forms of cardiovascular disease, heart disease, and stroke. Stress, eating high-fat foods, obesity, tobacco, and alcohol use are just some risk factors of developing heart disease.

- Recent research suggests that taking a small dose of aspirin daily may help prevent a heart attack (because aspirin inhibits platelet clumping).
- The length of all your blood vessels lined up is about 60,000 miles long! To put this in perspective, the Earth's circumference is 40,075.02 kilometers and 60,000 miles is around 96,000 km—so your blood vessels would go twice around the world and still have some to spare!

Maintaining Heart Health

- Eating healthfully; good nutrition.
- Fitness and exercise.
- Having a healthy lifestyle; don't drink, smoke, or do drugs.
- Lowering LDL cholesterol and high blood pressure.
- Reduce the fat, sodium, and calories in your diet.
- The total length of capillaries in an average adult human is approximately 25,000 mi (42,000 km).

CRITICAL-THINKING QUESTIONS

1. Why is the pressure in the pulmonary circulation lower than in the systemic circulation?
2. Why is the plateau phase so critical to cardiac muscle function?
3. How does the delay of the impulse at the atrioventricular node contribute to cardiac function?
4. How do gap junctions and intercalated discs aid in the contraction of the heart?
5. Why do the cardiac muscles cells demonstrate autorhythmicity?
6. Why does increasing EDV increase contractility?
7. Why is afterload important to cardiac function?
8. Arterioles are often referred to as resistance vessels. Why?
9. Cocaine use causes vasoconstriction. Is this likely to increase or decrease blood pressure, and why?
10. A blood vessel with a few smooth muscle fibers and connective tissue, and only a very thin tunica externa, conducts blood toward the heart. What type of vessel is this?
11. An obese patient comes to the clinic complaining of swollen feet and ankles, fatigue, shortness of breath, and often feeling "spaced out." She is a cashier in a grocery store, a job that requires her to stand all day. Outside of work, she engages in no physical activity. She confesses that, because of her weight, she finds even walking uncomfortable. Explain how the skeletal muscle pump might play a role in this patient's signs and symptoms.
12. A patient arrives at the emergency department with dangerously low blood pressure. The patient's blood colloid osmotic pressure is normal. How would you expect this situation to affect the patient's net filtration pressure?
13. All tissues, including malignant tumors, need a blood supply. Explain why drugs called angiogenesis inhibitors would be used in cancer treatment.

Chapter Fifteen

Lymphatic System and Immunity

The lymphatic system is a vascular system that contains capillaries, vessels, and lymph nodes (Fig. 15.1). The lymph capillaries pick up interstitial fluid lost by the circulatory system. The fluid known as lymph moves through the system and is returned to venous circulation. The lymphatic system also transports dietary fats from the gastrointestinal system. Small lymphatic structures called lacteals are located in the small intestine in structures called villi. Fats are broken down and packaged as structures known as chylomicrons. The fats then move through the system to the venous circulation. A large portion of the immune system resides in the lymphatic system as well. Lymph nodes containing white blood cells work to destroy pathogens.

Lymphatic Capillaries

Lymphatic capillaries are distributed throughout the interstitium. Lymphatic capillaries are not found in the central nervous system and bone marrow. They are also not resident in tissues without blood flow, such as the epidermis or cartilage. They are designed to allow one-way fluid flow into the capillary (Fig. 15.2). Lymphatic capillaries consist of overlapping simple squamous epithelium. They also form one-way valves. This arrangement allows for increased permeability and fluid movement toward the venous circulation.

Lymphatic Vessels

The lymphatic capillaries form larger structures called lymphatic vessels. The vessels have a similar structure to veins and contain three layers. The three layers consist of an inner endothelium, a middle smooth muscle layer, and an outer layer of thin fibrous connective tissue. Lymphatic vessels also contain

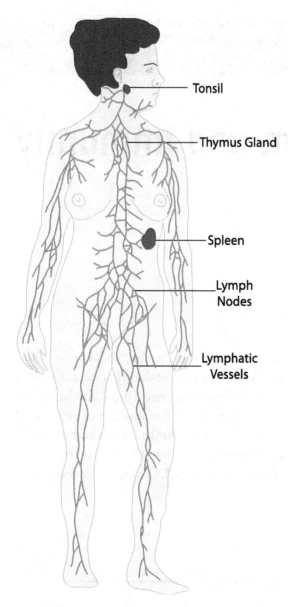

Figure 15.1 Lymphatic system.

valves to allow the one-way flow of blood. Smooth muscle contraction moves blood from one area separated by a valve to another. Some cells in the lymph vessel walls are capable of generating action potentials that cause the smooth muscle to contract. Skeletal muscle contraction also moves lymph fluid by means of generating pressure on the outside of the vessels, causing them to constrict. The valves allow only one-way flow, so lymph is moved toward the venous circulation. Lymph fluid also moves into vessels in the thoracic cavity as a result of dilation of lymph vessels, resulting in decreases in thoracic cavity pressure. The thoracic cavity expands during inspiration, causing a decrease in thoracic pressure. The vessels react by dilating and creating an area of lower pressure. Lymph fluid then moves toward the area of lower pressure.

Lymph Nodes

Lymph nodes are located throughout the lymphatic system. The lymphatic vessels connect with the nodes and fluid moves through them. There are numbers of nodes connected to a vessel so that lymph fluid moves from one node to another. Lymph nodes are small oval structures and are generally not felt during examinations unless enlarged or calcified (Fig. 15.3). Lymph nodes act as filters and work to remove pathogens such as bacteria and viruses. Although diffusely located throughout the body, lymph nodes tend to conglomerate in certain areas. These include the cervical, axillary, inguinal, popliteal, and mammary glands (Fig. 15.4). Lymph nodes consist of a dense connective tissue covering and a trabeculated internal structure. The nodes contain reticular connective tissue that forms an interconnected web-like structure. Vessels entering the nodes are known as afferent vessels. Likewise vessels exiting the nodes are known as efferent vessels. Lymph nodes consist of an outer cortex and an inner medulla. The cortex contains open areas called sinuses. The medulla contains medullary cords, which are branching structures of lymphatic tissue. Open areas called medullary sinuses are also present. Lymph nodes contain white blood cells called macrophages and lymphocytes. Macrophages are located in the sinuses and phagocytize bacteria and debris. Lymphocytes are located in germinal centers and when activated can move into the bloodstream.

The lymphatic vessels eventually form larger structures known as lymphatic trunks. The lymphatic trunks drain specific portions of the body. The subclavian trunks drain the upper extremities. The jugular trunks

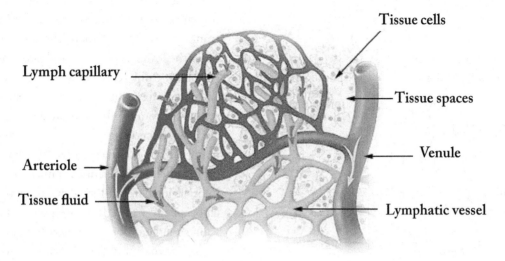

Figure 15.2 Lymphatic capillaries.

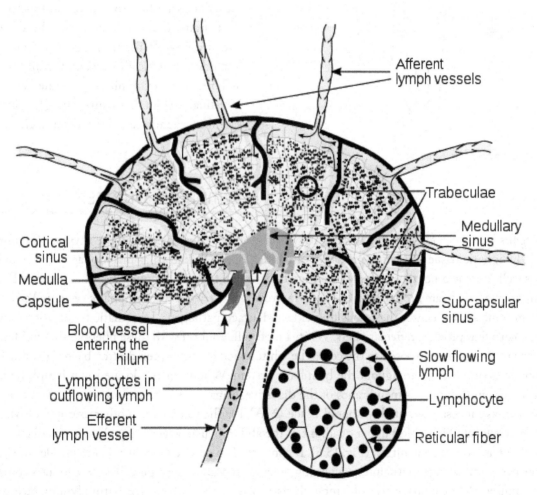

Figure 15.3 Lymph node.

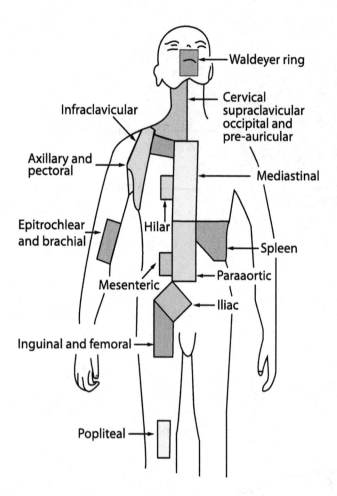

Figure 15.4 Lymph node locations.

drain the head and neck. The bronchomediastinal trunks drain the thoracic area. The intestinal trunks drain the abdomen. The lumbar trunks drain the lower extremities and pelvic area. The lymphatic trunks connect with larger structures called lymphatic ducts, which connect with the venous system at the subclavian veins. There are two ducts: the thoracic duct and right lymphatic duct. The jugular, subclavian, and bronchomediastinal trunks connect to the right internal jugular, right subclavian, or right brachiocephalic trunk. In some people, the three trunks merge to form the right lymphatic duct.

The remaining trunks connect with the thoracic duct. The drainage of lymph fluid is therefore asymmetrical with respect to the arrangement of the right lymphatic and thoracic duct. In other words, the right lymphatic duct drains the right side of the head, neck, and trunk while the thoracic duct drains the left side of the head, neck, and trunk as well as both lower extremities. In some cases the intestinal and lumbar trunks merge to form a sac-like structure called the cisterna chyli.

Lymphatic Organs

Two organs are associated with the lymphatic system. These are the spleen and the thymus. The organs contain lymphatic tissue consisting primarily of white blood cells known as macrophages and lymphocytes as well as some other types of cells. There are two general types of lymphocytes. These are the T- and B-lymphocytes. Both are produced in the bone marrow and carried to the lymphatic system. Activation of the immune system causes these cells to divide and attach pathogens. Lymphatic tissue also contains reticular cells that produce reticular fibers. White blood cells connect with these fibers so that fluid moving through the tissue is exposed to the cells. The white blood cells can then destroy bacteria and debris. Lymphatic tissue resides throughout the lymphatic system. When it is not located in a lymph node or organ such as in the mucous membranes of the digestive, urinary, respiratory and reproductive systems, it is known as mucosa associated lymphoid tissue (MALT). The tonsils are another example of MALT. The spleen is located in the left upper quadrant of the abdominal area, generally close to the diaphragm, and is about as large as an adult fist (Fig. 15.5). It consists of an outer connective tissue capsule. The inner portion has a trabeculated structure, containing areas of red and white pulp. The spleen also contains venous sinuses. White pulp consists of lymphatic tissue associated with arteries within lymphatic organs. Red pulp contains both white and red blood cells and is associated with veins. The splenic artery and vein enter and exit the spleen at the hilum. Blood flows into the spleen and through the trabeculated

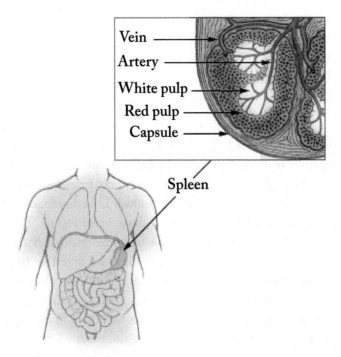

Figure 15.5 Spleen.

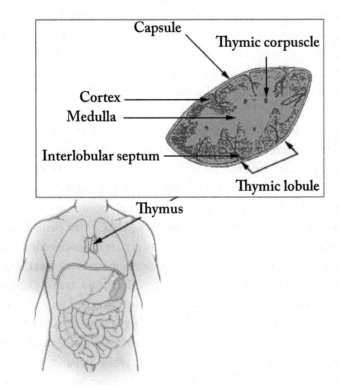

Figure 15.6 Thymus.

network. The cells in the spleen work to destroy pathogens. Lymphocytes in the spleen can react to pathogens and trigger the immune system. The spleen also acts as a blood reservoir.

The thymus is a gland located just deep to the sternum in the superior portion of the mediastinum (Fig. 15.6). Early in life the thymus is larger and decreases in size with age although it continues to produce white blood cells. The thymus has two lobes each surrounded by a connective tissue capsule. It contains an outer cortex and inner medulla. The internal region of the thymus is trabeculated and filled with lymphocytes. The thymus produces large numbers of T-lymphocytes that can travel to the blood.

Immunity

Our immune systems offer us protection against a world full of pathogens. Our immune systems work by providing two types of immunity: non-specific and specific. In non-specific immunity, our bodies present the same kinds of defense systems regardless of the type of pathogens. Non-specific immunity works much like a fence around your property. The fence does not differentiate between friend and foe. It keeps everyone out. The other type of immunity is known as specific immunity (defense). Specific defense produces an attack against a specific pathogen. This is much like having an attendant at the gate of the fence around your house. The attendant can identify potential foes and keep them out. Before birth the body inventories all of the cells and tissues of the body and classifies them as "self" cells. The presentation of non-self cells can then trigger the immune system.

Non-Specific Defense

Non-specific defense (innate immunity) consists of mechanisms that either keep pathogens out or destroy them regardless of their type. Non-specific defense includes mechanical barriers, chemical substances, cells, and inflammation. Mechanical barriers include the skin and mucous membranes. Besides presenting a physical barrier that stops pathogens, they also work to remove substances from the surface of membranes. Examples include the movement of mucus moving substances toward the digestive tract and tears washing substances from the eyes. Chemical substances work to destroy pathogens. These include enzymes, cytokines, and the complement system. For example, mucus from the respiratory tract moves toward the pharynx and esophagus where it is swallowed. Upon reaching the digestive tract, pathogens are destroyed by powerful digestive enzymes. Cytokines are a series of protein substances secreted by cells that work to destroy pathogens. Interferons are cytokines that bind to cells, causing them to produce substances that inhibit viral replication. One type of interferon can affect many types of viruses. Interferons can also activate other immune cells such as macrophages and natural killer cells. Some cytokines produce fever. Interleukin I (endogenous pyrogen) is a cytokine that acts as a pyrogen (raises body temperature). This cytokine is released in response to toxins or pathogens and causes an increase in body temperature. The complement system is a series of about 20 plasma proteins (Fig. 15.7). They include proteins that are named C1–C9 and factors B, D, P. They act much like the clotting cascade (see blood chapter), in that activation of the first complement protein causes the others to activate. There are two pathways in which to activate the complement system. The alternative pathway is activated by the presentation of a pathogen to the body. The C3 protein is normally inactivated by the body's cells; however, presentation of a non-self cell can cause it to remain active triggering a response. Complement system responses include inflammation, phagocytosis from white blood cells attracted to the area, and attacking non-self cells.

Inflammation is produced by facilitating the release of histamine from white blood cells called mast cells. Histamine promotes local vasodilation, increasing capillary permeability and bringing more blood to the area. Neutrophils and macrophages are attracted to activated complement proteins for phagocytosis of pathogens. Certain antibodies (we will cover antibodies later) called opsonins work with complement proteins to facilitate phagocytosis. This process is called opsonization. Some complement proteins (C5–C9) bind to cell membranes to form a membrane attack complex (MAC) that drills holes in cells, allowing substances to rush in and burst the cell. The classical pathway is part of specific defense that will be covered later in this chapter. Inflammation is characterized by swelling, redness, heat, and pain (tumor, rubor, calor, dolor). Inflammation is produced by tissue destruction from trauma, cuts, temperature, and chemicals. Inflammation causes an increased blood flow to the damaged area. Blood brings substances for repair and the stasis of blood in the area prevents further spread of pathogens. Inflammation is caused primarily by the release of histamine and heparin from mast cells (similar to basophils). Histamine promotes local vasodilation and capillary permeability, while heparin inhibits clotting. Phagocytes are also attracted to the area and remove debris. Neutrophils release substances that activate fibroblasts to begin to repair the area. Substances released by cells stimulate pain receptors in the tissue, causing the sensation of pain.

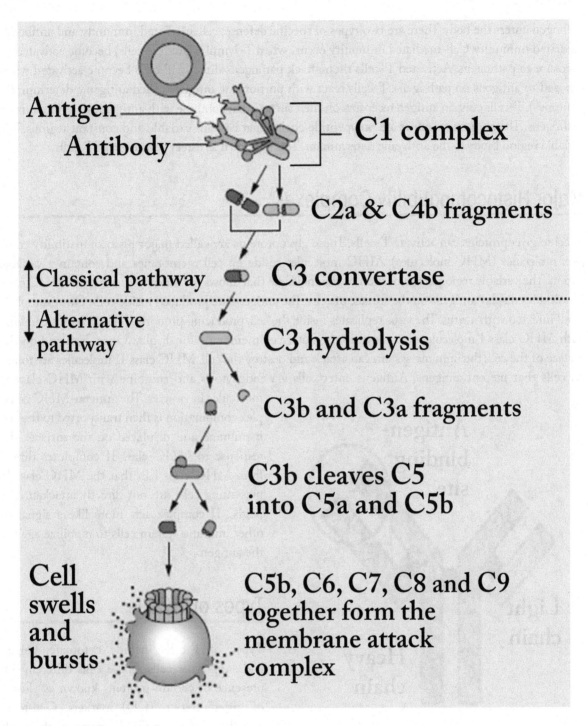

Figure 15.7 Complement system.

Specific Defense

Specific defense (sometimes called adaptive immunity) recognizes and coordinates attacks against specific pathogens. The system can also remember pathogens and produce a powerful response the next time a pathogen enters the body. There are two types of specific defense: cell-mediated immunity and antibody-mediated immunity. Cell-mediated immunity occurs when T-lymphocytes (T-cells) become activated by exposure to pathogens. Activated T-cells then attack pathogens directly. T-cells become activated when exposed to antigens on pathogens. T-cells react with portions of antigens called antigenic determinants (epitopes). T-cells contain antigen receptors on their surface that combine with antigenic determinants on pathogens. The antigen receptors are polypeptide chains that contain variable and constant regions. The variable region binds to the antigenic determinant. This is known as direct activation of T-cells.

Major Histocompatibility Complexes

Specific glycoproteins can activate T-cells. These glycoproteins are called major histocompatibility complex molecules (MHC molecules). MHC molecules reside on cell membranes and contain a variable region. The variable region is the portion of the molecule that allows for binding to antigens. MHC class I molecules display antigens on the surface of cells. The antigens are produced inside cells. One example is a cell infected with a virus. The virus replicates inside the cell, producing proteins. These proteins combine with MHC class I molecules that move to the outer cell membrane for display. Once displayed on the surface of the cell, the immune system can attack and destroy the cell. MHC class II molecules are found on cells that present antigens. Antigens enter cells via endocytosis and combine with MHC class II molecules in vesicles. The antigen-MHC complex combination is then transported to the cell membrane and displayed on the surface. The response to MHC class II complexes differs from MHC class I in that the MHC class II presenting cells are not directly attacked. The MHC II complex acts more like a signal to other immune system cells to mobilize against the antigen.

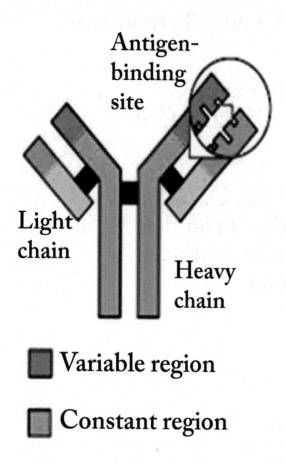

Figure 15.8 Antibody structure.

Types of T-Cells

Types of T-cells include cytotoxic, helper, and suppressor cells. These cells differ in the presence of certain proteins known as cluster of differentiation (CD) markers. Cytotoxic T-cells contain the CD8 protein in their cell membranes. Cytotoxic T-cells respond to MCH class I molecules. Helper T-cells have CD4 markers and respond to MCH class II molecules. T-cell activation typically requires co-stimulation in order to fully activate the cell. Co-stimulation involves a secondary binding to an antigen. Co-stimulation helps to ensure that the appropriate cell is attacked. Activated cytotoxic T-cells destroy pathogenic cells by either phagocytosis—release of a substance that drills holes in the infected cell called perforin, secreting a substance called a lymphotoxin that is toxic to the cell—or activating genes in the infected cell that tell it to destroy itself. The latter is known as apoptosis. Suppressor T-cells also develop from CD8 T-cells. Suppressor T-cells secrete substances known as suppression factors that suppress the action of T-cells and B-cells. These cells require a longer period of time for activation and help to protect against overactivation of the immune system. Helper T-cells contain the CD4 protein. Helper T-cells facilitate rapid mitosis of other T-cells, cause chemotaxis of macrophages to the infected area, help to activate B-cells, and stimulate natural killer cells.

B-Cells

The other major type of lymphocyte is the B-cell (B for bursa of fabricius of the chicken after where they were discovered). There are millions of B-cells in the body and each contains a specific set of antibodies. Some antibodies are present on the surface of the B-cell. Antigen-containing pathogens bind to antibodies, causing sensitization of the B-cells. Antigens are then displayed on MHC class II proteins on the surface of the B-cells. Helper T-cells complete the activation of B-cells by binding to the MHC class II proteins and secreting cytokines that stimulate the B-cells. Activated B-cells undergo rapid mitosis with some of the cells remaining immature memory cells. Activated B-cells produce and secrete antibodies. Antibodies consist of a pair of polypeptide chains called light chains connected to another pair of polypeptide chains called heavy chains. The chains are connected by disulfide bonds and contain both constant and variable segments. The base of the antibodies is formed by the constant segments of the heavy chains that help to identify the antibody. This region can also activate the complement system. The other end of the antibody contains the variable region. The antigen binding sites are located on this variable region. These sites can connect with antigens on pathogens

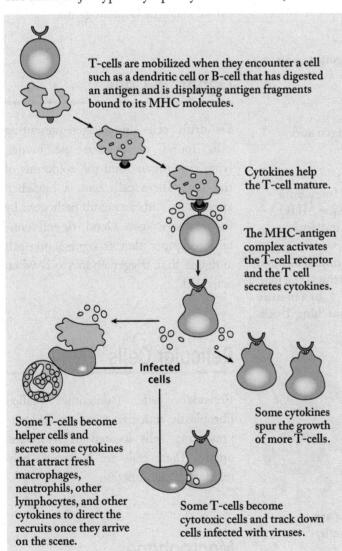

Figure 15.9 T-cell activation.

to form antibody-antigen complexes. Haptens are incomplete antigens and do not activate B-cells unless they combine with carrier molecules that act as complete antigens. A number of different effects result from antibody-antigen complexes. Antibodies can bind to cellular receptors and neutralize cells so they cannot enter other cells. Antibodies can cause agglutination or clumping of cells that attract phagocytes (opsonization). Antibodies can produce inflammation by stimulating basophils and activate the complement system. The activation of the complement system by antibody-antigen complexes is known as the classical pathway. There are a number of different types of antibodies. Antibodies are also known as immunoglobulins and can be organized into five categories. These include IgE (immunoglobulin E), IgG, IgM, IgD, and IgA. IgG is the largest category and accounts for 80% of all antibodies. IgG antibodies attack viruses and bacteria. IgE antibodies function in allergic reactions. They facilitate the release of histamine and heparin from basophils. IgD antibodies bind to antigens on the surface of B-cells and help in activation of B-cells. IgM antibodies work with IgG antibodies to form immune complexes. IgM antibodies are also resident in plasma as anti-A and anti-B antibodies. IgA is found in secretions such as tears, mucus, and saliva, and attacks pathogens.

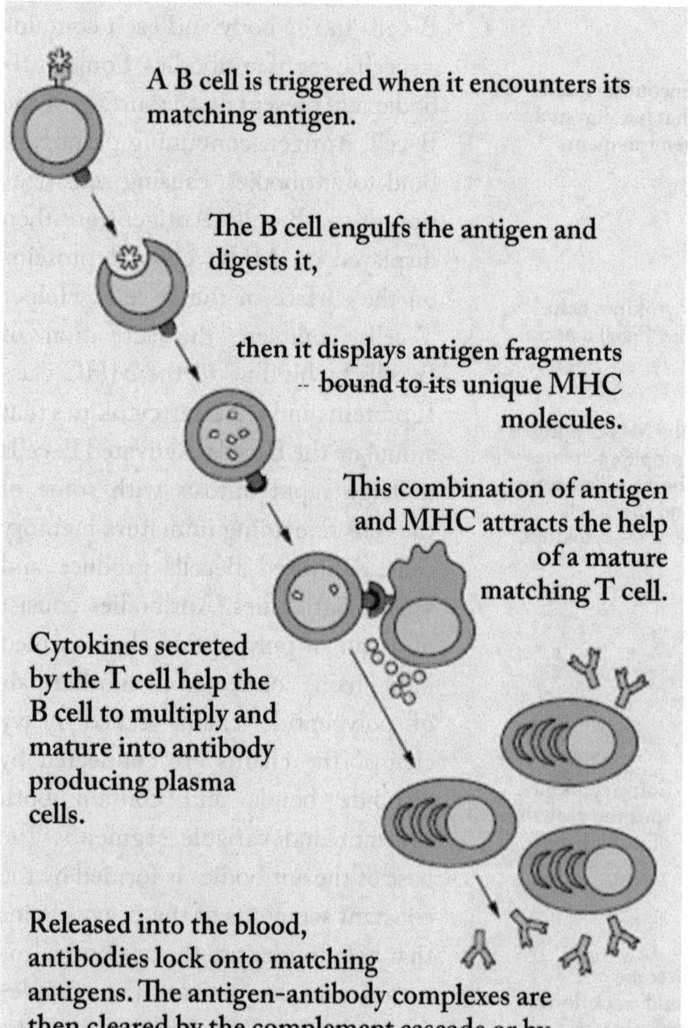

A B cell is triggered when it encounters its matching antigen.

The B cell engulfs the antigen and digests it,

then it displays antigen fragments bound to its unique MHC molecules.

This combination of antigen and MHC attracts the help of a mature matching T cell.

Cytokines secreted by the T cell help the B cell to multiply and mature into antibody producing plasma cells.

Released into the blood, antibodies lock onto matching antigens. The antigen-antibody complexes are then cleared by the complement cascade or by the liver and spleen.

Figure 15.10 B-cell activation.

Dendritic Cells

Dendritic cells are antigen-presenting cells found in mucous membranes, lymphatic organs, and the epidermis of the skin. These cells have a branched appearance and can engulf pathogens by way of endocytosis. Dendritic cells contain receptors that recognize non-self antigens that trigger endocytosis when activated.

Reticular Cells

Reticular cells (sometimes called fibroblastic reticular cells) are antigen-presenting cells located in lymphatic organs. These cells are known to help regulate T-cell function.

Macrophages

Macrophages develop from monocytes that have moved out of the blood.

They ingest pathogens by phagocytosis. They also clean up cellular debris, including dead neutrophils. Macrophages can display pathogenic antigens on their surface.

Primary and Secondary Immune Response

The first exposure to an antigen with activation of the immune system is known as the primary response. The immune system produces memory cells so that the next time the antigen is presented, the system is ready to respond. The second presentation of the same antigen to the immune system is known as the secondary response. The primary response takes longer to develop (from one to two weeks). During this time B-cells are producing antibodies, resulting in a gradual increase in antibodies. The immune system is also producing clones and memory cells that will be ready for the next presentation of the antigen. These memory cells can last for as long as 20 years. The secondary response is much faster. The second exposure to an antigen results in maturation of the memory cells and production of antibodies. Vaccinations rely primarily on the secondary response (Fig. 15.11).

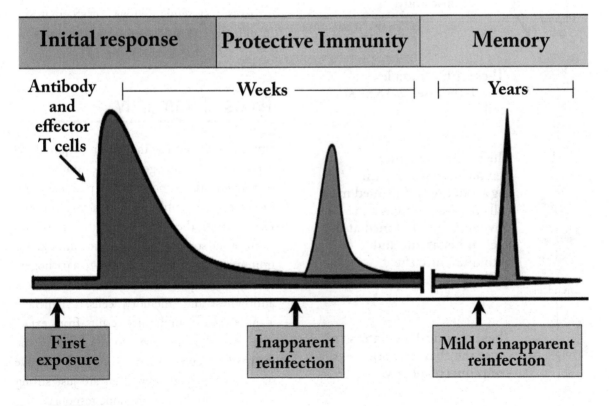

Figure 15.11 Immune response.

Allergies

Allergies are immune responses to non-pathological or inert substances. Allergic responses produce a large number of antibodies that can produce a number of adverse effects. Types of allergic responses are Type I anaphylactic, Type II cytotoxic reactions, Type III immune complex disorders, Type IV delayed hypersensitivity (Fig. 15.12). The type I reaction is also known as an immediate hypersensitivity reaction and can be life threatening. It is caused by an inherited tendency to overproduce the IgE antibodies in response to a specific antigen. The first exposure generally does not produce symptoms due to the time it takes for B-cell activation. However, the second exposure can be quite severe, producing large amounts of inflammation. The most severe reaction is known as anaphylaxis and produces a number of adverse effects within a very brief time period. These include hives, constriction of bronchioles, and peripheral vasodilation that can cause shock.

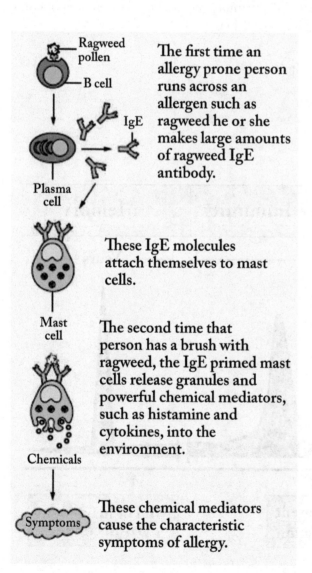

Figure 15.12 Allergic response.

Autoimmune Disorders

In some cases, the immune system reacts to self cells and tissues and produces an immune response to them. B-cells produce antibodies known as autoantibodies. Examples of autoimmune disorders include rheumatoid arthritis, systemic lupus erythematosus, insulin-dependent diabetes, and thyroidosis.

Types of Immunity

Immunity is either innate or acquired by exposure to a pathogen. Active immunity is developed after exposure to a pathogen and production of antibodies. Passive immunity results from the presentation of antibodies from other sources. Naturally acquired active immunity results from exposure to a pathogen. The immune system is activated and produces antibodies and memory cells. Artificially acquired active immunity results from exposure to pathogens given to the body in the form of vaccines. Vaccines contain inactive or attenuated pathogens that are just strong enough to produce an immune response.

Naturally acquired passive immunity occurs in utero with the passing of antibodies to the fetus from the mother. Antibodies are also passed to the infant through breast milk after birth. Artificially acquired passive immunity occurs when antibodies are given to a person who has a damaged immune system. Because of the short life span of antibodies, they must be injected periodically.

CRITICAL-THINKING QUESTIONS

1. Describe the flow of lymph from its origins in interstitial fluid to its emptying into the venous bloodstream.
2. Children born without a thymus gland must be kept in a germ-free environment if they are to survive. Explain why this is necessary.
3. In organ transplantation, the patient is often given an anti-lymphocyte serum to prevent rejection of the transplant. Explain why this serum prevents the rejection of the transplant but could also produce harmful side effects.
4. It is said: "No T cells, no immunity." Explain this statement.
5. Why is immune tolerance so important in the prevention of spontaneous abortions?
6. Why is it important that IgG can cross the placenta?
7. Which innate or adaptive defense would be used in the following situations?
    A. prevent debris from entering lower respiratory tract
    B. a cold (caused by a virus)
    C. keep mold from growing on skin
    D. rejection of transplanted organs
    E. removal of pus and debris

Section Five

Absorption and Excretion

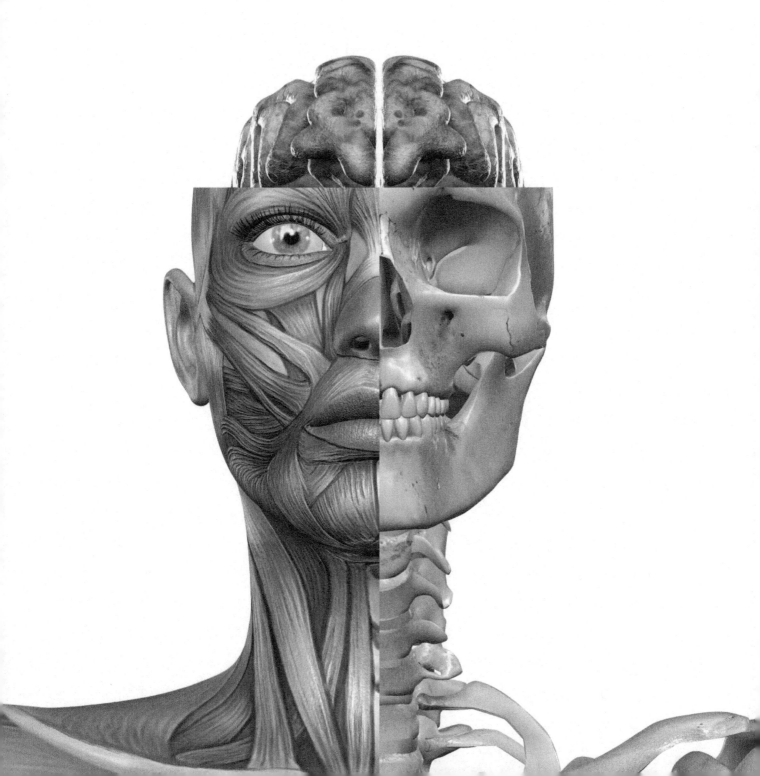

Chapter Sixteen

Respiratory System

The respiratory system supplies the body with oxygen and removes carbon dioxide. It consists of a series of structures that allow for the passage of air into the body and the exchanges of gases with the blood (Fig. 16.1). There are essentially two types of respiration. External respiration is the movement of gases into the body and blood. Cellular respiration is the use of oxygen and production of carbon dioxide by the cells. We will cover external respiration in the next two chapters. The respiratory system can be divided into the upper and lower respiratory systems. The upper respiratory system consists of the nose and nasal cavity, the sinuses, and the pharynx. The lower respiratory system consists of the larynx, trachea, bronchi, bronchioles, lungs, and alveoli. Air moves into the upper respiratory system through the nose at the nostrils or external nares and enters the nasal cavity. The nasal cavity is also known as the nasal vestibule and is lined with epithelium that contains hairs. The epithelium contains columnar and mucus-secreting goblet cells. Beneath the epithelium is a highly vascular area known as the lamina propria. The vascularization helps to provide heat and humidity to the air of the nasal cavity. The nasal cavity contains bony protuberances called conchae. There are superior, middle, and inferior conchae. The purpose of the conchae is to create a turbulent flow of air. This works to warm the air and to provide more contact with the nasal mucosa and hairs so that particles can be picked up by the mucosa. The turbulent air can also reach the upper nasal cavity, which contains sensory receptors for smell. The nasal cavity is divided into right and left portions by the nasal septum. The nasal septum is formed by two bones. The superior portion consists of the perpendicular plate of the ethmoid bone and the inferior portion consists of the vomer bone. The anterior portion of the nasal septum consists of cartilage. Located between the conchae are the superior, middle, and inferior meatuses, which are small grooves that allow air to flow between the nasal cavity, paranasal sinuses, and nasolacrimal ducts. The floor of the nasal cavity consists of the hard palate. The hard palate is formed by the maxilla (anterior) and palatine (posterior) bones. The hard palate separates the nasal and oral cavities. Just posterior to the hard palate is the soft palate and uvula. Air exits the nasal cavity to the nasopharynx by way of a passage known as the internal nares.

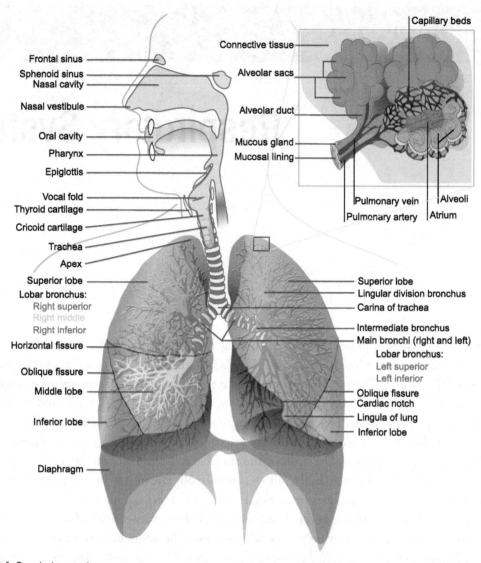

Figure 16.1. Respiratory system.

The Pharynx

Air passing through the internal nares enters the upper portion of the pharynx known as the nasopharynx. The nasopharynx begins posterior to the conchae and extends inferiorly to the soft palate. The soft palate raises to close off the nasopharynx during swallowing to prevent substances from moving into the nasopharynx. The nasopharynx is lined with ciliated pseudostratified columnar epithelium with goblet cells that secrete mucus. The cilia move substances through the nasopharynx so that they can be swallowed. The nasopharynx also contains connections from the Eustachian tubes. The pharyngeal tonsil (adenoid) is also located in the nasopharynx. Inferior to the nasopharynx is the oropharynx, which extends from the soft palate to the epiglottis. The oropharynx is a shared passageway for air and substances on their way to the digestive tract. The palatine and lingual tonsils are located in the oropharynx. The oropharynx is lined with stratified squamous epithelium. The most inferior portion of the pharynx is the laryngopharynx, which extends from the tip of the epiglottis to the larynx. The laryngopharynx is also a shared pathway with the digestive tract and is lined with stratified squamous epithelium.

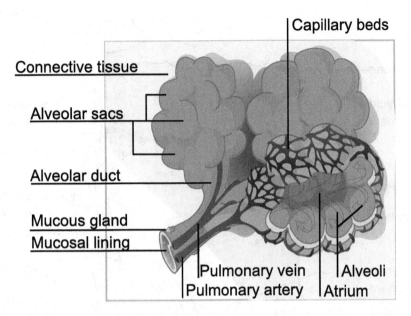

Figure 16.2. Alveolus.

The Larynx

The larynx begins at the base of the tongue and extends to the trachea (Fig. 16.2). The pharynx consists of nine cartilages that are interconnected with muscles and ligaments. The largest of the cartilages is the thyroid cartilage (Adam's apple). Inferior to the thyroid cartilage is the cricoid cartilage. The epiglottis is an elastic cartilage flap that closes during swallowing to keep substances from moving into the trachea and air passages. Other cartilages include the arytenoids, and corniculate and cuneiform cartilages. These cartilages are paired. The vocal cords reside in the larynx and consist of two pairs of ligaments that extend from the arytenoid to the thyroid cartilages. One set of ligaments (superior set) is called the false vocal cords. The inferior set is called the true vocal cords. When the vocal cords are relaxed, they form a triangular space called the glottis. The larynx is lined with pseudostratified columnar epithelium. The vocal cords are covered by a mucous membrane. Different pitches in the voice are produced by vibrations of the vocal cords. Vibration of smaller areas of the vocal cords results in higher pitches. Males typically have longer vocal cords than females have that result in lower pitches.

Trachea

Air travels from the larynx to the trachea. The trachea is a tubular structure consisting of dense connective tissue and rings of hyaline cartilage. The trachea is lined with ciliated pseudostratified columnar epithelium with goblet cells. The epithelium moves substances toward the larynx and esophagus for swallowing. The cartilage rings do not completely encircle the trachea but are open posteriorly. The posterior section of the trachea contains a ligament and smooth muscle known as the trachealis muscle.

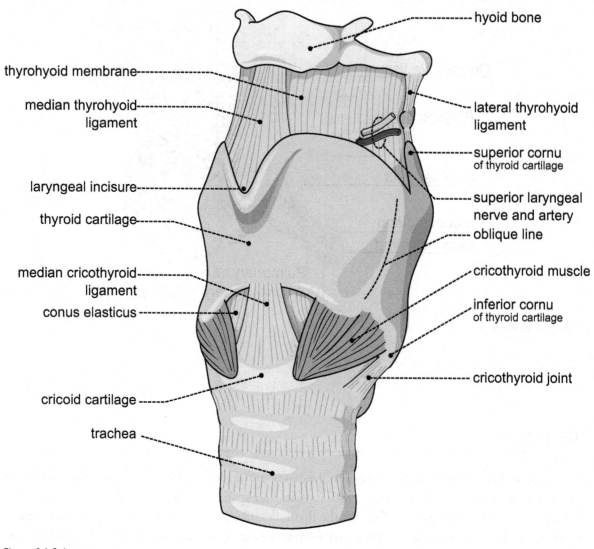

Figure 16.3. Larynx.

The trachealis muscle can contract and constrict the trachea. The trachea usually ends at about the level of the fifth thoracic segment. The inferior end of the trachea divides into right and left bronchi at an area known as the carina. The carina is the last tracheal cartilage and forms a cartilage division between the two bronchi.

Bronchial Tree

The trachea ends at the carina and divides into two tubular structures called the right and left primary bronchi. The bronchi then divide into smaller branches called secondary or lobar bronchi and then even smaller branches called tertiary or segmental bronchi. The structure of the bronchi is similar to the trachea with incomplete cartilage rings and smooth muscle. As the bronchi get smaller, there is less cartilage and more smooth muscle, until reaching the tertiary bronchi, which consist entirely of smooth muscle. The smooth muscle can constrict the bronchi and impede air passage. The bronchi continue to branch and form small bronchioles, which divide to form terminal bronchioles. The terminal bronchioles divide to form respiratory bronchioles that connect with alveolar ducts. The alveolar ducts give rise to alveoli. Alveoli are considered the functional unit of the lung and consist of small hollow areas for gas exchange. The alveolar ducts and alveoli are lined with simple squamous epithelium that allows for gas exchange. The cells of the simple squamous epithelium are called Type I pneumocytes. The alveoli also contain other cells known as type II pneumocytes. These cells secrete a substance known as surfactant that helps to decrease the surface tension in the alveoli. The lungs contain about 300 million alveoli.

The Lungs

The lungs are two cone-shaped structures residing in the thoracic cavity. The inferior portion of each lung reaches to the diaphragm. The superior portion extends about one inch above each clavicle. The right lung contains three lobes (superior, middle, and inferior) and is larger than the left lung, which contains two lobes (superior and inferior). The lobes are separated by fissures. The right lung includes a horizontal and oblique fissure, while the left lung contains only an oblique fissure. The medial surface of each lung contains an area known as the hilum where vessels enter and exit. The left lung also contains the cardiac notch, which is an indentation for the heart. The lungs are surrounded by two pleural membranes. The surface of each lung contains a visceral pleural membrane that closely adheres to the lung's surface. Lining the interior of the thoracic wall is the parietal pleural membrane. Both are serous membranes. A fluid known as pleural fluid is secreted by each membrane, which reduces friction and helps to hold the membranes together.

Inhalation and Exhalation

Inhalation and exhalation depends on changes in lung volume and air pressure. One cycle of inspiration and expiration is called a respiratory cycle. The movement of air into and out of the lungs is known as pulmonary ventilation. Air moves into the lungs and to the alveoli where oxygen and carbon dioxide diffuse between the alveoli and blood. It is important to maintain good airflow to the alveoli (alveolar ventilation) at all times. Air is a gas and gas moves via pressure gradients. Gas will move from areas of higher pressure to areas of lower pressure. Pressure in the lungs must be lower than atmospheric pressure for air to move into the lungs.

Boyle's Law
Boyle's law relates pressure and volume.
It can be represented by:
$P = 1/V$
P = pressure
V = volume

Molecules of a gas will move at random within an enclosed space producing pressure on the walls of the space. The same amount of gas in a smaller space will exert a greater pressure than when in a larger space. So increasing the volume will lower the pressure for a given temperature and vice versa.

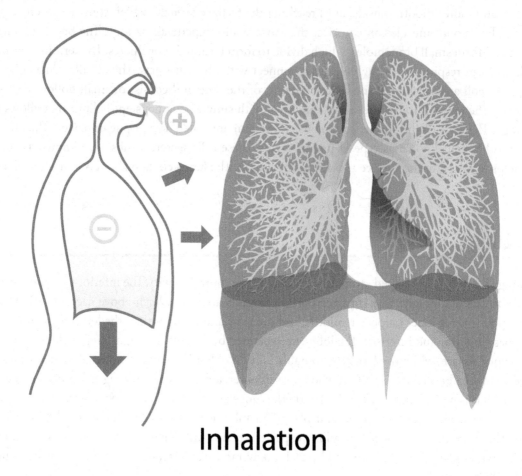

Figure 16.4. Inhalation.

This is just what happens during inhalation. The diaphragm contracts and pulls downward increasing the volume of the thoracic cavity. The external intercostals also contract and expand the ribcage. The increased volume decreases the pressure inside of the lungs and air flows from higher pressure outside the lungs to lower pressure inside the lungs (Fig. 16.4). Expansion of the thoracic cavity causes the lungs to expand because of the pleural cavity. The pleural membranes secrete a fluid that forms a bond between the membranes. The force of this bond produces a pressure that is about −4 mm Hg, or 4 mm Hg below atmospheric pressure. The lungs also contain elastic fibers. The elastic fibers create a force that opposes the force of the fluid bond between the pleural membranes. If the fluid bond did not exist the lungs would collapse to about 5% of their normal size.

Compliance

Compliance represents the lung's ability to expand. The more compliant the lung, the easier it will expand. As the lungs become less compliant, they require more force to expand. The tissue structure of the lungs, flexibility of the thoracic cage, and production of surfactant all affect compliance. For example, people

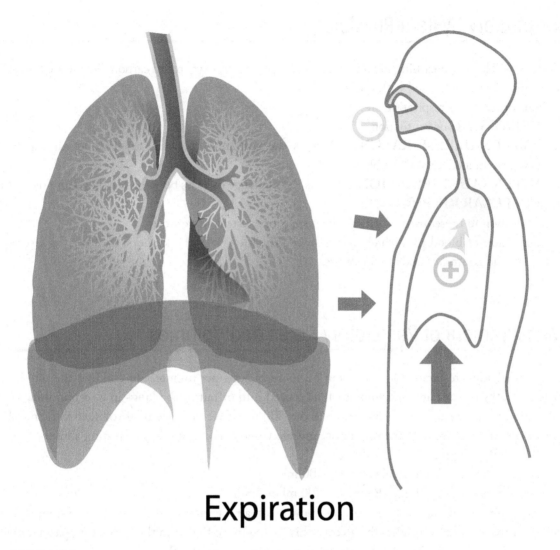

Expiration

Figure 16.5. Exhalation.

with emphysema have a decreased compliance. During exhalation the diaphragm relaxes, decreasing the volume of the thoracic cavity (Fig. 16.5). The elastic fibers of the lungs work to move the lungs back to their original shape and the pressure increases, moving air out of the lungs. Resting exhalation is considered a passive process.

Accessory Muscles of Respiration

Other muscles besides the diaphragm and external intercostals are involved in respiration when greater amounts of air need to be moved into the lungs. Muscles assisting in inhalation include the sternocleidomastoid, serratus anterior, pectoralis minor, and scalenes. Muscles assisting in exhalation include the internal intercostals, transverse thoracic, and abdominals.

Respiratory System Physiology

Next, we will look at breathing in more detail and explore how oxygen and carbon dioxide are transported in the blood and exchanged with the tissues.

Four processes of respiration are:
1. **BREATHING** or ventilation
2. **EXTERNAL RESPIRATION**, which is the exchange of gases (oxygen and carbon dioxide) between inhaled air and the blood.
3. **INTERNAL RESPIRATION**, which is the exchange of gases between the blood and tissue fluids.
4. **CELLULAR RESPIRATION**

In addition to these main processes, the respiratory system serves for:

Regulation of blood pH, which occurs in coordination with the kidneys, and as a defense against microbes, and control of body temperature.

Measurement of Respiratory Rates and Volumes

The normal adult respiratory rate is about 12 to 18 breaths per minute. For children the rate is about 18–20 breaths per minute. The volume of air moved into or out of the lungs in a resting inhalation or exhalation is known as tidal volume and is about 500 ml. If we multiply tidal volume by the number of breaths per minute, we have the respiratory minute volume. For example, in a resting adult:

RMV 5 Breaths/min 3 Tidal volume

If breaths/min 5 15 and tidal volume 5 500 ml

Respiratory minute volume (RMV) 5 7500 ml or 7.5 liters

The respiratory minute volume indicates how much air has entered the respiratory system. However, not all of the air inhaled reaches the alveoli. This is because of the air in the respiratory passages known as anatomic dead space. To calculate the amount of air reaching the alveoli:

Alveolar ventilation 5 breaths per minute 3 (Tidal volume 2 Anatomic dead space)

Anatomic dead space is usually 350 ml.

Example 1: Normal resting breathing

AV 5 12 3 (500 2 350)

AV 5 1800 ml or 1.8L

Example 2: Rapid shallow breathing

AV 5 20 3 (400 2 350)

AV 5 1000 ml or 1.0 L

Can you see that less air reaches the alveoli with rapid shallow breathing than in normal resting breathing?

Respiratory volumes can be measured with a device called a spirometer. Besides tidal volume other volumes can be measured, including inspiratory reserve volume and expiratory reserve volume. Inspiratory reserve volume (IRV) is the maximum amount of air that can be inhaled in addition to tidal volume. IRV is usually about 3300 ml in males and 1900 ml in females. Expiratory reserve volume (ERV) is the maximum amount of air that can be exhaled in addition to tidal volume. ERV is about 1000 ml. Residual volume (RV) is the amount of air remaining in the lungs after a maximal exhalation. RV is about 1200 ml in males and 1100 ml in females. Combining respiratory volumes gives us respiratory capacities.

These include vital capacity, inspiratory capacity, functional residual capacity, and total lung capacity. Vital capacity is the maximal amount of air that can move into and out of the lungs in a single breath. It is the sum of tidal volume, inspiratory reserve volume, and expiratory reserve volume. It is about 4800 ml in males and 3400 ml in females. Inspiratory capacity is the amount of air that can move into the lungs after resting inhalation and exhalation. Inspiratory capacity is the sum of tidal volume and inspiratory reserve volume. Functional residual capacity is the air remaining in the lungs after a resting inhalation and exhalation. Functional residual capacity is the sum of expiratory reserve volume and residual volume. Total lung capacity is the total volume of air in the lungs. It is the sum of vital capacity and residual volume. It is about 6000 ml in males and 4500 ml in females.

Gas Laws

Air moves into the lungs by means of changes in volume and pressure. Air is a combination of a number of gases. Air consists of nitrogen (78.6%), oxygen (20.9%), carbon dioxide (.04%), and a trace amount of other gases. Air produces an atmospheric pressure of 760 mm Hg and this pressure is produced by a combination of gases. Each gas produces a pressure that is proportional to its amount in the whole. This is known as Dalton's Law. The pressure each gas produces in the mixture of gases is known as the partial pressure of gas. We can represent the partial and total pressure of a gas, such as air, as follows:

P (nitrogen) + P (oxygen) + P (water vapor) + P (carbon dioxide) = P (air) = 760 mm Hg

For example, if oxygen produces 20.9% of the total pressure of air then 20.9% of 760 mm Hg is about 159 mm Hg. So the partial pressure of oxygen is 159 mm Hg. We can denote partial pressure as PO_2 or PCO_2. Partial pressure can be thought to be analogous to concentration. Henry's Law states that at a given temperature the amount of gas in a solution is directly proportional to the partial pressure of the gas. Gas, like other substances, follows a concentration gradient. We can say that gas follows a partial pressure gradient. For example, oxygen will move from a PO_2 of 100 mm Hg to a PO_2 of 80 mm Hg.

Gas Exchange

Air enters the respiratory tract and is warmed and humidified. It eventually reaches the alveoli and mixes with the air resident there. Thus alveolar air differs from atmospheric air. For example, alveolar air contains more carbon dioxide than atmospheric air. After reaching the alveoli, gases diffuse across the respiratory membrane and into the surrounding capillaries. The PO_2 of alveolar air is about 104 mm Hg and the PCO_2 is about 40 mm Hg. The PO_2 of deoxygenated blood is about 40 mm Hg and the PCO_2 is about 45 mm Hg. Oxygen and carbon dioxide both diffuse in opposite directions across the respiratory membrane. Oxygen diffuses from the alveolus to the blood (PO_2 of 104 mm Hg to PO_2 of 40 mm Hg) and carbon dioxide diffuses from the blood (PCO_2 of 45 mm Hg) to the alveolus (PCO_2 of 40 mm Hg). Other factors affecting the diffusion of gases include the solubility, the size of the concentration gradient, and the surface area and thickness of the respiratory membrane. The solubility of a gas in liquid is represented by the solubility coefficient. The solubility coefficient for oxygen is .024 and for carbon dioxide is .57. This means that carbon dioxide is much more soluble (or able to dissolve) in water than oxygen. Both oxygen and carbon dioxide are lipid soluble as well and can easily move across the respiratory membrane. Damage to the respiratory membrane tends to affect the diffusion of oxygen before affecting carbon dioxide due to the increased solubility of carbon dioxide. Internal oxygen levels can then decrease to dangerous levels. Giving supplemental oxygen helps to increase the concentration of oxygen and aid diffusion. The respiratory membrane's total area is about 70 square meters. Some diseases can adversely affect the respiratory membrane. These include emphysema and lung cancer. Emphysema creates large chambers within the lung that decrease the surface area of the respiratory membrane. Lung cancer produces tumors that decrease surface area as well. Partial pressure and the subsequent pressure gradient can change by increasing or decreasing alveolar ventilation. Breathing slowly and deeply lowers alveolar PCO_2 as more CO_2 exits the lungs with each breath. Oxygenated blood leaves the pulmonary circulation and enters the systemic circulation for distribution to the tissues. The PO_2 of oxygenated blood is 104 mm Hg and the PCO_2 is 40 mm Hg in the pulmonary circulation. The oxygenated blood mixes with blood from the bronchial veins, causing the PO_2 to decrease to 95 mm Hg. Blood leaving the pulmonary circulation and entering the systemic circulation has a PO_2 of 95 mm Hg. Oxygenated blood eventually reaches the tissues. The intracellular PO_2 is about 40 mm Hg and decreases to about 20 mm Hg in the cells. Oxygen then diffuses down its partial pressure gradient into the interstitium and cells. The blood is now deoxygenated with a PO_2 of 40 mm Hg. Carbon dioxide is produced in the cells as a byproduct of metabolism. Therefore the highest PCO_2 in the system is at the cells. The PCO_2 is about 46 mm Hg in the cells and about 45 mm Hg in the interstitium. The PCO_2 of oxygenated blood is about 40 mm Hg. Carbon dioxide diffuses from the interstitium to the blood. The resulting PCO_2 of deoxygenated blood leaving the tissues is then 45 mm Hg. Deoxygenated blood returns to the lungs where the alveolar PCO_2 is about 40 mm Hg. Carbon dioxide then diffuses to the alveoli and is expelled with each exhaled breath.

Carbon Dioxide Transport in the Blood

Carbon dioxide is transported in the blood by three mechanisms: carbon dioxide dissolved in plasma, carbon dioxide combining with hemoglobin, and storage of carbon dioxide in the bicarbonate ion. About 7% of the total carbon dioxide in blood is dissolved in plasma. Carbon dioxide also combines with hemoglobin to form a compound known as carbaminohemoglobin. About 23% of carbon dioxide is transported as carbaminohemoglobin. The majority of carbon dioxide (about 70%) is transported in the bicarbonate ion.

Carbon dioxide diffuses into red blood cells and encounters the enzyme carbonic anhydrase to form carbonic acid. Carbonic acid is an ionically bonded molecule that dissociates into bicarbonate and hydrogen ions. Bicarbonate ions diffuse out of the red blood cells into the plasma. In order to maintain ionic stability, chloride ions move into the red blood cell in what is called the chloride shift. The reaction is reversible with either the storage or release of carbon dioxide, depending on what is needed. For example, in areas of low PCO_2, such as in the alveoli, the reaction will work in the direction to release CO_2 for removal by the lungs. In areas of high PCO_2, such as in the tissues, the reaction will work in the direction to store CO_2 in the bicarbonate ion. The hydrogen ions will bind to hemoglobin or move into the blood plasma. Most of the hydrogen ions bind to hemoglobin, which acts as a buffer to help to maintain a narrow range of blood pH. Some hydrogen ions move into plasma, directly affecting the blood pH.

Respiratory Acidosis/Alkalosis

Because most of the carbon dioxide is transported by the bicarbonate ion with subsequent release of hydrogen ions, a buildup of carbon dioxide in the blood will produce a lower pH. This is known as respiratory acidosis and can result from the inability of the lungs to get rid of excess carbon dioxide, such as in diseases like emphysema or chronic bronchitis. You can generate a mild case of respiratory acidosis by simple holding your breath. The cells continue to produce carbon dioxide, but the lungs are not removing it through exhalation. Carbon dioxide builds up in the lungs, producing the hydrogen ion byproduct, and the blood begins to become acidic. Likewise you can produce a mild state of respiratory alkalosis by hyperventilating. In this case too much carbon dioxide is removed by the lungs and the hydrogen ion concentration subsequently decreases.

Oxygen Transport

Most of the oxygen transported in blood is bound to hemoglobin to form oxyhemoglobin. A small amount of oxygen is dissolved in plasma. Each hemoglobin molecule can bind with four oxygen molecules. Hemoglobin can also release oxygen to form deoxyhemoglobin. There are almost 300 million hemoglobin molecules in one red blood cell. The functional characteristics of hemoglobin are also variable and respond to changes in PO_2, pH, and temperature. The degree of oxygen binding to hemoglobin can be represented by what is known as an oxygen–hemoglobin saturation curve. If all of the hemoglobin is fully bound with oxygen molecules, then the saturation is 100%. By examining the saturation curve, we can see that in areas of lower PO_2, hemoglobin tends to release oxygen. In other words, we can say that hemoglobin decreases its affinity for oxygen binding. This makes sense because low areas of PO_2, such as in the tissues, require free oxygen to diffuse into the interstitium and supply the cells. For example, tissue PO_2 is about 40 mm Hg. At this PO_2, hemoglobin is about 75% saturated, which means that about 23% of the oxygen bound to hemoglobin was released. The remaining 75% of oxygen acts like a reserve in case PO_2 goes even lower. In the tissues, small changes in PO_2 can have a large affect on the release of oxygen. This helps to ensure that the tissues receive enough oxygen so they can function properly. Likewise in areas of higher PO_2, we see that hemoglobin is almost completely saturated (about 98%). This means that hemoglobin increases its affinity for oxygen binding. Again it makes sense that hemoglobin would bind with oxygen in areas such as the alveolar capillaries so that oxygen can be carried to the tissues. For example, alveolar PO_2 is about 104 mm Hg. By examining the curve, we can see that hemoglobin is 98% saturated. Even if PO_2 drops to 80 mm Hg, we see that hemoglobin is still 95.8% saturated. The functional characteristics of hemoglobin can change. Hemoglobin saturation is affected by changes in pH, temperature, and PCO_2. As pH decreases, more free hydrogen ions are bound to hemoglobin, changing its function and causing it to release oxygen more readily. In other words, we can say that in areas of lower pH, hemoglobin decreases its affinity for oxygen binding. Likewise areas of increased pH will increase hemoglobin's affinity for oxygen binding. This change in binding affinity for oxygen is known as the Bohr effect (Christian Bohr). Hemoglobin also changes with respect to carbon dioxide binding at the same time it changes for oxygen binding. As pH decreases, hemoglobin increases its affinity for carbon dioxide binding. Likewise in areas of increased pH, hemoglobin will decrease its affinity for carbon dioxide binding. These changes in hemoglobin function for carbon dioxide are known as the Haldane effect. An increase in PCO_2 has the same effect as a decrease in pH. Hemoglobin again decreases its affinity for oxygen binding. Release of hydrogen ions and bound carbon dioxide work to produce this effect. Increased temperatures also decrease hemoglobin's affinity for oxygen binding. Again hemoglobin works to release oxygen into the tissues during times of increased metabolism, such as with exercise or fever.

The changes in hemoglobin's function with decreases in pH, and increases in PCO_2 and temperature can be represented by a right shift in the saturation curve. Likewise in areas of higher pH, lower PCO_2 and decreased temperature hemoglobin resumes its normal function. We can say the saturation curve shifts to the left. For example, exercise will cause an increased metabolic demand in the tissues and subsequent need for oxygen supply and removal of carbon dioxide. PCO_2 increases in skeletal muscle tissue with a subsequent increase in hydrogen ion concentration, resulting in a lower pH. Byproducts of anaerobic metabolism such as lactic acid also work to decrease pH. The decrease in pH causes a right shift in the hemoglobin saturation curve. In other words, hemoglobin works to release oxygen in the tissues and pick up carbon dioxide more readily. At the lungs, however, pH is in the normal range. Hemoglobin then works to release carbon dioxide and bind with oxygen for transport to the tissues. The saturation curve then shifts to the right in skeletal muscle and then back to the left in the lungs.

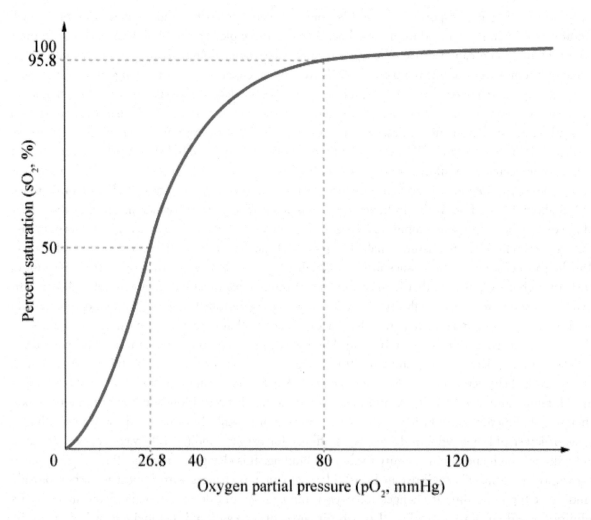

Figure 16.6. Oxygen-hemoglobin saturation curve.

Neural Control of Respiration

Neural control of respiration begins in the brainstem at the medulla oblongata and the pons. The medulla contains the medullary respiratory center. The medullary respiratory center consists of two groups of neurons called the dorsal and ventral respiratory groups. The dorsal respiratory group consists of two groups of neurons located in the posterior area of the medulla oblongata. This group is responsible primarily for contraction of the diaphragm for regulation of breathing rate. The neurons receive input from other parts of the brain and receptors that sense changes in concentrations of gases and pH.

The ventral respiratory group stimulates the external and internal intercostals and abdominal muscles. This group works to regulate breathing rhythm. The pons contains the pneumotaxic center (now called the pontine respiratory group). This center works with the centers in the medulla and helps to fine tune breathing rate and rhythm. The pneumotaxic center also receives input from other centers in the brain. The apneustic center also resides in the pons. The pneumotaxic center inhibits the apneustic center to help control exhalation. However, if damage to the brainstem occurs, the person can exhibit what is known as apneustic breathing. This consists of a very slow respiration rate with a deep inhalation held for ten to twenty seconds followed by shallow and brief exhalations that provide little pulmonary ventilation. All of the above respiratory centers innervate the phrenic and intercostals nerves.

For normal resting breathing, the following neural events occur. The dorsal respiratory group becomes active, causing contraction of the diaphragm and external intercostal muscles. Air moves into the lungs. The dorsal respiratory group is now inhibited, causing relaxation of the respiratory muscles and passive exhalation. For forceful breathing, both the dorsal and ventral respiratory groups are active, causing the respiratory muscles and accessory muscles to contract. Part of the ventral respiratory group that innervates the muscles of expiration is inhibited. Air moves into the lungs. The dorsal respiratory group is now inhibited, while the ventral respiratory group is activated. The muscles of inspiration relax, while the muscles of expiration contract. Air is expelled from the lungs.

Breathing is not entirely unconscious. We can decide to take in a deep breath or hold our breath. The cerebral cortex provides connections to the brainstem centers for breathing. The limbic system also affects breathing. For example, strong emotions elicited in the limbic system can speed up breathing. Chemoreceptors that sense changes in concentration of PO_2 and pH are involved in controlling respiration. These receptors, as well as the carotid and aortic bodies, are located in the medulla oblongata. The carotid bodies connect to the medulla via the glossopharyngeal nerve (CN IX). The aortic body connects to the medulla via the vagus nerve (CN X).

The medulla oblongata senses changes in pH by way of carbon dioxide diffusion. Increased blood levels of carbon dioxide result in an increased rate and depth of breathing. The respiratory centers are very sensitive to changes in PCO_2. Small increases in PCO_2 can cause large increases in respiratory rate. A greater-than-normal PCO_2 is called hypercapnia, while a lower-than-normal PCO_2 is called hypocapnia. Neural respiratory centers are also sensitive to changes in PO_2 but changes in PCO_2 account for the majority of respiratory regulation. If PO_2 levels decrease while PCO_2 levels remain normal, there will be a subsequent increase in respiration rate. A lower-than-normal PO_2 is called hypoxia. Small changes in PO_2 do not cause an appreciable stimulation of the respiratory centers. Another neural control mechanism is the Hering-Breuer reflex. This reflex is a protective mechanism and prevents over-inflation of the lungs. Stretch receptors on the walls of the bronchi and bronchioles send impulses to the vagus nerve to the medulla oblongata. The impulses inhibit the respiratory centers and produce exhalation.

Problems Associated with the Respiratory Tract and Breathing

The environment of the lung is very moist, which makes it a hospitable environment for bacteria. Many respiratory illnesses are the result of bacterial or viral infection of the lungs. Because we are constantly being exposed to harmful bacteria and viruses in our environment, our respiratory health can be adversely affected. There are a number of illnesses and diseases that can cause problems with breathing. Some are simple infections, and others are disorders that can be quite serious.

Carbon Monoxide Poisoning: caused when carbon monoxide binds to hemoglobin in place of oxygen. Carbon monoxide binds much tighter, without releasing, causing the hemoglobin to become unavailable to oxygen. The result can be fatal in a very short amount of time. Mild Symptoms: flu-like symptoms, dizziness, fatigue, headaches, nausea, and irregular breathing. Moderate Symptoms: chest pain, rapid heartbeat, difficulty thinking, blurred vision, shortness of breath, and unsteadiness. Severe Symptoms: seizures, palpitations, disorientation, irregular heartbeat, low blood pressure, coma, and death.

Pulmonary Embolism: blockage of the pulmonary artery (or one of its branches) by a blood clot, fat, air, or clumped tumor cells. By far the most common form of pulmonary embolism is a thromboembolism, which occurs when a blood clot, generally a venous thrombus, becomes dislodged from its site of formation and embolizes to the arterial blood supply of one of the lungs. Symptoms may include difficulty breathing, pain during breathing, and more rarely, circulatory instability and death. Treatment, usually, is with anticoagulant medication.

Upper Respiratory Tract Infections

The upper respiratory tract consists of our nasal cavities, pharynx, and larynx. Upper respiratory infections (URI) can spread from our nasal cavities to our sinuses, ears, and larynx. Sometimes a viral infection can lead to what is called a secondary bacterial infection. Strep throat is a primary bacterial infection and can lead to an upper respiratory infection that can be generalized or even systemic (affects the body as a whole). Antibiotics aren't used to treat viral infections, but are successful in treating most bacterial infections, including strep throat. The symptoms of strep throat can be a high fever, severe sore throat, white patches on a dark red throat, and stomachache.

Sinusitis

An infection of the cranial sinuses is called sinusitis. Only about 1–3% of URIs are accompanied by sinusitis. This sinus infection develops when nasal congestion blocks off the tiny openings that lead to the sinuses. Some symptoms are post nasal discharge, facial pain that worsens when bending forward, and sometimes even tooth pain can be a symptom. Successful treatment depends on restoring the proper drainage of the sinuses. Taking a hot shower or sleeping upright can be very helpful. Otherwise, using a spray decongestant or sometimes a prescribed antibiotic will be necessary.

Otitis Media

Otitis media in an infection of the middle ear. Even though the middle ear is not part of the respiratory tract, it is discussed here because it is often a complication seen in children who has a nasal infection. The infection can be spread by way of the auditory (Eustachian) tube that leads form the nasopharynx to the middle ear.

The main symptom is usually pain. Sometimes, though, vertigo, hearing loss, and dizziness may be present. Antibiotics can be prescribed and tubes are placed in the eardrum to prevent the buildup of pressure in the middle ear and the possibility of hearing loss.

Tonsillitis

Tonsillitis occurs when the tonsils become swollen and inflamed. The tonsils, located in the posterior wall of the nasopharynx, are often referred to as adenoids. If you suffer from tonsillitis frequently and breathing becomes difficult, they can be removed surgically in a procedure called a tonsillectomy.

Laryngitis

An infection of the larynx is called laryngitis. It is accompanied by hoarseness and being unable to speak in an audible voice. Usually, laryngitis disappears with treatment of the URI. Persistent hoarseness without a URI is a warning sign of cancer, and should be checked into by your physician.

Lower Respiratory Tract Disorders

Lower respiratory tract disorders include infections, restrictive pulmonary disorders, obstructive pulmonary disorders, and lung cancer.

Acute Bronchitis

An infection that is located in the primary and secondary bronchi is called bronchitis. Most of the time, it is preceded by a viral URI that led to a secondary bacterial infection. Usually, a nonproductive cough turns into a deep cough that will expectorate mucus and sometimes pus.

Pneumonia

A bacterial or viral infection in the lungs where the bronchi and the alveoli fill with a thick fluid. Usually it is preceded by influenza. Symptoms of pneumonia include high fever and chills, with headache and chest pain. Pneumonia can be located in several lobules of the lung and obviously, the more lobules involved, the more serious the infection. It can be caused by a bacterium that is usually held in check, but due to stress or reduced immunity, has gained the upper hand.

Restrictive Pulmonary Disorders

Pulmonary Fibrosis

Vital capacity is reduced in these types of disorders because the lungs have lost their elasticity. Inhaling particles such as sand, asbestos, coal dust, or fiberglass can lead to pulmonary fibrosis, a condition where fibrous tissue builds up in the lungs. This makes it so our lungs cannot inflate properly and are always tending toward deflation.

Asthma

Asthma is a respiratory disease of the bronchi and bronchioles. The symptoms include wheezing, shortness of breath, and sometimes a cough that will expel mucus. The airways are very sensitive to irritants, which can include pollen, dust, animal dander, and tobacco. Even being out in cold air can be an irritant. When exposed to an irritant, the smooth muscle in the bronchioles undergoes spasms. Most asthma patients have at least some degree of bronchial inflammation that reduces the diameter of the airways and contributes to the seriousness of the attack.

Respiratory Distress Syndrome

At birth the pressure needed to expand the lungs requires high inspiratory pressure. In the presence of normal surfactant levels, the lungs retain as much as 40% of the residual volume after the first breath and thereafter will only require far lower inspiratory pressures. In the case of deficiency of surfactant, the lungs will collapse between breaths; this makes the infant work hard and each breath is as hard as the first breath. If this goes on further, the pulmonary capillary membranes become more permeable, letting in fibrin-rich fluids between the alveolar spaces that in turn form a hyaline membrane. The hyaline membrane is a barrier to gas exchange, which then causes hypoxemia and carbon dioxide retention that in turn will further impair surfactant production.

Type two alveolar cells produce surfactant and do not develop until the 25th to the 28th week of gestation; in this, respiratory distress syndrome is one of the most common respiratory diseases in premature infants. Furthermore, surfactant deficiency and pulmonary immaturity together lead to alveolar collapse. Predisposing factors that contribute to poorly functioning type II alveolar cells in premature babies are found in males, white infants, infants of mothers with diabetes, precipitous deliveries, and caesarean section performed before the 38th week of gestation. Surfactant synthesis is influenced by hormones, ranging from insulin to cortisol. Insulin inhibits surfactant production, explaining why infants of mothers with diabetes type 1 are at risk of development of respiratory distress syndrome.

Cortisol can speed up maturation of type II cells and therefore production of surfactant. Finally, babies delivered by caesarean section are at greater risk of developing respiratory distress syndrome, because cortisol production is reduced in the absence of the stress that happens during vaginal delivery. Cortisol increases in times of high stress and helps in the maturation of type II cells of the alveoli that cause surfactant.

Sleep Apnea

Sleep apnea or sleep apnoea is a sleep disorder characterized by pauses in breathing during sleep. These episodes, called apneas (literally, "without breath"), each last long enough so one or more breaths are missed, and occur repeatedly throughout sleep. The standard definition of any apneic event includes a minimum 10-second interval between breaths, with either a neurological arousal (3-second or greater shift in EEG frequency, measured at C3, C4, O1, or O2), or a blood oxygen desaturation of 3–4 percent or greater, or both arousal and desaturation. Sleep apnea is diagnosed with an overnight sleep test called polysomnogram. One method of treating central sleep apnea is with a special kind of CPAP, APAP, or VPAP machine with a Spontaneous Time (ST) feature. This machine forces the wearer to breathe a constant number of breaths per minute. CPAP, or continuous positive airway pressure, uses a controlled air compressor to generate an airstream at a constant pressure. This pressure is prescribed by the patient's physician, based on an overnight test or titration.

Nutrition for COPD (Chronic Obstructive Pulmonary Disease) Patients

Nutrition is particularly important for ventilator-dependent patients. When metabolizing macronutrients, carbon dioxide and water are produced. The respiratory quotient (RQ) is a ratio of produced carbon dioxide to amount consumed. Carbohydrate metabolism produces the most amount of carbon dioxide, so it has the highest RQ. Fats produce the least amount of carbon dioxide. Protein has a slightly higher RQ ratio. It is recommended that patients with COPD not exceed a 1.0 RQ. Lowering carbohydrates and supplementing fat or protein in the diet might not result in maintaining the desired outcome because excess amounts of fat or protein may also result in a respiratory quotient higher than 1.0.

Cystic Fibrosis

This disease is most common in Caucasians and will happen to 1 in every 2,500 people. It is most known for its effects on the respiratory tract although it does affect other systems as well. The respiratory passages become clogged with a thick mucus that is difficult to expel even with vigorous coughing. Breathing becomes difficult and affected individuals run the risk of choking to death on their own secretions unless strenuous effort is made to clear the lungs multiple times every day. Victims frequently will die in their 20s of pneumonia. All of us secrete mucus by certain cells in the epithelium that line the respiratory passage ways. In normal cases the cells also secrete a watery fluid that will dilute the mucus, making it easier to pass through the airways. In cystic fibrosis that secretion of watery fluid is impaired. This makes the mucus thicker and difficult to clear from the passageways. A recent discovery found that cystic fibrosis is caused by a defect in a type of chloride protein found in apical membranes of epithelial cells in the respiratory system and elsewhere. This defect directly impedes chlorine ion transport, which will then indirectly affect the transport of potassium ions. This causes the epithelium to not create its osmotic gradient necessary for water secretion. It has been known for a long time that cystic fibrosis is caused by a recessive gene inheritance. This gene codes for a portion of the chloride channel protein, which can malfunction in a variety of ways, each with specific treatment required.

CRITICAL-THINKING QUESTIONS

1. If a person sustains an injury to the epiglottis, what would be the physiological result?
2. Compare and contrast the conducting and respiratory zones.
3. Why are the pleurae not damaged during normal breathing?
4. Describe what is meant by the term "lung compliance."
5. What is respiratory rate and how is it controlled?
6. A smoker develops damage to several alveoli that then can no longer function. How does this affect gas exchange?
7. Describe the relationship between the partial pressure of oxygen and the binding of oxygen to hemoglobin.
8. Describe three ways in which carbon dioxide can be transported.
9. Describe the neural factors involved in increasing ventilation during exercise.
10. What is the major mechanism that results in acclimatization?

Chapter Seventeen

Digestive System

Which organ is the most important organ in the body? Most people would say the heart or the brain, completely overlooking the gastrointestinal tract (GI tract). Though definitely not the most attractive organs in the body, they are certainly among the most important. The 30+-foot-long tube that goes from the mouth to the anus is responsible for the many different body functions that will be reviewed in this chapter. The GI tract is imperative for our well-being and our life-long health. A non-functioning or poorly functioning GI tract can be the source of many chronic health problems that can interfere with your quality of life. In many instances the death of a person begins in the intestines. The old saying "you are what you eat" perhaps would be more accurate if worded "you are what you absorb and digest."

The gastrointestinal system is responsible for the breakdown and absorption of various foods and liquids needed to sustain life. Many different organs have essential roles in the digestion of food, from the mechanical disrupting by the teeth to the creation of bile (an emulsifier) by the liver. Bile production of the liver plays an important role in digestion, from being stored and concentrated in the gallbladder during fasting stages to being discharged to the small intestine. In order to understand the interactions of the different components, we shall follow the food on its journey through the human body. During digestion, two main processes occur at the same time:

<u>Mechanical Digestion</u>: larger pieces of food get broken down into smaller pieces while being prepared for chemical digestion. Mechanical digestion starts in the mouth and continues into the stomach.

<u>Chemical Digestion</u>: starts in the mouth and continues into the intestines. Several different enzymes break down macromolecules into smaller molecules that can be absorbed. The GI tract starts with the mouth and proceeds to the esophagus, stomach, small intestine (duodenum, jejunum, ileum), and then to the large intestine (colon), rectum, and terminates at the anus. You could probably say the human body is just like a big donut. The GI tract is the donut hole. We will also be discussing the pancreas and liver, and accessory organs of the gastrointestinal system that contribute materials to the small intestine.

The Alimentary Canal

The digestive system can be thought of as a long tube extending from the mouth to the anus with some accessory organs attached. The alimentary canal consists of the mouth, esophagus, stomach, duodenum, jejunum, ileum, cecum, colon, rectum, and anus. The accessory organs include the tongue, teeth, salivary glands, pancreas, liver, and gallbladder (Fig. 17.1).

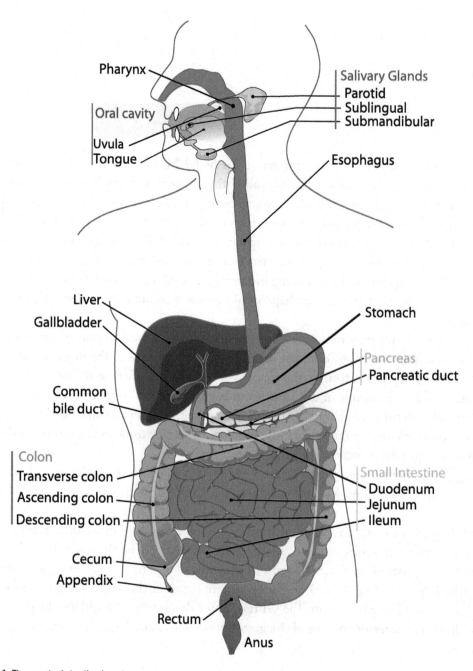

Figure 17.1. The gastrointestinal system.

There are some important tissue similarities that are consistent throughout the alimentary canal. The canal contains four layers of tissue. The outer portion consists of a serous membrane called the serosa. This layer secretes a slimy serous fluid that helps reduce friction so the organs can slide over one another. Deep to this layer is a muscle layer called the muscularis layer. This consists of two layers of smooth muscle, the longitudinal muscle and bands of circular muscle. There is also skeletal muscle under voluntary control in the mouth, pharynx, and portions of the upper esophagus. The next layer is known as the submucosa. The submucosa consists of connective tissue that helps to hold the mucosa and muscularis layers together. It also contains blood and lymphatic vessels and nerves. Deep to the muscularis layer is a mucous membrane called the mucosa. The mucosa consists of three layers: an epithelial layer, a layer of connective tissue called the lamina propria, and a smooth muscle layer. The epithelial layer of the mucosa consists of stratified squamous in the mouth, pharynx, esophagus, and anal canal and columnar epithelium in the stomach and intestines. The thin smooth muscle layer of the mucosa helps to produce folds in the membrane to increase the surface area. Two membranes surround the digestive system. We already mentioned the serosal layer and this layer in the abdominal cavity is known as the visceral peritoneum. There is also a membrane, known as the parietal peritoneum, lining the abdominal cavity. The peritoneum consists of simple squamous epithelium and connective tissue. The peritoneum also creates folds of tissue that extend between some of the organs. The greater omentum is one such fold that lies over the transverse colon and small intestines. The greater omentum also attaches to the greater curvature of the stomach. It contains blood vessels, lymphatic vessels, nerves, and adipose tissue. The lesser omentum is much smaller and connects to the lesser curvature of the stomach. It has the same structure as the greater omentum. The falciform ligament extends from the liver to the anterior abdominal wall. The mesentery connects the small intestine to the posterior abdominal wall. The mesocolon connects the large intestine to the posterior abdominal wall as well.

The Enteric Nervous System

The digestive system contains a complex system of nerves called the enteric nervous system. The nerves form plexi that reside in the digestive tract walls and connect with higher nervous system centers in the central and autonomic nervous systems. Many of the actions of the enteric nervous system occur within the system. However, both sympathetic and parasympathetic divisions of the autonomic nervous system also have an effect on digestion. The neurons in the enteric nervous system include sensory neurons that sense changes in chemical concentration and mechanical deformation of the tract. There are also motor neurons that help to control smooth muscle contraction and glandular secretions. Lastly, interneurons located in the enteric nervous system interconnect other neurons. The enteric nervous system provides a good deal of control over the GI tract without influence from other parts of the nervous system. It produces reflex actions, controls secretions, and smooth muscle contraction as well as blood flow.

Mouth

The mouth or oral cavity is bordered by the lips anteriorly, cheeks laterally, tongue inferiorly, and the hard and soft palate superiorly. The mouth is lined with a mucous membrane. The lips have an outer membrane consisting of skin and an inner mucous membrane. The upper lip contains a groove at the midline known as the philtrum. The cheeks also contain an outer skin and inner mucous membrane. The buccinator muscle lies between the membranes. The hard palate consists of the two palatine bones and the palatine processes of the maxilla. The soft palate is a muscular structure that forms two arches with the uvula in the midline. The arches form openings to the oropharynx called the fauces. The tongue consists of skeletal muscles covered by a mucous membrane. The tongue is connected externally by a series of muscles including the genioglossus and hyoglossus. Parts of the tongue include a root, tip, and body. The superior surface of the tongue contains papillae. There are several types of papillae including vallate, fungiform, circumvallate, and filiform. The fungiform papillae contain taste buds. A section of mucous membrane connecting the tongue to the floor of the mouth is known as the lingual frenulum.

Salivary Glands

There are three pairs of salivary glands. These are the parotid, submandibular, and sublingual glands. The salivary glands produce about 1L of saliva per day. In addition to the large glands, small glands on the insides of the cheeks called buccal glands also secrete saliva (about 5% of the total volume per day), which helps to keep the mouth moist. The salivary glands are considered exocrine glands, which secrete their substances into tubes. The parotid glands are the largest salivary glands. They are also somewhat superficial and lie between the skin and masseter muscles just anterior and inferior to the ear. These glands secrete a serous fluid containing enzymes. The secretions travel by way of ducts (parotid or Stenson ducts) that pierce the buccinator muscles and empty into the oral cavity. The submandibular glands are located just inferior to the angle of the mandible. They generate both serous and mucous secretions and are known as compound glands. The ducts of these glands (known as Wharton ducts) open into the floor of the mouth near the frenulum. The sublingual glands are located just under the mucous membrane of the floor of the mouth. The sublingual glands are the smallest salivary glands. They contain a series of up to 20 ducts (Rivinus ducts) that drain into floor of the mouth. These glands secrete only mucus.

The Teeth

A typical tooth consists of a crown, neck, and a root. The crown is the visible portion. The root is a cone-shaped process that lies below the gumline and forms a joint with the alveolar process of the mandible or maxilla. The neck is the portion surrounded by the gums. The crown is covered by enamel, which is a hard substance that protects the teeth. Deep to the enamel is the dentin. In the crown, dentin is covered by enamel. In the root, the dentin is covered by cementum. The deepest portion of the tooth consists of a pulp cavity and root canal that contain blood vessels and nerves (Fig. 17.2). The baby teeth are also known as deciduous teeth. There are 20 of these, which are gradually replaced by the permanent teeth, which number 32.

The deciduous teeth:
4 central incisors
4 lateral incisors
4 canines
4 first molars
4 second molars

The secondary teeth:
4 central incisors
4 lateral incisors
4 canines
4 first premolars (bicuspids)
4 second premolars (bicuspids)
4 first molars
4 second molars
4 wisdom teeth

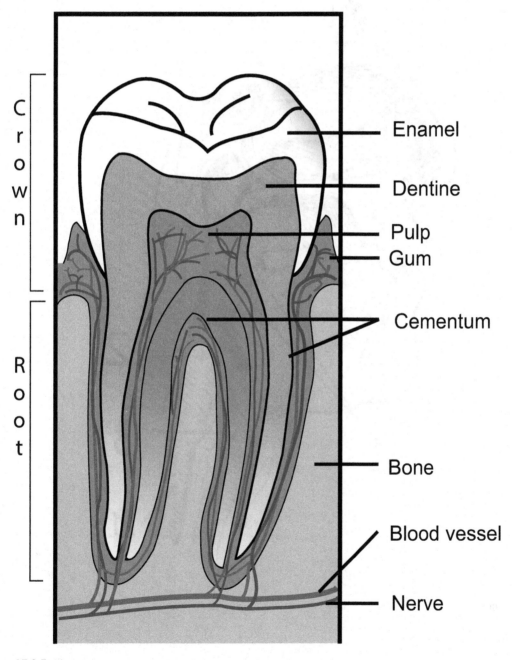

Figure 17.2. Tooth.

The Pharynx

The pharynx is a shared passageway for the respiratory and digestive systems. Food is chewed up and rolled in what is known as a bolus and pushed to the back of the mouth where it enters the pharynx for swallowing (Fig. 17.3). You may wish to review the anatomy of the pharynx in the respiratory anatomy chapter.

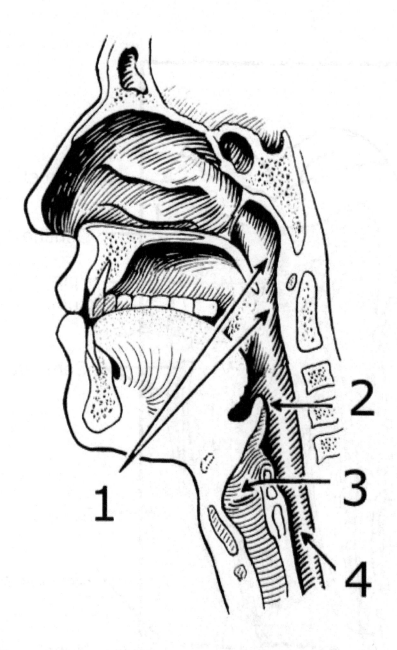

Figure 17.3 Pharynx.
1. Pharynx
2. Epiglottis
3. Larynx
4. Esophagus

The Esophagus

The esophagus is a muscular tube extending from the pharynx to the stomach (Fig. 17.4). It lies posterior to the trachea. The upper portion of the esophagus contains voluntary skeletal muscle. The middle and lower sections contain involuntary smooth muscle. The esophagus has two circular sphincter muscles. The upper esophageal sphincter keeps the esophagus closed during breathing to keep air from moving into the digestive tract. The lower esophageal sphincter (cardiac sphincter) is located at the inferior end of the esophagus where it pierces the diaphragm at the esophageal hiatus. The lower esophageal sphincter remains closed until swallowing occurs. In some cases the diaphragm is weakened near the hiatus and the sphincter enlarges. This is known as a hiatal hernia and can allow contents of the stomach to enter the esophagus and cause gastric reflux.

The Stomach

The esophagus empties into the stomach, which is a curved pouch-like organ. There are four major divisions of the stomach: the cardiac region, body, fundus, and pylorus. The cardiac region is the superior portion just after the esophagus. The fundus is an upward bulge that is located on the left side. The body is the central portion and the pylorus the inferior portion (Figs. 17.5, 17.6).

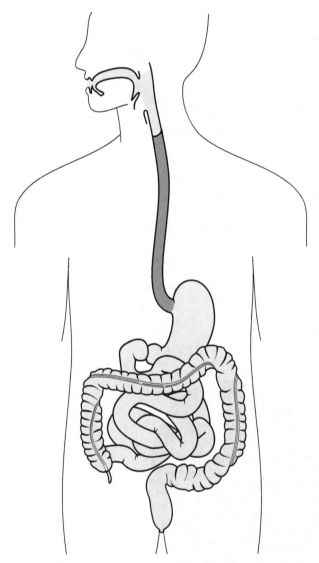

Figure 17.4. Esophagus.

The outer portion of the stomach contains a concave and convex curve. The concave curve is called the lesser curvature, while the convex curve is the greater curvature. The stomach contains two sphincters, one at each opening. The lower esophageal (cardiac) sphincter allows substances to enter while the pyloric sphincter allows substances to exit. The stomach has three smooth muscle layers. The outer layer runs longitudinally across the stomach. The middle layer is a circular layer that produces constriction of the stomach. The internal layer is an oblique muscle layer. The stomach is lined with simple columnar epithelium. The epithelium contains tube-like openings called gastric pits. Secretions from gastric glands flow through the pits to the inside of the stomach. There are also a good deal of mucus-secreting cells that secrete and alkaline mucus to help to protect the lining. The inner membrane creates folds called rugae. The rugae increase the surface area and help in mixing the contents of the stomach. The gastric glands consist of mucus-secreting cells, parietal cells, and chief cells. Parietal cells secrete hydrochloric acid and intrinsic factor. Chief cells secrete a precursor enzyme called pepsinogen. Pepsinogen combines

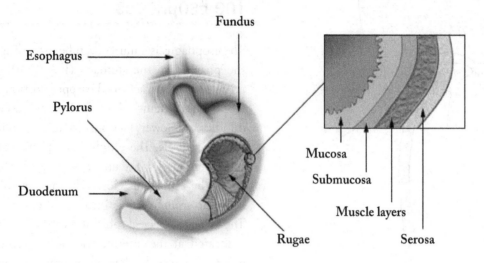

Figure 17.5. Stomach.

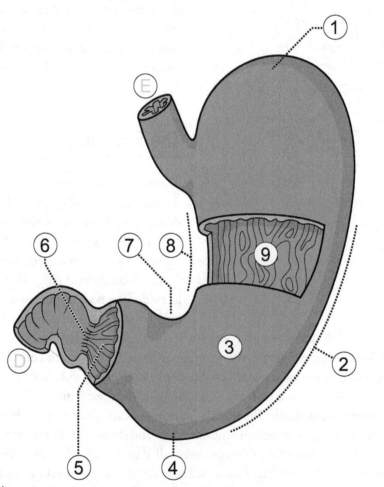

Figure 17.6. Stomach.
1. fundus; 2. greater curvature; 3. Body; 4. inferior aspect; 5. pyloric antrum; 6. pyloric canal; 7. angular notch; 8. lesser curvature; 9. rugal folds; E. esophagus; D. duodenum (bulbus and part of descending part)

with hydrochloric acid to become an active form known as pepsin. Pepsin digests proteins. There are also endocrine cells that secrete hormones that help to control digestion. An example is intrinsic factor that combines with vitamin B12 to help in absorption. Parietal cells contain a proton pump that helps to produce hydrochloric acid. Carbon dioxide and water enter the parietal cell and combine with the enzyme carbonic anhydrase to form carbonic acid. The carbonic acid dissociates into bicarbonate and hydrogen ions. The hydrogen ions are actively transported out of the cell and into the stomach by a transport protein (proton pump). Bicarbonate ions flow down their concentration gradient and are exchanged for chloride ions. The parietal cells are capable of creating a very acidic pH of 2.0 within the stomach. The combination of all of the stomach secretions is known as gastric juice. Food enters the stomach and combines with gastric juice to form a pasty substance called chyme. Chyme then leaves the stomach by way of the pyloric sphincter and enters the duodenum.

Stomach Movements

The stomach can mix substances and move them through. Both of these movements result from smooth muscle contractions. Weaker peristaltic waves help to mix the stomach contents, while stronger waves move the contents toward the pylorus. Emptying of the stomach is controlled so that it occurs at a rate that allows for adequate digestion of substances. Vomiting is a reverse peristaltic action of the stomach. The vomiting center in the medulla oblongata is sensitive to stimuli such as toxins and rapid body movements. The vomiting center also receives cortical input so certain thoughts can cause vomiting.

The Small Intestine

The small intestine consists of three parts. The proximal section is the duodenum, which is followed by the jejunum and the ileum. The duodenum begins at the pylorus and extends about 10 inches. It becomes the jejunum at its distal curve. The jejunum extends for about 8 feet before gradually becoming the ileum. There is no anatomical separation between jejunum and ileum. The small intestine is built for absorption with a large surface area. The inside of the small intestine consists of circular folds called plicae circulares. The plicae also contain numerous finger-like projections known as villi. The villi contain blood vessels and a lymphatic system tubule called a lacteal. The intestine is lined with cilia containing epithelium. The epithelial membrane resembles a brush and is sometimes referred to as a brush border. Enzymes and mucus are secreted by the cells lining the intestine. The membrane also contains intestinal crypts (crypts of Lieberkuhn) that are areas of rapid mitosis. The crypts help the intestinal membrane to renew itself as old cells are pushed out of the villi as they are replaced by new cells.

The Large Intestine

The large intestine begins at a pouch called the cecum. The junction between the ileum and cecum occurs at a smooth muscle sphincter in the cecum known as the ileocecal valve or sphincter. The diameter of the large intestine (2.5 inches) is much larger than the small intestine (1 inch). The cecum contains a

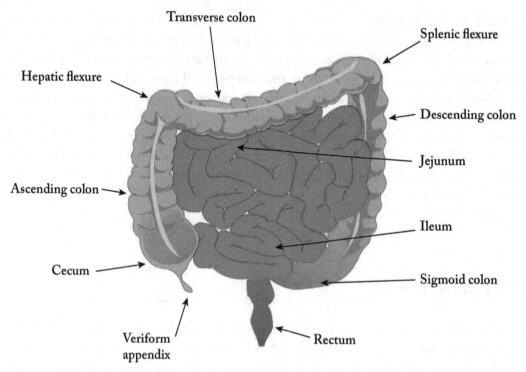

Figure 17.7. Intestines and colon.

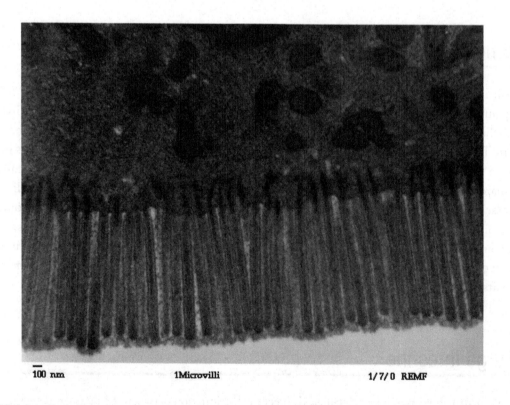

Figure 17.8. Intestinal villi.

finger-like projection (8–10cm) called the vermiform appendix. The function of the appendix is not known, but it may be an area for breeding intestinal bacteria (intestinal flora). Extending vertically from the cecum is the first segment of the colon, known as the ascending colon. The ascending colon takes a left turn at the liver (hepatic flexure) and continues horizontally as the transverse colon. The transverse colon takes a downward turn at the spleen (splenic flexure) and continues as the descending colon. As the descending colon extends beyond the iliac crest it becomes the sigmoid colon, which is S-shaped. The sigmoid colon becomes the rectum, which runs about 7–8 inches long. The last inch or so of the rectum is known as the anal canal. The anal canal contains vertical folds called anal columns. The anal columns contain blood vessels. The rectum contains two sphincter muscles; an internal and an external sphincter. The internal sphincter consists of smooth muscle, while the external sphincter is striated muscle. The opening of the anal canal is called the anus. The large intestine contains numerous mucus-secreting glands. Along the outside of the colon are bands of smooth muscles called taenia coli that run longitudinally. There are also rings of smooth muscle that divide the colon into pouch-like structures called haustra.

The Liver

The liver is located in the right upper quadrant of the abdominal cavity close to the diaphragm. The liver consists of four lobes: right, left, quadrate, and caudate. The right and left lobes are separated by the falciform ligament. The lobes are further divided into lobules by blood vessels and connective tissue. Tributaries of the hepatic vein extend into each lobule. Hepatic plates radiate outward from the central region of the lobules. Bile ducts and interlobular arteries are located on the outer regions of the plates. Smaller bile vessels called canaliculi permeate the plates and collect bile from the hepatic cells. Sinusoids containing white blood cells called Kupffer cells are located between the plates. These cells help phagocytize bacteria and debris. The small bile ducts merge into one large duct known as the hepatic duct. The hepatic duct merges with the cystic duct emerging from the gallbladder to form the common bile duct. The common bile duct carries bile to the duodenum. The common bile duct merges with the pancreatic duct just before entering the duodenum. The liver performs many functions and is considered a vital organ. Its functions include detoxifying the blood; producing bile; metabolism of carbohydrates, fats, and proteins; storing iron, blood, and vitamins; recycling red blood cells; and producing plasma proteins. Bile is secreted by the liver and stored in the gallbladder. Bile contains bile salts that are formed from cholesterol. Bile works to break down fat by emulsification and eliminates products from the breakdown of red blood cells. The gallbladder contains an outer serous membrane as well as a smooth muscle layer and inner mucous membrane. The inside of the gallbladder contains rugae much like the stomach. The gallbladder is about 3–4 inches long. In some cases bile can precipitate and form gallstones. The gallbladder can become inflamed in a condition known as cholecystitis.

The Pancreas

The pancreas has a dual endocrine and exocrine role. We investigated the endocrine role in the endocrine system chapter. The exocrine portion consists of compound acinar glands. These are branching duct structures containing clusters of cells that secrete substances into the ducts. The smaller ducts merge

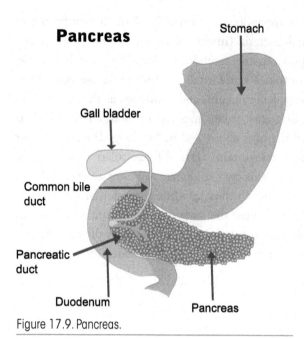

Figure 17.9. Pancreas.

with the larger pancreatic duct. The pancreatic duct merges with the common bile duct at an area in the duodenum known as the hepatopancreatic ampulla. The hepatopancreatic ampulla is encircled by smooth muscle that forms the hepatopancreatic sphincter. The exocrine glands secrete digestive enzymes. The endocrine cells are called alpha and beta cells. The alpha cells secrete glucagon and the beta cells secrete insulin. The pancreas consists of a body, a head, and a tail. It is located in the curve of the duodenum.

Digestive System Physiology

In order to supply the body with the substances it needs, the digestive system must perform a number of processes. These include breaking the substances down by mechanical and chemical digestion, absorption, assimilation, and elimination of waste products. Mechanical digestion begins in the mouth with chewing, or mastication. The teeth physically break down food and the mandible moves to provide the necessary force. The muscles of mastication include the masseter and temporalis muscles. The front teeth, known as the incisors, work to tear the food, while the molars work to grind it into smaller pieces. The final goal of mastication is to increase the surface area of the food so that the digestive enzymes can continue to break it down. Before swallowing food, the mouth forms it into an oval bolus. The bolus is pushed to the pharynx by the tongue. Chemical digestion also begins in the mouth. The parotid and submandibular salivary glands contain an enzyme called salivary amylase. This enzyme begins carbohydrate digestion. Salivary amylase essentially breaks complex sugars(polysaccharides) into disaccharides such as maltose and isomaltose. The structure of the mouth does not allow for much absorption, however, some chemicals can be absorbed. These include nitroglycerin and some vitamins.

Swallowing

There are essentially three phases to swallowing (deglutition). These are a voluntary phase followed by pharyngeal and esophageal phases. The first phase consists of pushing the bolus of food toward the back of the mouth to the pharynx. During the second phase, or pharyngeal phase, the uvula of the soft palate raises to block the nasopharynx and a series of pharyngeal muscles, known as the superior pharyngeal constrictor muscles, contract. The action of the constrictor muscles pushes the bolus toward the esophagus. The epiglottis closes to block the passageway to the larynx and the larynx is elevated. One can view the elevation of the larynx during swallowing by observing the thyroid cartilage elevate during swallowing. During the last phase, or esophageal phase, the smooth muscles of the esophagus create a peristaltic wave that pushes the food to the stomach. The lower esophageal sphincter relaxes to allow food to pass into the stomach.

Digestion in the Stomach

Once food reaches the stomach, it mixes with the secretions of the stomach. The stomach contains cells that produce secretions that continue the chemical digestion of food. The parietal cells secrete hydrochloric acid and intrinsic factor, and chief cells secrete pepsinogen. Mucus-secreting cells that secrete an alkaline mucus help protect the stomach lining. G-cells secrete gastrin and enterochromaffin cells secrete histamine. Pepsinogen combines with hydrochloric acid to produce pepsin. Pepsin is an enzyme that digests proteins by breaking peptide bonds. Intrinsic factor combines with vitamin B12 so that it can be absorbed. Gastrin and histamine help to regulate secretions. The combination of all of the gastric secretions is known as gastric juice. Gastric juice combines with food to produce a pasty substance called chyme. There are three phases of gastric secretions. These are the cephalic, gastric, and gastrointestinal phases. The cephalic phase is a preparatory phase for the arrival of food. Thoughts and smells can initiate the cephalic phase. During this phase, the parasympathetic nervous system sends impulses to the stomach by way of the vagus nerve. The postganglionic parasympathetic neurons secrete acetylcholine that stimulates the production of stomach secretions. Gastrin is secreted in response to parasympathetic stimulation and promotes the release of more hydrochloric acid. Gastrin also causes the release of histamine, which also promotes the release of hydrochloric acid. The action of the parasympathetics' causing secretion of hydrochloric acid and gastrin, which in turn causes more release of hydrochloric acid, constitutes a positive feedback mechanism. This is one of the few positive feedback mechanisms in the body. The hormone somatostatin also plays a role in the regulation of digestion. We have seen somatostatin in the endocrine system as it played a role in the regulation of growth hormone. There are 2 types of somatostatin secreted in the body. The type SS-14 is produced by the hypothalamus and pancreas, while type SS-28 is produced by the intestines. This type of somatostatin has an inhibitory effect on gastrointestinal secretions including the hormones gastrin and secretin. The parasympathetic nervous system can facilitate digestion by inhibiting the release of somatostatin type SS-28.

The gastric phase is characterized by food entering the stomach. Stimuli such as distention of the stomach and the presence of proteins promote additional secretions of gastric juice. The stimuli activate the parasympathetic nervous system to send impulses to the stomach, causing a series of events similar to those in the cephalic phase. The regulation of gastric secretions is also affected by the pH of the stomach. If the pH gets lower than 2.0, stomach secretions are inhibited. During the gastrointestinal phase, stomach secretions are inhibited when chyme exits the stomach and reaches the duodenum. This occurs most when chyme is acidic. Digestive enzymes work better with alkaline conditions. Acid chyme entering the duodenum causes the release of the hormone secretin, which has an inhibitory effect on gastric secretions. This is known as the enterogastric reflex. Fatty chyme causes the release of another hormone called cholecystokinin, which has an inhibiting effect on gastric secretions.

Absorption in the Stomach

Since digestion is usually not completed in the stomach, there is little absorption there. Water, alcohol, and lipid-soluble drugs are absorbed in the stomach.

Disorders of the Stomach

Disorders of the stomach are common. There can be a lot of different causes with a variety of symptoms. The strength of the lining of the stomach needs a careful balance of acid and mucus. If there is not enough mucus in the stomach, ulcers, abdominal pain, indigestion, heartburn, nausea, and vomiting could all be caused by the extra acid. Erosions, ulcers, and tumors can cause bleeding. When blood is in the stomach, it starts the digestive process and turns black. When this happens, the person can have black stool or vomit. Some ulcers can bleed very slowly so the person won't recognize the loss of blood. Over time, the iron in your body will run out, which in turn, will cause anemia. There isn't a known diet to prevent against getting ulcers. A balanced, healthy diet is always recommended. Smoking can also be a cause of problems in the stomach. Tobacco increases acid production and damages the lining of the stomach. It is not proven that stress alone can cause an ulcer.

Digestion in the Duodenum

Chyme entering the duodenum combines with a number of secretions. Many of these secretions come from the pancreas and gallbladder. The inner membrane of the duodenum contains many mucus-secreting cells that help to protect the membrane from acidic chyme. The pancreas secretes its exocrine secretions into the pancreatic duct. The pancreatic duct combines with the common bile duct at an area in the duodenum known as the hepatopancreatic ampulla. Fatty chyme entering the duodenum stimulates the secretion of cholecystokinin (CCK). CCK is a hormone that travels to the gallbladder to stimulate the release of bile. Bile works to break down fats through emulsification. Fatty chyme also stimulates the release of secretin. Besides inhibiting gastric secretions, secretin targets the pancreas to facilitate the release of bicarbonate ions that help to neutralize acidic chyme. Bicarbonate ions are produced by the columnar epithelium that lines pancreatic ducts. CCK released by the duodenum stimulates the pancreas to secrete a variety of enzymes from pancreatic acinar cells. These include trypsin, chymotrypsin, and carboxypeptidase for digesting proteins. The enzymes are secreted in an inactive form and activated by the presence of certain peptides. CCK also stimulates the release of pancreatic lipase for fat digestion, pancreatic amylase for carbohydrate digestion, and nucleases for breaking down DNA into nucleotides. Pancreatic secretions are also stimulated by the parasympathetic nervous system but to a lesser degree than by secretin and CCK.

Digestion in the Jejunum and Ileum

The walls of the jejunum and ileum contain cells with additional enzymes for performing the final stages of digestion. These enzymes include peptidases for breaking peptide bonds, disaccharidases for breaking carbohydrates into monosaccharides, and lipases for breaking down lipids. The ileum also contains lymphatic nodules called Peyer's patches that help to fight infection.

Enzyme	Produced In	Site of Release	pH Level
Carbohydrate Digestion:			
Salivary amylase	Salivary Glands	Mouth	Neutral
Pancreatic amylase	Pancreas	Small Intestine	Basic
Maltase	Small Intestine	Small Intestine	Basic
Protein Digestion:			
Pepsin	Gastric glands	Stomach	Acidic
Trypsin	Pancreas	Small Intestine	Basic
Peptidases	Small Intestine	Small Intestine	Basic
Nucleic Acid Digestion:			
Nuclease	Pancreas	Small Intestine	Basic
Nucleosidases	Pancreas	Small Intestine	Basic
Fat Digestion:			
Lipase	Pancreas	Small Intestine	Basic

Figure 17.10. Enzymes in the digestive system.

Absorption in the Small Intestine

Most digestion is completed by the time substances reach the jejunum and the majority of substances are absorbed by the time they reach the end of the ileum. The inner membrane of the small intestine is built for absorption with a very large surface area produced by the plicae circulares and intestinal villi.

Digested carbohydrates (monosaccharides) are absorbed in the small intestine by way of active and passive transport proteins. For example, glucose and galactose move into intestinal villi cells by way of symporters powered by the sodium gradient. Other monosaccharides are transported by facilitated diffusion. The monosaccharides then move out of the villi and into capillaries for transport to other areas of the body. Digested proteins (amino acids and dipeptides) also enter villi cells by way of a sodium symporter or facilitated diffusion. Once inside the cells, additional enzymes break down dipeptides and tripeptides into amino acids. The amino acids then exit the cell and are carried to the liver by the hepatic portal system. The amino acids are then either reconfigured or released into the bloodstream for use by other body tissues. Digested lipids (glycerol and fatty acids) enter the villi cells by simple diffusion. Once inside they are reassembled to form triglycerides by the smooth endoplasmic reticulum and packaged by the Golgi apparatus into packages called chylomicrons. The chylomicrons then leave the cell and enter the lacteals of the villi. The lacteals are extensions of the lymphatic system. Chylomicrons then travel through the lymph back to the venous blood. Lymph fluid containing large amounts of fat is called chyle. Chylomicrons end up in the liver where they are used for energy, converted to other molecules, or stored. Excess lipids are also stored in adipose tissue for later use. The liver converts lipids by combining them with proteins to form lipoproteins. Lipoproteins are named for the amount of protein and lipid within them. Very low density lipoproteins (VLDL) contain about 92% lipid and 8% protein. Low density lipoproteins (LDL) contain

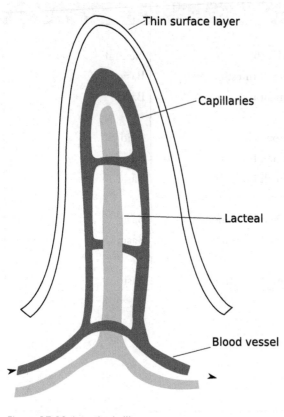

Figure 17.11. Intestinal villus.

about 75% lipid and 25% protein. High density lipoproteins (HDL) contain about 55% lipid and 45% protein.

Digestion and Absorption in the Large Intestine

By the time material has reached the large intestine digestion, and absorption are primarily complete. The colon secretes mucus that holds feces together. There are also resident bacteria known as intestinal flora. The bacteria help to break down undigested substances. Substances move through the small intestine by slow peristaltic movements. In the large intestine, substances move in what are called mass movements. Mass movements are stimulated by parasympathetic impulses, stretch of the colon walls, and impulses from the enteric nervous system. The mass movements move large sections of the colon (up to 20 cm), resulting in the movement of feces toward the rectum. Water and electrolytes are absorbed in the large intestine with the resulting material called feces. When feces enter the rectum, it triggers the defecation reflex. Distention of the rectum sends sensory impulses to the conus medullaris of the spinal cord. The resultant motor impulses produce peristaltic contractions in the colon and cause the internal anal sphincter muscle to relax. Impulses also travel to the cerebrum where the urge to defecate is sensed. This allows for voluntary control of the external anal sphincter. Relaxation of the external anal sphincter results in defecation. Defecation can be produced voluntarily by holding the breath and bearing down. The increases the intra-abdominal pressure and moves feces toward the rectum, triggering the defecation reflex.

Pancreas, Liver, and Gallbladder

The pancreas, liver, and gallbladder are essential for digestion. The pancreas produces enzymes that help digest proteins, fats, and carbohydrates; the liver produces bile that helps the body absorb fat; and the gallbladder stores the bile until it is needed. The enzymes and bile travel through special channels called ducts and into the small intestine where they help break down the food.

Pancreas

The pancreas is located posterior to the stomach and in close association with the duodenum. In humans, the pancreas is a 6- to 10-inch elongated organ in the abdomen located retroperitoneal. It is often described as having three regions: a head, body, and tail. The pancreatic head abuts the second part of the duodenum, while the tail extends toward the spleen. The pancreatic duct runs the length of the pancreas and empties into the second part of the duodenum at the ampulla of Vater. The common bile duct commonly joins the pancreatic duct at or near this point. The pancreas is supplied arterially by the pancreaticoduodenal arteries, themselves branches of the superior mesenteric artery of the hepatic artery (branch of celiac trunk from the abdominal aorta). The superior mesenteric artery provides the inferior pancreaticoduodenal arteries, while the gastroduodenal artery (one of the terminal branches of the hepatic artery) provides the superior pancreaticoduodenal artery. Venous drainage is via the pancreatic duodenal veins, which end up in the portal vein. The splenic vein passes posterior to the pancreas but is said to not drain the pancreas itself. The portal vein is formed by the union of the superior mesenteric vein and splenic vein posterior to the body of the pancreas. In some people (as many as 40%) the inferior mesenteric vein also joins with the splenic vein behind the pancreas; in others it simply joins with the superior mesenteric vein instead. The function of the pancreas is to produce enzymes that break down all categories of digestible foods (exocrine pancreas) and secrete hormones that affect carbohydrates metabolism (endocrine pancreas).

The pancreas is composed of pancreatic exocrine cells, whose ducts are arranged in clusters called acini (singular "acinus"). The cells are filled with secretory granules containing the precursor digestive enzymes (mainly trypsinogen, chymotrypsinogen, pancreatic lipase, and amylase) that are secreted into the lumen of the acinus. These granules are termed "zymogen granules" ("zymogen" referring to the inactive precursor enzymes). It is important to synthesize inactive enzymes in the pancreas to avoid auto degradation, which can lead to pancreatitis.

The pancreas is near the liver, and is the main source of enzymes for digesting fats (lipids) and proteins—the intestinal walls have enzymes that will digest polysaccharides. Pancreatic secretions from ductal cells contain bicarbonate ions and are alkaline in order to neutralize the acidic chyme that the stomach churns out. Control of the exocrine function of the pancreas are via gastrin, cholecystokinin, and secretin, which are hormones secreted by cells in the stomach and duodenum, in response to distension and/or food and which cause secretion of pancreatic juices.

The two major proteases that the pancreas secretes are trypsinogen and chymotrypsinogen. These zymogens are inactivated forms of trypsin and chymotrypsin. Once released in the intestine, the enzyme enterokinase present in the intestinal mucosa activates trypsinogen by cleaving it to form trypsin. The free trypsin then cleaves the rest of the trypsinogen and chymotrypsinogen to their active forms. Pancreatic secretions accumulate in intralobular ducts that drain the main pancreatic duct, which drains directly into the duodenum. Due to the importance of its enzyme contents, injuring the pancreas is a very dangerous situation. A puncture of the pancreas tends to require careful medical intervention.

Scattered among the acini are the endocrine cells of the pancreas, in groups called the islets of Langerhans. They are Insulin-producing beta cells (50–80% of the islet cells), Glucagon-releasing alpha cells (15–20%), Somatostatin-producing delta cells (3–10%), and Pancreatic polypeptide-containing PP cells (remaining %). The islets are a compact collection of endocrine cells arranged in clusters and cords and are crisscrossed by a dense network of capillaries. The capillaries of the islets are lined by layers of endocrine cells in direct contact with vessels, and most endocrine cells are in direct contact with blood vessels, by either cytoplasmic processes or by direct apposition.

Liver

The liver is an organ in all vertebrates, including humans. It plays a major role in metabolism and has a number of functions in the body, including glycogen storage, plasma protein synthesis, and drug detoxification. It also produces bile, which is important in digestion. It performs and regulates a wide variety of high-volume biochemical reactions that require specialized tissues. The liver normally weighs between 1.3 and 3.0 kilograms and is a soft, pinkish-brown boomerang-shaped organ. It is the second largest organ (the largest being the skin) and the largest gland within the human body. Its anatomical position in the body is immediately under the diaphragm on the right side of the upper abdomen. The liver lies on the right side of the stomach and makes a kind of bed for the gallbladder.

The liver is supplied by two main blood vessels on its right lobe: the hepatic artery and the portal vein. The hepatic artery normally comes off the celiac trunk. The portal vein brings venous blood from the spleen, pancreas, and small intestine, so that the liver can process the nutrients and byproducts of food digestion. The hepatic veins drain directly into the inferior vena cava. The bile produced in the liver is collected in bile canaliculi, which merge from bile ducts. These eventually drain into the right and left hepatic ducts, which in turn merge to form the common hepatic duct. The cystic duct (from the gallbladder) joins with the common hepatic duct to form the common bile duct. Bile can either drain directly into the duodenum via the common bile duct or be temporarily stored in the gallbladder via the cystic duct. The common bile duct and the pancreatic duct enter the duodenum together at the ampulla of Vater. The branchings of the bile ducts resemble those of a tree, and indeed term "biliary tree" is commonly used in this setting.

The liver is among the few internal human organs capable of natural regeneration of lost tissue: as little as 25% of remaining liver can regenerate into a whole liver again. This is predominantly due to hepatocytes acting as unipotential stem cells. There is also some evidence of bio-potential stem cells, called oval cells, which can differentiate into either hepatocytes or cholangiocytes (cells that line bile ducts).

The various functions of the liver are carried out by the liver cells or hepatocytes.
- The liver produces and excretes bile requires for dissolving fats. Some of the bile drains directly into the duodenum, and some is stored in the gallbladder.
- The liver performs several roles in carbohydrate metabolism:
  - gluconeogenesis (the formation of glucose from certain amino acids, lactate, or glycerol)
  - glycogenolysis (the formation of glucose from glycogen)
  - glycogenesis (the formation of glycogen from glucose)
  - the breakdown of insulin and other hormones
  - the liver is responsible for the mainstay of protein metabolism
- The liver also performs several roles in lipid metabolism:
  - cholesterol synthesis
  - the production of triglycerides (fats)
- The liver produces coagulation factors I (fibrinogen), II (prothrombin), V, VII, IX, X, and XI, as well as protein C, protein S, and antithrombin.
- The liver breaks down hemoglobin, creating metabolites that are added to bile as pigment.
- The liver breaks down toxic substances and most medicinal products in a process called drug metabolism. This sometimes results in toxication, when the metabolite is more toxic than its precursor.
- The liver converts ammonia to urea.
- The liver stores a multitude of substances, including glucose in the form of glycogen, vitamin B12, iron, and copper.

- In the first trimester, a fetus's liver is the main site of red blood cell production. By the 32nd week of gestation, the bone marrow has almost completely taken over that task.

Gallbladder

The gallbladder is a pear-shaped organ that stores about 50 ml of bile (or "gall") until the body needs it for digestion. The gallbladder is about 7–10 cm long in humans and is dark green in appearance due to its contents (bile), not its tissue. It is connected to the liver and the duodenum by biliary tract. The gallbladder is connected to the main bile duct through the gallbladder duct (cystic duct). The main biliary tract runs from the liver to the duodenum, and the cystic duct is effectively a cul de sac, serving as entrance and exit to the gallbladder. The surface marking of the gallbladder is the intersection of the midclavicular line (MCL) and the transpyloric plane, at the tip of the ninth rib. The blood supply is by the cystic artery and vein, which runs parallel to the cystic duct. The cystic artery is highly variable, and this is of clinical relevance since it must be clipped and cut during a cholecystectomy.

The gallbladder has an epithelial lining characterized by recesses called Aschoff's recesses, which are pouches inside the lining. Under the epithelium is a layer of connective tissue, followed by a muscular wall that contracts in response to cholecystokinin, a peptide hormone secreted in the duodenum.

The gallbladder stores bile, which is released when food containing fat enters the digestive tract, stimulating the secretion of cholecystokinin (CCK). The bile emulsifies fats and neutralizes acids in partly digested food. After being stored in the gallbladder, the bile becomes more concentrated than when it left the liver, increasing its potency and intensifying its effect in fats.

Conditions Affecting the Esophagus

There are two different types of conditions that may affect the esophagus. The first type is called congenital, meaning a person is born with it. The second type is called non-congenital, meaning the person develops it after birth. Some examples of these are:

Tracheoesophageal Fistula and Esophageal Atresia

Both of these conditions are congenital. In *tracheoesophageal fistula* there is a connection between the esophagus and the windpipe (trachea) where there shouldn't be one. In *Esophageal atresia* the esophagus of a newborn does not connect to the stomach but comes to a dead end right before the stomach. Both conditions require corrective surgery and are usually detected right after the baby is born. In some cases, it can be detected before the baby is born.

Esophagitis

Esophagitis is inflammation of the esophagus and is a non-congenital condition. Esophagitis can be caused by certain medications or by infections. It can also be caused by gastroesophageal reflux disease (GERD), a condition where the esophageal sphincter allows the acidic contents of the stomach to move back up into the esophagus. Gastroesophageal reflux disease can be treated with medications, but it can also be corrected by changing what you eat.

Conditions Affecting the Stomach and Intestines

Everyone has experienced constipation or diarrhea in their lifetime. With constipation, the contents of the large intestines don't move along fast enough and waste material stays in the large intestines too long. All water is extracted out of the waste and it becomes hard. With diarrhea you get the opposite reaction. Waste moves along too fast and the large intestines can't absorb the water before the waste is pushed through. Common flora bacteria assist in the prevention of many serious problems. Here are some more examples of common stomach and intestinal disorders:

Appendicitis

Appendicitis is the inflammation of the appendix, the finger-like pouch that extends from the cecum. The most common symptoms are abdominal pain, loss of appetite, fever, and vomiting. Kids and teenagers are the most common victims of appendicitis, which must be corrected by surgery. While mild cases may resolve without treatment, most require removal of the inflamed appendix, either by laparotomy or laparoscopy. Untreated, mortality is high, due mainly to peritonitis and shock.

Celiac Disease

Celiac disease is a disorder in which a person's digestive system is damaged by the response of the immune system to a protein called gluten, which is found in rye, wheat, and barley, common ingredients in foods like breakfast cereal and pizza crust. People who have celiac disease experience abdominal pain, diarrhea, bloating, exhaustion, and depression when they eat foods with gluten in them. They also have difficulty digesting their food. Celiac disease runs in families and becomes active after some sort of stress, like viral infections or surgery. The symptoms can be managed by following a gluten-free diet. Doctors can diagnose this condition by taking a full medical history or with a blood test.

Diverticulitis

Develops from diverticulosis, which involves the formation of pouches (diverticula) on the outside of the colon. Diverticulitis results if one of these diverticula becomes inflamed. In complicated diverticulitis, bacteria may subsequently infect the outside of the colon if an inflamed diverticula bursts open. If the infection spreads to the lining of the abdominal cavity (peritoneum), this can cause a potentially fatal peritonitis. Sometimes inflamed diverticula can cause narrowing of the bowel, leading to an obstruction. Also, the affected part of the colon could adhere to the bladder or other organ in the pelvic cavity, causing a fistula, or abnormal communication, between the colon and an adjacent organ.

Gastritis and Peptic Ulcers

Usually the stomach and the duodenum are resistant to irritation because of the strong acids produced by the stomach. But sometimes a bacterium called Helicobacter pylori, or the chronic use of drugs or certain medications, weakens the mucous layer that coats the stomach and the duodenum, allowing acid to get through the sensitive lining beneath. This can cause gastritis, an irritation and inflammation of the lining of the stomach, or peptic ulcers, which are holes or sores that form in the lining of the stomach and duodenum and cause pain and bleeding. Medications are the best way to treat this condition.

Gastrointestinal Infections

Gastrointestinal infections can be caused by bacteria such as Campylobacter, Salmonella, E. coli, or Shigella. They can also be caused by viruses or by intestinal parasites like amebiasis and Giardiasis. The most common symptoms of gastrointestinal infections are abdominal pain and cramps, diarrhea, and vomiting. These conditions usually go away on their own and don't need medical attention.

Inflammatory Bowel Disease

Inflammatory bowel disease is the chronic inflammation of the intestines, which usually affect older kids, teens, and adults. The two major types are ulcerative colitis and Crohn's disease; indeterminate colitis occurs in 10–15% of patients. Ulcerative colitis usually affects just the rectum and large intestine, while Crohn's disease can affect the whole gastrointestinal tract from mouth to anus, along with some other parts of the body. Patients with these diseases also suffer from extraintestinal symptoms including joint pain and red eye, which can signal a flare of the disease. These diseases are treated with medications and if necessary, intravenous or IV feeding, or in the more serious cases, surgery to remove the damaged areas of the intestines.

Polyp

A polyp is an abnormal growth of tissue (tumor) projecting from a mucous membrane. If it is attached to the surface by a narrow elongated stalk, it is said to be pedunculated. If no stalk is present, it is said to be sessile. Polyps are commonly found in the colon, stomach, nose, urinary bladder, and uterus. They may also occur elsewhere in the body where mucous membranes exist, like the cervix and small intestine.

Disorders of the Pancreas, Liver, and Gallbladder

Disorders of the pancreas, liver, and gallbladder affect the ability to produce enzymes and acids that aid in digestion. Examples of these disorders are:

Cystic Fibrosis

Cystic fibrosis is a chronic, inherited illness where the production of abnormally thick mucus blocks the duct or passageways in the pancreas and prevents the digestive fluids from entering the intestines, making it difficult for the person with the disorder to digest protein and fats, which cause important nutrients to pass through without being digested. People with this disorder take supplements and digestive enzymes to help manage their digestive problems.

Hepatitis

Hepatitis is a viral condition that inflames a person's liver, which can cause it to lose its ability to function. Viral hepatitis, like hepatitis A, B, and C, is extremely contagious. Hepatitis A, which is a mild form of hepatitis, can be treated at home, but more serious cases that involve liver damage might require hospitalization.

Cholecystitis

Acute or chronic inflammation if the gallbladder causes abdominal pain. Ninety percent of cases of acute cholecystitis are caused by the presence of gallstones. The actual inflammation is due to secondary infection with bacteria of an obstructed gallbladder, with the obstruction caused by the gallstones. Gallbladder conditions are very rare in kids and teenagers but can occur when the kid or teenager has sickle cell anemia or in kids being treated with medications over a long term.

Cholestasis

Cholestasis is the blockage in the supply of bile into the digestive tract. It can be "intrahepatic" (the obstruction is in the liver) or "extrahepatic" (outside the liver). It can lead to jaundice and is identified by the presence of elevated bilirubin level that is mainly conjugated.

Biliary Colic

This is when a gallstone blocks either the common bile duct or the duct leading into it from the gallbladder. This condition causes severe pain in the right upper abdomen and sometimes through to the upper back. It is described by many doctors as the most severe pain in existence, amongst childbirth and a heart attack. Other symptoms are nausea and vomiting, diarrhea, bleeding caused by continual vomiting, and dehydration caused by the nausea and diarrhea. Another more serious complication is total blockage of the bile duct, which leads to jaundice, which if it is not corrected naturally or by surgical procedure, can cause fatal liver damage. The only long-term solution is the removal of the gallbladder.

Gastrointestinal Dysfunctions

As we age, the amount of digestive enzymes produced by the body drops way down. This leads to decreased and slower digestion, slower absorption of nutrients, and increased accumulation of fecal matter in the intestinal tract. Undigested food material and metabolic waste can also build up due to slow elimination, starting of a series of health problems.

When digestion slows, it turns the intestines into a toxic environment. Helpful organisms cannot live in toxic environments. When the beneficial organisms die, they are replaced by harmful organisms, such as yeasts and parasites, the most common being *Candida albicans*. This leads to changes in the intestinal wall, which produce *leaky gut syndrome*, which allows many toxic chemicals to be introduced into the bloodstream. As a result the entire toxic load of the body is increased, which causes a bigger burden on the liver, kidneys, and other body organs. When this happens, the organs that are normally used for eliminating waste from and supplying nutrients to the GI tract becomes a large dumpsite for waste. This problem is made worse by the use of junk food, prescriptions, over-the-counter medications, antibiotics, and a diet that is too low in fiber. Most people never even think about their GI tract. We are all concerned about what the outside of our body looks like, but we completely ignore the inside. Because our bodies a very resilient,. deterioration of the digestive system can go on for years with no symptoms or side effects. When symptoms finally do appear, they are usually very non-specific; they include decreased energy, headaches, diarrhea, constipation, heartburn, and acid reflux. Over the years these symptoms become more serious; they include asthma, food allergies, arthritis, and cancer.

Poor digestion, poor absorption, and bacterial imbalance can be traced to a lot of chronic conditions. Every organ in the body receives nutrients from the GI tract. If the GI tract is malfunctioning, then the whole body suffers. It is possible to return good health to your GI tract by improving digestion, consuming the right amount of fiber, and cutting out junk food and refined sugars. You can improve the function of the intestines by taking fiber supplements and vitamins (especially B12 and vitamin K). Some doctors suggest herbal or vitamin enemas to cleanse and relieve constipation and to help stimulate *peristaltic movement*, which will help to move the bowels.

Irritable Bowel Syndrome

Irritable Bowel Syndrome (IBS) is a disorder with symptoms that are most commonly bloating, abdominal pain, cramping, constipation, and diarrhea. IBS causes a lot of pain and discomfort. It does not cause permanent damage to the intestines and does not lead to serious diseases such as cancer. Most of the people affected with IBS can control their symptoms with stress management, diet, and prescription medication. For others IBS can be debilitating; they may be unable to go to work, travel, attend social events, or leave home for even short periods of time.

About 20 percent of the adult population has some symptoms of IBS, making it one of the most common intestinal disorders diagnosed by physicians. It is more common in men than women and in about 50 percent of people affected, it starts at about age 35.

Researchers have not found out what exactly causes IBS. One idea is that people with IBS have a large intestine (colon) that is sensitive to certain foods and stress. The immune system may also be involved. It has also been reported that *serotonin* is linked with normal GI functioning. Ninety-five percent of the body's serotonin is located in the GI tract (the other 5 percent is in the brain). People with IBS have diminished receptor activity, causing abnormal levels of serotonin in the GI tract. Because of this, IBS patients experience problems with bowel movement, mobility, and sensation; they have more sensitive pain receptors in their GI tract. Many IBS patients suffer from depression and anxiety, which can make symptoms worse. There is no cure for IBS, but medications are an important part of relieving symptoms. Fiber supplements or laxatives are helpful for constipation. Antidiarrheals such as Imodium can help with diarrhea. An antispasmodic is commonly prescribed for colon muscle spasms. Antidepressants and pain medication are also commonly prescribed.

Gastrointestinal Stromal Tumor

Gastrointestinal Stromal Tumor, or GIST, is an uncommon type of cancer in the GI tract (esophagus, stomach, small intestine, and colon). These types of cancers begin in the connective tissue like fat, muscles, nerves, and cartilage. GIST originates in the stroma cells. Stroma cells are strung along the GI tract and are part of the system that helps the body to know when to move food through the digestive system. Over half of GISTs occur in the stomach. Most cases occur in people between the ages of forty and eighty, but can also show up in a person of any age.

All GISTs of any size or location have the ability to spread. Even if a GIST is removed, it can reappear in the same area, or may even spread outside of the GI tract. In the early stages, GIST is hard to diagnose because in the early stages symptoms cannot be recognized. In the later stages, a person can have vague abdominal pain, vomiting, abdominal bleeding that shows up in stool or vomit, low blood counts causing anemia, and an early feeling of being full that causes a decrease in appetite. GIST is now recognized as an aggressive cancer that is able to spread to other parts of the body. People who have been diagnosed with GIST should get treatment as soon as possible.

Food Allergies

Food allergies occur when the immune system thinks that a certain protein in any kind of food is a foreign object and will try to fight against it. Only about eight percent of children and two percent of adults have a food allergy. A person can be allergic to any kind of food, but the most common food allergies are from nuts, cow's milk, eggs, soy, fish, and shellfish. Most people who have a food allergy are allergic to fewer than four different foods.

The most common signs of food allergies are hives, swelling, itchy skin, itchiness, tingling or swelling in the mouth, coughing, trouble breathing, diarrhea, and vomiting. The two most common chronic illness that are associated with food allergies are eczema and asthma. Food allergies can be fatal if they cause the reaction called anaphylaxis. This reaction makes it hard for the person to breathe. This can be treated by an epinephrine injection.

GERD, Heartburn, Acid Reflux

GERD, or Gastroesophageal Reflux Disease, occurs when the lower esophageal sphincter is not able to close properly. When this happens, contents from the stomach called reflux leak back into the esophagus and the stomach. When the stomach refluxes, stomach acid touches the lining of the esophagus and causes it to have a burning feeling in the throat or the chest. This is what heartburn is. When you taste the fluid in the back of your throat, it is called acid indigestion. It is common for a person to get occasional heartburn, but when it occurs more than twice a week it can be considered GERD. GERD can occur in people of all ages including infants. Some symptoms of GERD include having a pain in your chest, hoarseness, having trouble swallowing, or having the feeling of food being stuck in your throat. The main symptoms are having persistent heartburn and acid regurgitation. GERD can also cause bad breath and a dry cough. No one knows why people get GERD. Some things that could contribute to GERD are alcohol use, pregnancy, being overweight, and smoking. Certain foods might also contribute like citrus fruits; caffeine; spicy, fatty, and dried foods; and also mint flavorings. Over-the-counter antacids or medications that help stop acid production and help the muscles empty the stomach are commonly used to treat GERD.

Constipation

Not everyone is on the same schedule for having a bowel movement. Depending on the person, a "normal" schedule can range anywhere from three times a day to three times a week. If you start having bowel movements less frequently than your own personal schedule, then you might be getting the signs of constipation. Constipation is when you have trouble having bowel movements. The stool is very hard, making it hard to pass and causing a person to strain. You may even feel like you have to have a bowel movement even after you have already had one.

When you digest food, the waste products go through your intestines by the muscles contracting. When in the large intestine, most of the water and salt from the waste products are reabsorbed because they are needed by the body for our everyday functions. You can become constipated if too much water is absorbed, or if waste products move too slowly. Not getting enough fluids, a low-fiber diet, age, not being physically active, depression, stress, and pregnancy can all be causes of constipation. Medications and narcotics can also cause a person to get constipated. Chronic constipation may be a symptom of a liver problem such as a urea cycle disorder.

The best way for a person to treat constipation is to make sure that you are getting enough fluids as well as fiber in your diet. By doing this, the bulk of your stool is increased and also makes the stool softer so that it can move through your intestines more easily. Being more active and increasing your daily exercise also helps keep you regulated.

Hemorrhoids

Hemorrhoids (also known as haemorrhoids, emerods, or piles) are varicosities or swelling and inflammation of veins in the rectum and anus. Two of the most common types of hemorrhoids are external and internal hemorrhoids.

External hemorrhoids are those that occur outside of the anal verge (the distal end of the anal canal). They are sometimes painful, and can be accompanied by swelling and irritation. Itching, although often thought to be a symptom from external hemorrhoids, is more commonly due to skin irritation. If the vein ruptures and a blood clot develops, the hemorrhoid becomes a thrombosed hemorrhoid.

Internal hemorrhoids are those that occur inside the rectum. As this area lacks pain sensory receptors, internal hemorrhoids are usually not painful and most people are not aware that they have them. Internal hemorrhoids, however, may bleed when irritated. Untreated internal hemorrhoids can lead to two severe forms of hemorrhoids: prolapsed and strangulated hemorrhoids.

Prolapsed hemorrhoids are internal hemorrhoids that are so distended that they are pushed outside of the anus. If the anal sphincter muscle goes into spasm and traps a prolapsed hemorrhoid outside of the anal opening, the supply of blood is cut off, and the hemorrhoid becomes a strangulated hemorrhoid.

Bleeding in the Gastrointestinal Tract

Bleeding in the gastrointestinal tract doesn't always mean you have a disease, it's usually a symptom of a digestive problem. The cause of the bleeding may not be that serious; it could be something that can be cured or controlled such as hemorrhoids. However, locating the source of the bleeding is very important. The gastrointestinal tract contains many important organs like the esophagus, stomach, small intestine, large intestine or colon, rectum, and anus. Bleeding can come from one or more of these area from a small ulcer in the stomach or a large surface like the inflammation of the colon. Sometimes a person doesn't even know they are bleeding. When this happens, it is called hidden, or occult bleeding. Simple tests can detect hidden blood in the stool.

What Causes Bleeding in the Digestive Tract

Esophageal bleeding may be caused by Mallory-Weiss syndrome, which is a tear in the esophagus. Mallory-Weiss syndrome is usually caused by excessive vomiting or may be caused by childbirth, a hiatal hernia, or increased pressure in the abdomen caused by coughing. Various medications can cause stomach ulcers or inflammations. Medications containing aspirin or alcohol, and various other medications (mainly those used for arthritis) are some examples of these.

Benign tumors or cancer of the stomach may also cause bleeding. These disorders don't usually produce massive bleeding. The most common source of bleeding usually occurs from ulcers in the duodenum. Researchers believe that these ulcers are caused by excessive stomach acid and a bacterium called Helicobacter Pylori. In the lower digestive tract, the most common sources of bleeding are the large intestine and the rectum. Hemorrhoids are the most common cause of bleeding in the digestive tract. Hemorrhoids are enlarged veins in the anal area, which produce bright red blood that you see in the toilet or on the toilet paper.

How to Recognize Bleeding in the Digestive Tract

The signs of bleeding in the digestive tract vary depending on the site and severity of the bleeding. If the blood is coming from the rectum, it would be bright red blood. If it was coming from higher up in the colon or from the small intestine, the blood would be darker. When the blood is coming from the stomach, esophagus, or the duodenum, the stool will be black and tarry. If the bleeding is hidden, or occult, a person may not notice changes in the stool color. If extensive bleeding occurs, a person may feel dizzy, faint, weak, short of breath, have diarrhea, or cramps/abdominal pain. Shock can also occur along with rapid pulse, drop in blood pressure, and difficulty urinating. Fatigue, lethargy, and pallor from anemia will settle in if the bleeding is slow. Anemia is when the blood's iron-rich substance, hemoglobin, is diminished.

Common Causes of Bleeding in the Digestive Tract

- Hemorrhoids
- Gastritis (inflammation)
- Inflammation (ulcerative colitis)
- Colorectal Polyps
- Colorectal Cancer
- Duodenal Ulcer
- Enlarged Veins
- Esophagitis (inflammation of the esophagus)
- Mallory-Weiss Syndrome
- Ulcers

How Bleeding in the Digestive Tract Is Diagnosed

To diagnose bleeding in the digestive tract, the bleeding must be located and a complete history and physical are very important. The following are some of the procedures that diagnose the cause of bleeding.

Endoscopy

An endoscopy is a common diagnostic technique that allows direct viewing of the bleeding site. Since the endoscope can detect lesions and confirm the absence or presence of bleeding, doctors often use this method to diagnose acute bleeding; the endoscope can also be used to treat the cause of bleeding.

The endoscope is a flexible instrument that can be inserted through the mouth or rectum. The instrument allows the doctors to see inside the esophagus, stomach, duodenum (esophagoduodenoscopy), sigmoid colon (sigmoidoscopy), and rectum (rectoscopy) to collect small samples of tissues, take pictures, and stop the bleeding. There is a new procedure out using a long endoscope that can be inserted during surgery to locate a source of bleeding in the small intestine.

Capsule Endoscopy

Capsule endoscopy helps doctors to see and examine the lining of the middle part of the gastrointestinal tract, which includes the three parts of the small intestine (duodenum, jejunum, ileum). The capsule is a small pill-sized video camera called an endoscope. It has its own lens and light that transfers the images to a monitor so the doctor can view them outside of the body. This process is also referred to as small bowel endoscopy, capsule endoscopy, or wireless endoscopy. The most common reason for doing a capsule endoscopy is to look for the causes of bleeding that is coming from the small intestine. It is also able to help detect ulcers, tumors, and Crohn's disease.

Angiography

Angiography is a technique that uses dye to highlight blood vessels. This procedure is used when the patient is bleeding badly enough that it allows the dye to leak out of the blood vessels and identifies the bleeding site. In some situations, angiography allows the patient to have medication injections that may stop the bleeding.

Radionuclide Scanning

Radionuclide scanning is a non-invasive screening technique used for locating sites of acute bleeding, especially in the lower GI tract. This procedure injects small amounts of radioactive material that either attach to the person's red blood cells or are suspended in the blood. Special pictures are taken that allow doctors to see the blood escaping. Barium x-rays, angiography, and radionuclide scans can be used to locate sites of chronic occult bleeding.

CRITICAL-THINKING QUESTIONS

1. Explain how secretin and cholecystokinin work together to create a proper environment in the small intestine for the digestion of carbohydrates, fats, and protein fragments.
2. Explain why many people have a bowel movement shortly after a meal.
3. What symptoms might you expect if gallstones completely filled the gallbladder?
4. Explain the role of gastrin in providing a proper digestive environment in the stomach.
5. Explain why neither rapid emptying of the stomach nor slow emptying of the stomach is desirable.
6. Relate the importance of mechanical and chemical digestion to absorption of food.
7. What role does each of the following play in normal functioning of the gastrointestinal tract?
    A. parietal cells
    B. constrictor muscles of pharynx

C. lower esophageal sphincter
   D. gallbladder
   E. gastrin
   F. bile
   G. trypsin
   H. hepatocytes
   I. gastric inhibitory peptide
   J. mucus in the small intestine
8. Why might cutting the vagus nerve help someone who has a peptic ulcer?

Chapter Eighteen

Urinary System

The urinary system is a group of organs in the body concerned with filtering excess fluid and other substances from the bloodstream. The substances are filtered from the body in the form of urine. Urine is a liquid produced by the kidneys, collected in the bladder, and excreted through the urethra. Urine is used to extract excess minerals or vitamins as well as blood corpuscles from the body. The urinary organs include the kidneys, ureters, bladder, and urethra. The urinary system works with the other systems of the body to help maintain homeostasis. The kidneys are the main organs of homeostasis because they maintain the acid–base balance and the water–salt balance of the blood. The kidney's appearance is deceivingly simple and yet they perform extremely complex physiology. The urinary system consists of the kidneys, ureters, urinary bladder, and urethra. The urinary system can be thought of as a kind of purification system for the blood. The system functions to maintain fluid, electrolyte, and pH balance and remove toxins from the blood. We can gain a good deal of insight into how the kidneys work by examining the inputs and outputs. Blood flows into the kidney and urine and blood flow out. So the kidneys must somehow make urine from blood. The blood enters via the renal artery and exits via the renal vein. The urine exits by way of the ureters and flows to the bladder, urethra, and out of the body. One of the simplest ways to make urine from blood is to filter the blood. Filtration is actually the first stage of urine formation. Filters work by the movement of substances from areas of higher to lower pressure across a filtration membrane. The filtration membrane sorts substances based on size. One could think of it as being filled with holes. Smaller substances pass through the holes, while larger substances do not. Smaller substances that are filtered include water, electrolytes, and glucose.

There would be a problem if filtration were the only mechanism of urine formation. Our bodies need many of the filtered substances and they would be lost in the urine. So there must be some other mechanisms that help to maintain the balance. Fortunately, there are and these include tubular reabsorption and secretion. Tubular reabsorption and secretion work together to reclaim substances like water, glucose, and electrolytes after they have been filtered. Tubular reabsorption employs a number of mechanisms in order to move filtered substances back into the blood. Tubular secretion also uses a number of mechanisms to move substances from the blood to the urine. Besides reclaiming filtered substances, both of these processes fine tune electrolyte, water, and pH balance.

Functions of the Urinary System

One of the major functions of the urinary system is the process of excretion. Excretion is the process of eliminating waste products of metabolism and other materials that are of no use. The urinary system maintains an appropriate fluid volume by regulating the amount of water that is excreted in the urine. Other aspects of its function include regulating the concentrations of various electrolytes in the body fluids and maintaining normal pH of the blood. Several body organs carry out excretion, but the kidneys are the most important excretory organ. The primary function of the kidneys is to maintain a stable internal environment (homeostasis) for optimal cell and tissue metabolism. They do this by separating urea, mineral salts, toxins, and other waste products from the blood. They also do the job of conserving water, salts, and electrolytes. At least one kidney must function properly for life to be maintained. Six important roles of the kidneys are:

1. Regulation of plasma ionic composition. Ions such as sodium, potassium, calcium, magnesium, chloride, bicarbonate, and phosphates are regulated by the amount that the kidney excretes.
2. Regulation of plasma osmolarity. The kidneys regulate osmolarity because they have direct control over how many ions and how much water a person excretes.
3. Regulation of plasma volume. Your kidneys are so important they even have an affect on your blood pressure. The kidneys control plasma volume by controlling how much water a person excretes. The plasma volume has a direct effect on the total blood volume, which has a direct affect on your blood pressure. Salt (NaCl) will cause osmosis, the diffusion of water into the blood, to happen.
4. Regulation of plasma hydrogen ion concentration (pH). The kidneys partner up with the lungs and they together control the pH. The kidneys have a major role because they control the amount of bicarbonate excreted or held onto. The kidneys help maintain the blood pH mainly by excreting hydrogen ions and reabsorbing bicarbonate ions as needed.
5. Removal of metabolic waste products and foreign substances from the plasma. One of the most important things the kidneys excrete is nitrogenous waste. As the liver breaks down amino acids, it also releases ammonia. The liver then quickly combines that ammonia with carbon dioxide, creating urea, which is the primary nitrogenous end product of metabolism in humans. The liver turns the ammonia into urea because it is much less toxic. We can also excrete some ammonia, creatinine, and uric acid. The creatinine comes from the metabolic breakdown of creatine phosphate (a high-energy phosphate in muscles).
   Uric acid comes from the breakdown of nucleotides. Uric acid is insoluble and too much uric acid in the blood will build up and form crystals that can collect in the joints and cause gout.
6. Secretion of hormones The endocrine system has assistance from the kidneys when releasing hormones. Renin is released by the kidneys. Renin leads to the secretion of aldosterone, which is released from the adrenal cortex. Aldosterone promotes the kidneys to reabsorb the sodium (Na+) ions. The kidneys also secrete erythropoietin when the blood doesn't have the capacity to carry oxygen. Erythropoietin stimulates red blood cell production. Vitamin D from the skin is also activated with help from the kidneys. Calcium (Ca+) absorption from the digestive tract is promoted by vitamin D.

Kidneys

The kidneys are paired organs located behind the peritoneal membrane (retroperitoneal) (Fig. 18.1). They are bean shaped and about the size of an adult fist. They are located laterally in the flank area, from about the level of the twelfth thoracic vertebra to the third lumbar vertebra. A layer of adipose tissue called perirenal fat surrounds each kidney. The outer layer of the kidneys consists of a layer of fibrous connective tissue called the renal capsule. The kidneys are partially held in place by connective tissue called renal fascia that connect to the outer portion of the peritoneum. Each kidney has an indentation called a hilum that opens to a renal sinus where the renal artery, vein, and ureters enter and exit the kidney. The inside of the kidney is divided into an outer cortical region (cortex) and an inner medulla. The medulla contains conical structures called renal pyramids. Areas of the cortex called renal columns extend between the pyramids. The distal tip of the pyramid ends at the renal papilla. The renal papillae connect with minor calyces. The minor calyces combine to form larger major calyces that combine to form the renal pelvis that extends to form the ureter. A renal artery supplies each kidney with blood. The renal artery branches off the abdominal aorta and extends into the hilum of the kidney. The renal artery forms smaller branches called segmental arteries that form smaller interlobar arteries. The interlobar arteries flow through the renal columns and branch to form arcuate arteries, which are located between the cortex and medulla. The arcuate arteries branch to form interlobular arteries that travel in the cortex. The interlobular arteries then branch to form afferent arterioles that supply the functional unity of the kidney, known as the nephron. Exiting the nephron is the efferent arteriole. The efferent arterioles of nephrons feed the interlobular veins that empty into the arcuate veins. The arcuate veins empty into the interlobular veins, which empty into the renal vein. The kidneys also have a nerve supply. The renal nerves, which are sympathetic postganglionic neurons from the celiac plexus and inferior splanchnic nerves, innervate the kidneys.

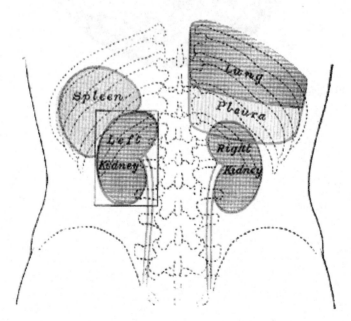

Figure 18.1. Location of the kidneys.

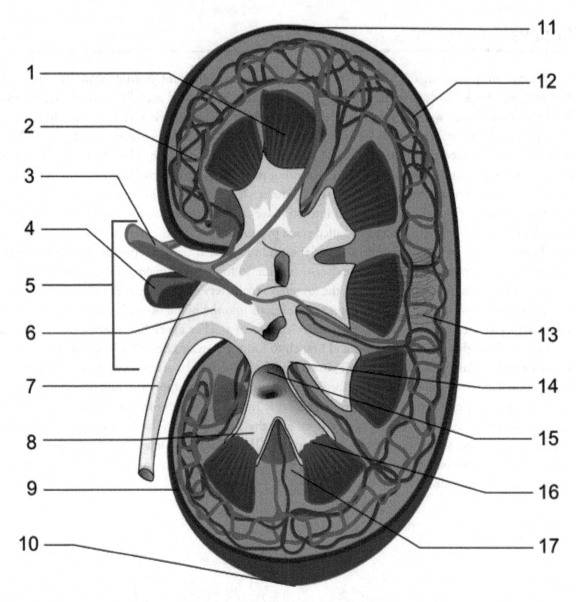

Figure 18.2. Structures of the kidney:
1. Renal pyramid
2. Interlobar artery
3. Renal artery
4. Renal vein
5. Renal hilum
6. Renal pelvis
7. Ureter
8. Minor calyx
9. Renal capsule
10. Inferior renal capsule
11. Superior renal capsule
12. Interlobar vein
13. Nephron
14. Minor calyx
15. Major calyx
16. Renal papilla
17. Renal column

The Nephron

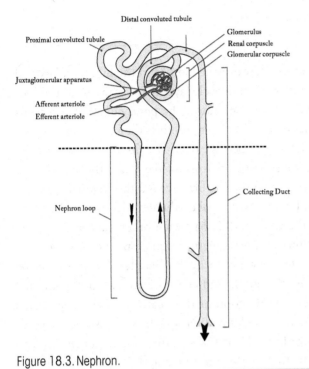

Figure 18.3. Nephron.

The functional unit of the kidney is a microscopic structure known as the nephron (Fig. 18.3). There are over one million nephrons in one kidney. Some nephrons lie near the medulla and are called juxtamedullary nephrons. These nephrons extend deep into the medulla. Other nephrons reside in the cortex and extend only minimally into the medulla. These are known as cortical nephrons. The nephron consists of a renal tubule and a renal corpuscle. The renal corpuscle is a spherical structure that consists of a capillary network called the glomerulus, surrounded by a fibrous capsule called the glomerular capsule (Bowman's capsule). The capillary network is fed by an afferent arteriole. Blood exiting the nephron flows through the efferent arteriole and peritubular capillaries that surround the nephron. The glomerulus and glomerular capsule is where filtration occurs. The glomerular capsule consists of two layers. The outer layer, or parietal layer, consists of simple squamous epithelium. The inner layer, or visceral layer, contains special cells called podocytes. Between the podocytes and the capillaries is a thin basement membrane. The podocytes surround the glomerular capillaries. Small openings between the podocytes, called filtration slits, act as holes in a filter. The glomerular capillaries also contain small openings called fenestrae. The combined action of these structures is to act as a filter. The filtrate from the glomerular capsule flows through the first part of the renal tubule, known as the proximal convoluted tubule. The renal tubule is lined with cuboidal epithelium and a basement membrane, with the exception of the ascending limb of the nephron loop. The renal tubule performs reabsorption and secretion of substances. Substances are selectively moved from the tubule to the peritubular capillaries or from the peritubular capillaries to the tubule. Fluid then moves through the nephron loop (loop of Henle). The nephron loop has a descending and an ascending limb, each with different tissue characteristics. The ascending limb is lined with simple squamous epithelium in the lower section that again becomes simple cuboidal epithelium in the thick section. Surrounding the nephron loop are capillaries known as vasa recta. After flowing through the nephron loop, the fluid flows through the distal convoluted tubule and past the juxtaglomerular apparatus, which consists of a group of cells that reside between the distal convoluted tubule and afferent arteriole. The juxtaglomerular apparatus monitors blood-solute concentration as well as the concentration of the urine. It aids in regulating fluid volume and blood pressure. Fluid (urine) then drains from the distal convoluted tubule into the collecting duct. Urine from many nephrons drains into one collecting duct. The collecting duct merges with other collecting ducts at the renal papilla. Urine flows from the renal papilla to the minor calyces, which combine to form major calyces, which combine to form the renal pelvis. Urine now flows into the ureter and to the urinary bladder, urethra, and out of the body.

Ureters, Bladder, Urethra

The ureters carry the urine from the kidney to the bladder. They exit the kidneys at the hilum and extend inferiorly and medially to the bladder. The ureters have a smooth muscle layer that is capable of producing peristaltic contractions that occur once every two to three minutes. The parasympathetic nervous system increases these contractions and the sympathetic nervous system inhibits them.

The urinary bladder is a hollow organ that resides in the pelvic cavity. The ureters connect at the posterolateral surface. The urethra carries the urine from the bladder out of the body. The area on the inside of the bladder between the two ureter connections and the urethra is called the trigone.

The urinary bladder and ureters are internally lined with transitional epithelium. Transitional epithelium is a special kind of epithelium that allows for the cells to slide past each other during distension of the bladder. The bladder also has a thick smooth muscle layer sometimes called the detrusor muscle. Contraction of the detrusor muscle increases the internal pressure of the bladder and causes urine to be expelled. Male bladders contain an area of smooth muscle and elastic tissue called the internal urinary sphincter. This area is not present in females. The function of this structure is to keep semen from entering the urinary bladder during intercourse. Both males and females have an external urinary sphincter located in the urethra that controls the flow of urine. The male urethra consists of three parts. The prostatic urethra exits the bladder and extends to the inferior prostate gland. It then becomes the membranous urethra until it enters the penis, where it becomes the penile urethra.

Micturition

Urine continuously flows from the kidney to the bladder. The bladder acts as a reservoir for urine and can store up to 1 liter. At about 300 ml, the urge to urinate becomes evident. Once the wall is stretched, the micturition reflex is stimulated. Stretch of the bladder sends impulses to sensory neurons in the pelvic nerves to the sacral segments of the spinal cord. Micturition is under parasympathetic control and parasympathetic impulses cause the bladder to contract. The motor impulses for micturition originate in a micturition center in the pons. The center also receives input from the cerebral cortex (so one can decide whether or not to micturate). Contraction of the bladder increases the internal pressure pushing urine into the urethra. The micturition reflex is an involuntary reflex in infants. Voluntary control of the reflex does not occur until around age 2–3 years.

Urinary Physiology: Glomerular Filtration

Recall that the structures of the glomerulus and glomerular capsule act together as a filter. They filter the blood, removing water and substances small enough to fit through the filtration slits in the glomerular capsule. Larger substances, such as blood cells and plasma proteins, remain in the blood. We can describe the amount of blood that passes through the kidneys as a percentage of cardiac output. This is known as renal fraction. For example, the renal fraction is between 12% and 30% of total cardiac output. It normally averages about 20%, which works out to a rate of blood flow of about 1.1 liters per minute. Also if blood consists of about 55% plasma, then we can determine the amount of plasma flowing through the kidney in one minute, which is about 616 ml. This is known as the renal

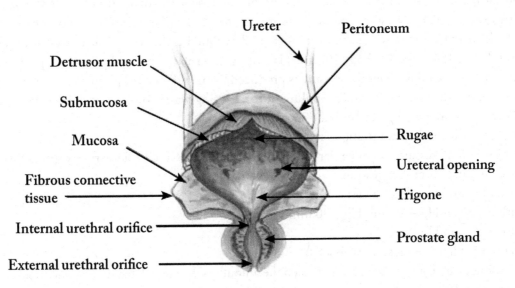
Figure 18.4. Urinary bladder.

plasma flow rate. The portion of the plasma that is filtered produces a filtrate known as the filtration fraction. The filtration fraction is about 20% of the plasma flowing into the filter. This works out to 616,ml 3 .2 or about 123 ml per minute. So the kidneys produce about 123 ml of filtrate per minute. This is called the glomerular filtration rate. That means that in one day the kidneys produce (123 3 60 3 24) 177 L per day. We usually round this up to about 180 L per day. That's a lot of filtrate. We obviously don't urinate 180 L per day, so much of that filtrate is reabsorbed via tubular reabsorption. We usually produce about 1,L to 2,L of urine per day. That means that only about 1% of the filtrate actually becomes urine. The rest is reabsorbed. The glomerular filter consists of an input (glomerular capillaries), an output (glomerular capsule), and a filtration membrane (podocytes, fenestrae, basement membrane). The filtration membrane is very permeable and allows small substances of less than 7 nm through. These include water, glucose, and electrolytes. Larger substances, such as plasma, proteins, and cells, do not pass through the membrane. In order to move substances through the filter, there must be a pressure gradient. Substances must move from an area of higher pressure to lower pressure. The pressure gradient is called filtration pressure or net filtration pressure. Net filtration pressure is directly proportional to the glomerular filtration rate. So if for some reason net filtration increases or decreases, so does glomerular filtration rate, and so does the amount of filtrate produced. Net filtration pressure is the combination of a series of pressures that exist in the renal corpuscle. These include glomerular capillary hydrostatic pressure, glomerular capsular hydrostatic pressure, and colloid osmotic pressure.

Glomerular capillary hydrostatic pressure is the blood pressure inside the capillaries. It is usually about 50 mm Hg and must be greater that the pressure inside the glomerular capsule, known as glomerular capsular hydrostatic pressure. The glomerular capillary hydrostatic pressure is controlled in part by the diameter of the afferent and efferent arterioles. The efferent arterioles have a smaller diameter than the afferent arterioles have. The smaller diameter works to decrease blood flow through the efferent arterioles, increasing the pressure inside the glomerular capillaries. Changing the diameter of the afferent and efferent arterioles changes the glomerular capillary hydrostatic pressure. For example, increasing the diameter of the afferent arteriole or decreasing the diameter of the efferent arteriole increases the capillary pressure. The pressure inside the glomerular capsule is called the glomerular capsular hydrostatic pressure. This pressure is created by fluid inside the capsule as well as downstream in the tubules. It is usually about 10 mm Hg. The glomerular capsular hydrostatic pressure works against filtration. Colloid osmotic pressure is produced by the presence of plasma proteins in the urine, called colloids. The colloids produce a pulling force, causing water to move back into the glomerular capillaries. This pressure is usually about 30 mm Hg.

We can calculate the net filtration pressure by the following:

Net Filtration Pressure = Glomerular capillary hydrostatic pressure – Glomerular capsular hydrostatic pressure – Colloid osmotic pressure

If we plug in the normal values:

NFP = 50 mm Hg – 10 mm Hg – 30 mm Hg

NFP = 10 mm Hg

So the net filtration pressure is about 10 mm Hg.

The kidneys are constantly working to keep the amount of filtrate relatively constant, despite changes in mean arterial pressure. For example, as systemic blood pressure increases, the afferent arterioles vasoconstrict, keeping the glomerular capillary hydrostatic pressure constant. Likewise when blood pressure decreases, the afferent arteriole dilates. The sympathetic nervous system affects the afferent arteriole by causing it to vasoconstrict under intense sympathetic activity, such as when exercising strenuously or when in shock. Mild to moderate changes in sympathetic activity have little effect on glomerular filtration rate. The prostaglandin E2 (PGE2) works to counteract the effects of intense sympathetic stimulation by promoting vasodilation. Colloid osmotic pressure can also be affected by changes in the filtration membrane. If the kidneys become damaged, such as with glomerular nephritis, the filtration membrane allows the passage of plasma proteins (a condition known as proteinuria). Loss of plasma proteins causes a decrease in systemic colloid osmotic pressure. Fluid then moves from the plasma into the interstitium, causing systemic edema.

Juxtaglomerular Apparatus

The juxtaglomerular apparatus is a group of cells residing at the junction of the afferent arteriole and distal ascending limb of the nephron loop. The juxtaglomerular apparatus consists of two different types of cells. Juxtaglomerular cells are located on the afferent arteriole side. These are granulated cells containing renin. These cells secrete renin in response to decreases in blood pressure and decreased urine-solute concentration. Macula densa cells are located on the nephron loop side. These cells secrete nitric oxide in response to changes in urine solute concentration. Renin is secreted by the juxtaglomerular cells in response to a decrease in systemic blood pressure lower than 80 mm Hg. The cells act as a kind of mechanoreceptor that monitors the stretch of the afferent arteriole. The cells secrete renin in response to decreased stretch. Renin is also secreted in response to feedback from the macula densa cells when the macula densa cells sense decreases in solute concentration of urine.

Renin activates the renin-angiotensin mechanism (see cardiovascular chapter). Renin acts like an enzyme to allow the conversion of angiotensinogen to angiotensin I. Angiotensin I is converted to angiotensin II by the angiotensin converting enzyme (ACE) located in capillary endothelium (especially in the lungs).

Angiotensin II produces systemic vasoconstriction, causes sodium reabsorption in the kidney tubules both directly and through the action of aldosterone, and stimulates the hypothalamus to release antidiuretic hormone (ADH). The combined actions of angiotensin II work to retain fluid volume and consequently raise blood pressure. Fluid volume increases by way of increased reabsorption of sodium. Water follows sodium from the kidney tubules to the interstitium and blood, thereby increasing blood volume and blood pressure. Angiotensin II also promotes vasoconstriction. Systemic vasoconstriction produces increased peripheral resistance, thereby increasing blood pressure. Vasoconstriction of the afferent arteriole also works to decrease glomerular capillary hydrostatic pressure and subsequent net filtration pressure. The resulting glomerular filtration rate then decreases as well, and fluid volume is conserved. The macula densa cells also work to control glomerular filtration by a process known as tubuloglomerular feedback. The macula densa cells secrete varying amounts of nitric oxide that work to vasodilate the afferent arteriole. When filtrate concentration decreases, as when the filter is producing too much fluid, the secretion of nitric oxide decreases, causing vasoconstriction of the afferent arteriole. The resulting vasoconstriction reduced glomerular capillary hydrostatic pressure and subsequent filtration rate.

Tubular Reabsorption

Tubular reabsorption occurs in the proximal and distal convoluted tubule. It works in conjunction with tubular secretion and involves moving filtered substances back into the blood. The processes involved in tubular reabsorption include diffusion, active transport, osmosis, and facilitated diffusion. The tubules are surrounded by peritubular capillaries. The walls of the tubules also have some special characteristics. The cells of the kidney tubules form two membranes. On the tubule side, cells form the basolateral membrane. On the opposite side, close to the interstitium, cells form the apical (sometimes called luminal) membrane. In order to move substances from the tubules to the blood, they must move through the apical and basolateral membranes of the tubules as well as the peritubular capillary endothelium. Some substances move passively (powered only by concentration gradients), while others move via special transport proteins using ATP. The walls of the tubules are somewhat permeable and allow movement of some substances by way of concentration gradients between the cells of the tubules. The electrolytes calcium, magnesium, potassium, and some sodium move this way.

There is a large sodium gradient in the kidney tubules. Sodium concentration is high in the tubules and lower in the surrounding interstitium and blood.

Sodium-Glucose Symporter

Some cells in the apical membrane of the tubules contain special transport proteins called symporters. One such symporter transports sodium and glucose. This symporter is powered by the sodium gradient. Sodium moves through the protein from the tubule to inside the cell. The transport protein also takes glucose along for the ride by transporting it to the inside of the cell as well. Once inside the cells, sodium is removed from the cell by traveling through the basolateral membrane by the action of the sodium-potassium pump. Sodium travels to the interstitium and blood. Glucose is also removed from the cell by way of a transport protein to the interstitium and blood. Since so much sodium is reabsorbed, water is also reabsorbed by osmosis. The tubules contain special water-containing regions called aquaporins that aid in water reabsorption. Since the reabsorption of many substances relies on the number of transport proteins in cells, there is a transport maximum for each of these substances. When the maximum is reached, additional substances flow into the urine. An example of this is with the increased blood glucose levels found in diabetes. Glucose passes through the glomerular filter and is reabsorbed in the tubules. If the amount of glucose exceeds the number of transporters, then the excess glucose spills out in the urine in a condition known as glucosuria. The apical membrane also contains numerous carrier proteins similar to the sodium glucose symporter. For example, symporters use the sodium gradient to transport amino acids. The amino acids are transported into the interstitium at the basolateral membrane by way of a second transport protein. Other such symporters include those for sodium and chloride, and sodium and potassium. The symporters of the apical membrane work with transport proteins in the basolateral membrane to maintain the concentration gradients for sodium and potassium.

Tubular Secretion

Tubular secretion involves the movement of substances from the blood and interstitium to the kidney tubules. Unlike tubular reabsorption that moves substances in order to maintain fluid and electrolyte balance, tubular secretion works primarily to eliminate toxic substances or byproducts of metabolism. Tubular secretion can involve active or passive transport. An example of passive transport is the sodium-hydrogen antiporter. This transport protein uses the sodium gradient to move sodium from the tubule to inside the apical membrane cell, while at the same time moving excess hydrogen ions out of the cell and into the lumen of the tubule. The secreted hydrogen ions result from water and bicarbonate combining in the cell. The bicarbonate and sodium travel into the cell by way of a symporter located in the basolateral membrane, and move into the peritubular capillaries. The secretion of hydrogen ions plays an important role in maintaining acid–base balance. Other examples of secreted substances include ammonia, potassium, penicillin, and para-aminohippuric acid. These substances are normally not produced in the body.

Action of Aldosterone

Aldosterone is secreted by the adrenal cortex in response to activation of the renin-angiotensin mechanism. Aldosterone targets receptors on cells in the apical membrane that transport sodium. When aldosterone attaches to these receptors, the cells respond by increasing sodium transport (increasing tubular reabsorption of sodium). At the same time they also stimulate the tubular secretion of potassium.

Atrial Natriuretic Hormone

Atrial natriuretic hormone (ANH) is secreted by the wall of the right atrium in the heart in response to atrial stretch. ANH has the opposite action of aldosterone and inhibits sodium and water reabsorption in the kidney tubules.

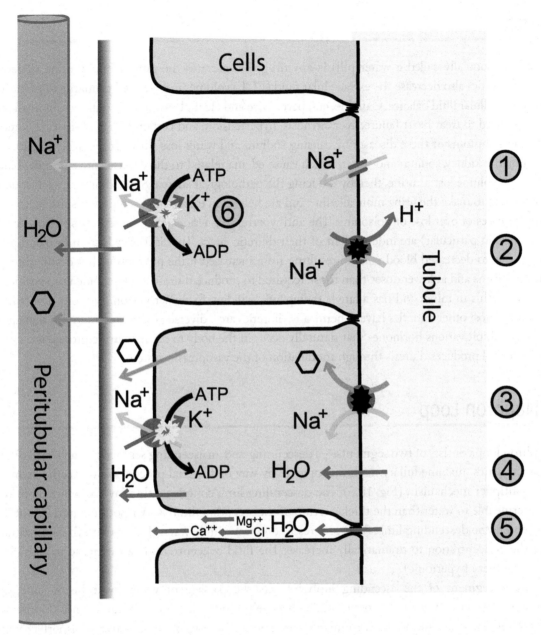

Figure 18.5. Tubular reabsorption and secretion. 1. Sodium passively transported; 2. Sodium-hydrogen antiporter; 3. Sodium-glucose symporter; 4. Water reabsorption by osmosis; 5. Passive diffusion of water, magnesium, chloride, and calcium.

Hypernatremia

An increase in plasma sodium levels above normal is hypernatremia. Sodium is the primary solute in the extracellular fluid. Sodium levels have a major role in osmolarity regulation. For excitable cells, the electrochemical gradient for sodium across the plasma membrane is critical for life. Water retention and an increased blood pressure usually are signs of hypernatremia. If the plasma sodium levels are below normal, it is called hyponatremia. Signs of this are low plasma volume and hypotension.

Diuretics

A diuretic (colloquially called a water pill) is any drug that elevates the rate of bodily urine excretion (diuresis). Diuretics also decrease the extracellular fluid (ECF) volume, and are used primarily to produce a negative extracellular fluid balance. Caffeine, cranberry juice and alcohol are all weak diuretics. In medicine, diuretics are used to treat heart failure, liver cirrhosis, hypertension, and certain kidney diseases. Diuretics alleviate the symptoms of these diseases by causing sodium and water loss through the urine. As urine is produced by the kidney, sodium and water, which cause edema related to the disease, move into the blood to replace the volume lost as urine, thereby reducing the pathological edema. Some diuretics, such as acetazolamide, help to make the urine more alkaline and are helpful in increasing excretion of substances such as aspirin in cases of overdose or poisoning. The antihypertensive actions of some diuretics (thiazides and loop diuretics, in particular) are independent of their diuretic effect. That is, the reduction in blood pressure is not due to decreased blood volume resulting from increased urine production, but occurs through other mechanisms and at lower doses than those required to produce diuresis. Indapamide was specifically designed with this in mind, and has a larger therapeutic window for hypertension (without pronounced diuresis) than most other diuretics have. Chemically, diuretics are a diverse group of compounds that either stimulate or inhibit various hormones that naturally occur in the body to regulate urine production by the kidneys. Alcohol produces diuresis through modulation of the vasopressin system.

The Nephron Loop

The nephron loop consists of two segments—a descending and an ascending segment—each with different characteristics. Juxtamedullary nephrons operate by way of a special process known as the countercurrent multiplier mechanism (Fig. 18.6). The descending limb contains a thin layer of epithelium that is more permeable to water than the thick portion of the ascending limb is. An isotonic fluid (about 300 mOsm) enters the descending limb. As it progresses down the limb, water diffuses into the interstitium, causing the concentration to dramatically increase. The fluid concentration can increase to as high as 1200 mOsm (very hypertonic).

The thick segment of the ascending limb inhibits the passage of water by diffusion and contains a series of active transport proteins that selectively move substances. Sodium and chloride are moved out of the ascending limb and into the interstitium by way of these active transport proteins. As fluid moves up the ascending limb, the concentration decreases. A 100 mOsm hypotonic solution exits the ascending limb and enters the distal convoluted tubule. The countercurrent consists of the current of water moving in one direction and the current of sodium chloride moving in the opposite direction. The high salt gradient is maintained by the active transport of sodium and chloride in the ascending limb. Urea also diffuses into the descending limb, adding to the increased concentration (Fig. 18.7).

Renal Clearance

Renal clearance is used to determine kidney function. Renal clearance is the volume of blood plasma from which a substance is completely removed in one minute. Renal clearance reflects the 3 processes of urine formation: glomerular filtration, tubular reabsorption, and tubular secretion. Substances passing though the glomerulus are added to the substances that are moving from the peritubular capillaries into the

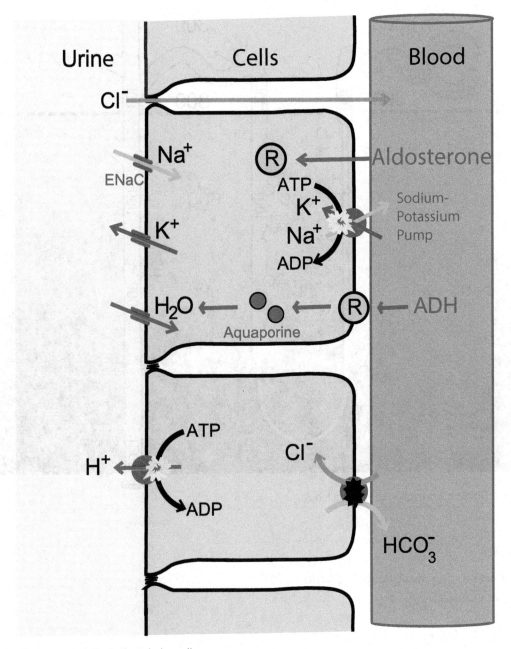

Figure 18.6. Transport proteins in the tubular cells.

tubules by way of secretion. The amount of the substance removed by tubular reabsorption is subtracted. However, this method is not practical. We can use an indirect method to determine renal clearance that includes the rate of urine output and the concentration of the substance in blood plasma and urine. For example, let's say we are determining the renal clearance for a substance X. We know the concentration of X in the urine is 5.0 mg/ml and the concentration of X is .3 mg/ml in the plasma. We also know the rate of urine output equals 2 ml/min. The renal clearance can be determined by the following:

Renal Clearance (C) = UV/P

U = concentration of substance in urine

V = rate of urine output

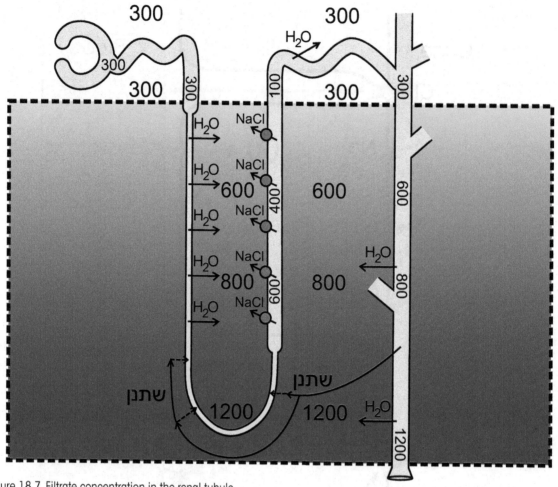

Figure 18.7. Filtrate concentration in the renal tubule.

P= concentration of substance in plasma
For our example:
C = (5.0)(2)/.3
C = 33.33 ml/minute
This can be interpreted as 33.33 ml of blood plasma is cleared of substance X every minute.

Glomerular Filtration Rate

We can use a substance that is not reabsorbed or secreted to determine the glomerular filtration rate. A substance that meets these criteria is inulin, a polysaccharide contained in artichokes and garlic. Glomerular filtration rate can be determined by injecting inulin and then measuring the urine ouput as well as the concentrations of inulin in the blood plasma and urine. Since inulin is not secreted or reabsorbed by the kidney tubules, the renal clearance is equal to the glomerular filtration rate. Thus we just need to calculate the renal clearance for inulin:

C = UV/P
U = 25 mg/ml
V = 2 ml/min
P = .4 mg/ml
C = (25)(2)/.4
C = 125 ml/min
Glomerular filtration rate (GFR) = C
GFR = 125 ml/min

Urine Composition

Adults produce about 1 to 2 liters of urine daily. Pathologies such as diabetes or some medications can produce a larger urine output known as polyuria. A urine output of less than 500 ml/day is known as oliguria and an output less than 100 ml is known as anuria. Urine is the final product of the kidney. It is mostly water (95%) with a few other solutes, including nitrogenous wastes, electrolytes, pigments, and toxins. It can also contain abnormal substances including glucose, albumin, bile, and acetone. Urine is usually clear or straw colored. An abnormal color may indicate presence of blood, bile, bacteria, drugs, food pigments, or high-solute concentration. Urine will become cloudy after standing, due to a buildup of bacteria. Pus from problems such as kidney infections will also make urine cloudy. Urine has a slight odor. It will develop an ammonia odor after standing, due the breakdown of urea. An acetone odor may indicate diabetes. The pH of urine varies between 4.6 and 8.0. The specific gravity is between 1.001 and 1.035.

Diseases of the Kidney

Diabetic nephropathy (nephropatia diabetica), also known as Kimmelstiel-Wilson syndrome and intercapillary glomerulonephritis, is a progressive kidney disease caused by angiopathy of capillaries in the kidney glomeruli. It is characterized by nodular glomerulosclerosis. It is due to longstanding diabetes mellitus, and is a prime cause for dialysis in many Western countries.

Kidney stones, also known as nephrolithiasis, urolithiasis, or renal calculi, are solid accretions (crystals) of dissolved minerals in urine found inside the kidneys or ureters. They vary in size from as small as a grain of sand to as large as a golf ball. Kidney stones typically leave the body in the urine stream; if they grow relatively large before passing (on the order of millimeters), obstruction of a ureter and distention with urine can cause severe pain most commonly felt in the flank, lower abdomen, and groin. Kidney stones are unrelated to gallstones.

Pyelonephritis: When an infection of the renal pelvis and calices, called pyelitis, spreads to involve the rest of the kidney as well, the result is pyelonephritis. It usually results from the spread of fecal bacterium Escherichia coli from the anal region superiorly through the urinary tract. In severe cases, the kidney swells and scars, abscesses form, and the renal pelvis fills with pus. Left untreated, the infected kidney may be severely damaged, but administration of antibiotics usually achieve a total cure.

Glomerulonephritis: Inflammation of the glomerular can be caused by immunologic abnormalities, drugs or toxins, vascular disorders, and systemic diseases. Glomerulonephritis can be acute, chronic, or progressive. Two major changes in the urine are distinctive of glomerulonephritis: hematuria and proteinuria with albumin as the major protein. There is also a decrease in urine as there is a decrease in GFR (glomerular filtration rate). Renal failure is associated with oliguria (less than 400 ml of urine output per day).

Renal failure: Uremia is a syndrome of renal failure and includes elevated blood urea and creatinine levels. Acute renal failure can be reversed if diagnosed early. Acute renal failure can be caused by severe hypotension or severe glomerular disease. Diagnostic tests include BUN and plasma creatinine level tests. It is considered to be chronic renal failure if the decline of renal function to less than 25%.

Diabetes insipidus: This is caused by the deficiency of or decrease of ADH. The person with diabetes insipidus (DI) has the inability to concentrate their urine in water restriction; in turn they will void up 3 to 20 liters/day. There are two forms of DI: neurogenic and nephrogenic. In nephrogenic DI, the kidneys do not respond to ADH. Usually nephrogenic DI is characterized by the impairment of the urine-concentrating capability of the kidney along with concentration of water. The cause may be a genetic trait, electrolyte disorder, or side effect of drugs such as lithium. Neurogenic DI is usually caused by head injury near the hypophyseal tract.

Urinary tract infections (UTIs): The second most common type of bacterial infections seen by health care providers is UTIs. Out of all the bacteria that colonize and cause urinary tract infections, the big gun is *Escherichia coli*. In the hospital, indwelling catheters and straight catheterizing predispose the opportunity for urinary tract infections. In females there are three stages in life that predispose urinary tract infections: menarche, manipulation between intercourse, and menopause. However, a small percentage of men and children will get urinary tract infections. In men it is usually due to the prostate gland growth, which usually occurs in older men. In children, it can occur in 3% to 5% of girls, where it may be the result of onset of toilet training, and in 1% of boys, and more commonly in uncircumcised boys than in circumcised ones. Some predispositions for getting urinary tract infection are family history and urinary tract anomalies. In neonates, urinary tract infections are most common when bacteremia is present.

Dialysis and Kidney Transplant

Generally, humans can live normally with just one kidney. Only when the amount of functioning kidney tissue is greatly diminished will renal failure develop. If renal function is impaired, various forms of medications are used, while others are contraindicated. Provided that treatment is begun early, it may be possible to reverse chronic kidney failure due to diabetes or high blood pressure. If creatinine clearance (a measure of renal function) has fallen very low (end-stage renal failure), or if the renal dysfunction leads to severe symptoms, dialysis is commenced. Dialysis is a medical procedure, performed in various different forms, where the blood is filtered outside of the body.

Kidney transplantation is the only cure for end-stage renal failure; dialysis is a supportive treatment, a form of "buying time" to bridge the inevitable wait for a suitable organ. The first successful kidney transplant was announced on March 4, 1954 at Peter Bent Brigham Hospital in Boston. The surgery was performed by Dr. Joseph E. Murray, who was awarded the Nobel Prize in Medicine in 1990 for this feat. There are two types of kidney transplants: living donor transplant and a cadaveric (deceased donor) transplant. When a kidney from a living donor, usually a blood relative, is transplanted into the patient's body, the donor's blood group and tissue type must be judged compatible with the patient's, and extensive medical tests are done to determine the health of the donor. Before a cadaveric donor's organs can be transplanted, a series of medical tests have to be done to determine if the organs are healthy. Also, in some countries, the family of the donor must give its consent for the organ donation. In both cases, the recipient of the new organ needs to take drugs to suppress their immune system to help prevent their body from rejecting the new kidney.

CRITICAL-THINKING QUESTIONS

1. What is suggested by the presence of white blood cells found in the urine?
2. Both diabetes mellitus and diabetes insipidus produce large urine volumes, but how would other characteristics of the urine differ between the two diseases?
3. Why are females more likely to contract bladder infections than males are?
4. What anatomical structures provide protection to the kidney?
5. How does the renal portal system differ from the hypothalamo–hypophyseal and digestive portal systems?
6. Explain what happens to Na^+ concentration in the nephron when GFR increases.
7. If you want the kidney to excrete more Na^+ in the urine, what do you want the blood flow to do?
8. What organs produce which hormones or enzymes in the renin–angiotensin system?
9. PTH affects absorption and reabsorption of what?
10. Why is ADH also called vasopressin?
11. How can glucose be a diuretic?
12. How does lack of protein in the blood cause edema?
13. Which three electrolytes are most closely regulated by the kidney?

Chapter Nineteen

Water, Electrolyte, and Acid–Base Balance

Fluids are an important constituent of tissues, membranes, and many other structures of the body, as well as play an important role in the body's chemistry. The human body is more than 70% water. Fluid levels must be maintained in order for the body to function properly. The same can be said for the millions of charged atoms called electrolytes. Our bodies run on the concentration gradients of electrolytes. Since electrolytes are carried by fluids, the balance of both fluids and electrolytes is of extreme importance in keeping our bodies alive. When we describe a solution in terms of concentration, we are defining the amount of solute in the solution. In many cases, the electrolytes represent the solute. Fluids also move across membranes by way of osmosis. For example, if we were losing water on one side of a membrane, we would say that this area is becoming hypotonic. Water would then move across the membrane by osmosis. Also, if we lost electrolytes from the same area, water would also be drawn across the membrane by osmosis. There are many electrolytes in the body. The most important of these, however, are sodium, potassium, and calcium, so you really want to know about these and how they are regulated. Acid–base balance is also important, as the blood is kept at a narrow range of pH. If it becomes too acidic, a condition called acidosis develops. Likewise, if the blood becomes too basic, a condition called alkalosis develops. Both acidosis and alkalosis can cause severe problems.

Fluids

The normal adult human can have as much as 75% water by weight. The amount of water depends on the proportion of body tissues contributing to the overall composition. The body loses water content with aging and can drop to about 45% water in the elderly. In physiology, fluid is explained in terms of two compartments. These are the compartment inside the cells or intracellular compartment (ICF) and the compartment outside the cells or extracellular compartment (ECF). If we were to generalize fluid and electrolyte problems, we could say that substances lost from the ECF are compensated for to a point. If the gain or loss exceeds the body's ability to compensate, then the gain or loss affects the cells, which causes a larger problem. The total fluid in a normal adult is about 40 liters. If we were to measure the fluid in the ICF and ECF, we would see that the ECF contains about 37% of the total fluid by volume. The ICF contains about 63%. The ICF consists of the cytoplasm of the cells. The ECF consists of a number of areas, including interstitial fluid, plasma, lymphatic fluid, and transcellular fluids. These include the cerebral spinal fluid, aqueous and vitreous humors, synovial fluid, and glandular secretions. The mechanisms allowing fluid movements are filtration and osmosis. The amount of fluid intake must equal the amount of fluid lost in order to maintain what is known as fluid balance. Normally about 1.5 liters of fluid is gained and lost per day. Fluid is gained by ingesting moist food (750 ml) and beverages (1500 ml) as well as through cellular metabolism (250 ml). Water is a byproduct of many metabolic processes of the body including aerobic metabolism. Fluid is lost through the urine (1500 ml), sweating (150 ml), feces (150 ml), and respiration (700 ml).

Fluid Regulation

The first method of fluid regulation is the thirst mechanism. Osmoreceptors in the hypothalamus sense an increase in solute concentration and stimulate the thirst center. A loss of as little as 1–2% of body fluid can stimulate thirst. Upon drinking, the resultant stretch of the stomach inhibits the thirst center. Water is then absorbed in the stomach and intestines. Solute concentration subsequently decreases.

Certain hormones have a powerful influence on regulating fluid volume. These include antidiuretic hormone, aldosterone, and atrial natriuretic peptide. Antidiuretic hormone (ADH) is secreted by the posterior pituitary gland in response to increased solute concentration, as sensed by osmoreceptors in the hypothalamus. ADH then travels through the blood to the kidneys, where it causes an increase in tubular permeability, particularly in the distal convoluted tubule of the nephron. When this occurs, water moves from the tubule to the interstitium and eventually into the blood. The primary effect is to conserve fluid volume. ADH also stimulates the thirst center in the hypothalamus. Aldosterone is a hormone secreted by the adrenal cortex in response to activation of the renin-angiotensin system (RA system). The RA system is triggered by the secretion of renin from the juxtaglomerular cells of the juxtaglomerular apparatus in the nephron of the kidney. Aldosterone targets cells in the kidney tubules, causing them to increase their permeability to sodium. When this occurs, sodium is reabsorbed and water follows by osmosis. Water moves into the interstitium and blood and is conserved. Atrial natriuretic peptide (ANP) is secreted by the right atrium of the heart in response to an increase in atrial stretch. Atrial stretch results from an increase in blood volume. The increase in plasma volume relates to an increase in fluid volume. ANP also inhibits the thirst mechanism.

Fluid Regulation Problems

When more is water is lost than gained, the result is dehydration. Water is first lost from the ECF, causing the osmotic pressure to rise due to a rise in solute concentration. If compensatory mechanisms cannot restore water balance, then the increase is osmotic pressure in the ECF causes a subsequent movement of water out of the ICF. If water loss continues, the person will experience thirst, dizziness, weakness, mental confusion, delirium, and coma. Water loss results from a decrease in fluid intake, vomiting, diarrhea, severe sweating known as profuse diaphoresis, or profuse urination from diseases such as diabetes. Dehydration can cause general hypovolemia, which can result in low cardiac output, electrolyte imbalances, and acid–base abnormalities. Treatment for dehydration includes restoring intake of fluids and electrolytes, either orally or through an intravenous route. Water intoxication or hyperhydration occurs when fluid intake exceeds water loss. It is a rare occurrence in adults and is more likely to occur in newborns who are given dilute formula or water. Newborns do not have a fully developed system for decreasing fluid volume. Water is gained first in the ECF, which causes the compartment to become hypotonic. This causes subsequent loss of fluid from the ICF. Water intoxication can be severe and result in muscle cramping, convulsions, confusion, coma, and brain edema.

Electrolytes

Like fluids, sodium balance depends on the intake versus excretion of sodium. A normal adult human has a sodium intake of about 1.1 to 3.4 g of sodium per day. Sodium enters the body through the digestive system epithelium. Sodium is excreted through the kidneys and skin via sweating. The kidneys are the main regulators of sodium in the body. Sodium is regulated by aldosterone. As discussed in the urinary system chapter, aldosterone increases the reabsorption of sodium in the kidney tubules. As stated earlier, sodium and water are often transported together as water moves by osmosis. This helps to keep the sodium concentration constant. For example, ingesting a large amount of sodium causes a subsequent increase in water absorption via osmosis. The additional water ends up in plasma and increases blood volume. The increase in blood volume results in an increase in blood pressure. This is why people with hypertension are told to limit their sodium intake.

When sodium concentration is reduced to below 130 mEq/L, a state of hyponatremia exists. Hyponatremia can result from prolonged and severe sweating, vomiting, diarrhea, renal disease, and a condition of the adrenal gland called Addison's disease. In hyponatremia, the ECF becomes hypotonic, causing swelling of the ICF. Symptoms include muscle spasms, postural blood pressure changes, nausea, vomiting, convulsions, confusion, and coma. Treatment for hyponatremia ranges from water restriction in mild cases to the administration of sodium orally or intravenously in more severe cases. When sodium concentration exceeds 145 mEq/L, a state of hypernatremia exists. Hypernatremia results from severe uncorrected diabetes insipidus/mellitus, severely high sodium intake, lack of fluid intake, diarrhea, heart disease, or renal failure. The signs and symptoms of hypernatremia include thirst, disorientation, lethargy, and central nervous system problems. Hypernatremia is treated with a hypotonic solution, which lowers sodium concentration. Most of the potassium in the body is located in the ICF (98%). Potassium enters the body through the digestive system and is excreted primarily by the kidney. Potassium is regulated mainly by aldosterone. Aldosterone causes secretion of potassium as well as reabsorption of sodium. High concentration of potassium in blood plasma stimulates the release of aldosterone. If potassium levels drop below 3.5 mEq/L, a state of hypokalemia exists. The causes of hypokalemia include Cushing's disease (affects the adrenal cortex causing increased aldosterone levels resulting potassium loss), potassium-wasting diuretics, increased urine output, gastric suctioning (without potassium replacement), and vomiting. Signs and symptoms of hypokalemia include muscle weakness, paralysis, atrial or ventricular arrhythmias, and respiratory problems. Potassium disorders can be very dangerous and result in a life-threatening condition. Hypokalemia can result in peaked T-waves on ECG, ventricular dysrhythmias, cardiac arrest, muscle weakness, failure of respiratory muscles, intermittent diarrhea, and intestinal colic. If potassium levels exceed 5.5 mEq/L, a state of hyperkalemia exists. This condition is very dangerous and life threatening. Causes of hyperkalemia include kidney disease, vomiting, diarrhea, potassium-sparing diuretics, extensive tissue damage, severe infections, and Cushing's syndrome. The treatment for

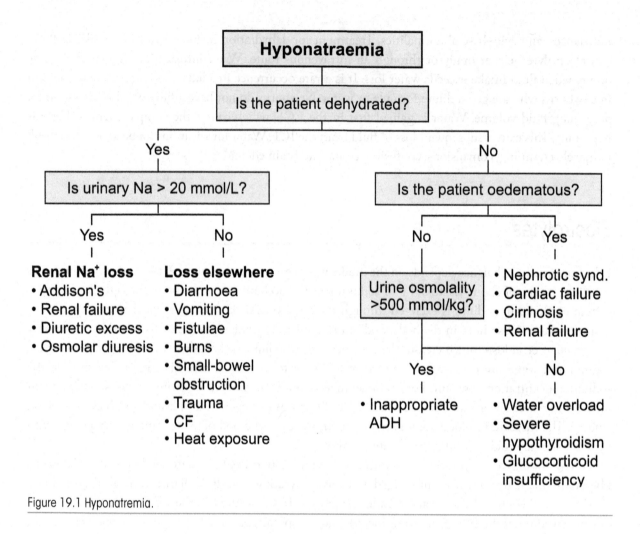

Figure 19.1 Hyponatremia.

hyperkalemia ranges from dietary restriction of potassium in mild cases to intravenous administration of calcium gluconate to correct cardiac problems along with dialysis to remove excess potassium. The underlying problem must also be corrected.

Calcium Balance

A typical adult human has about 1–2 kg of calcium. We have seen calcium play an important role in skeletal and cardiac muscle contraction as well as in the transmission of nervous system impulses. There is a dynamic interplay between calcium in blood and bone. Calcium can be deposited or removed from bone when necessary. Calcium enters the body through the digestive tract and is excreted in the kidneys with a small portion excreted in bile. A typical adult needs about .8 to 1.2 g/day of calcium. Calcium is regulated by the hormones calcitonin and parathyroid hormone. Calcitonin stimulates osteoblasts that remove calcium from blood and deposit it into bone. Calcitonin also inhibits osteoclastic activity. Osteoclasts remove calcium from bone so that it is available in the blood. Parathyroid hormone (PTH) has the opposite effect of calcitonin. PTH increases osteoclastic activity and decreases osteoblastic activity.

Both calcitonin and PTH work together to maintain calcium balance. When calcium levels decrease to lower than 4 mEq/L, a state of hypocalcemia exists. Hypocalcemia results from hypoparathyroidism, which produces a low level of parathyroid hormone, vitamin D deficiency, and renal failure. Signs and symptoms include muscle spasms, convulsions, and cardiac arrhythmias. When calcium levels exceed 11 mEq/L, a state of hypercalcemia exists. This can result from hyperparathyroidism and some cancers. Signs and symptoms include confusion, fatigue, arrhythmias, and calcification of the soft tissues of the body.

Magnesium Balance

Magnesium is needed for a number of metabolic reactions, including phosphorylation of glucose, and in muscle contraction. Magnesium is reabsorbed in the proximal convoluted tubule. Excess magnesium is excreted in the urine. The typical adult needs about .3–.4 g/day of magnesium.

Phosphate Balance

Phosphates are stored in the skeleton and used for phosphorylation of ADP. Phosphates are reabsorbed in kidney tubules and excreted in the urine. The typical adult needs about .8–1.2 g/day of phosphate.

Chloride Balance

Chlorides are the most numerous negative electrolytes in the body. Chloride ions are absorbed in the digestive tract and co-transported with sodium ions. Chloride ions are reabsorbed in the kidney tubules. The typical adult requires about 1.7–5.1 g/day of chloride.

Acid–Base Balance

The body maintains a narrow range of pH of the blood that is between 7.35 and 7.45. It must maintain this pH despite the constant release of acidic substances from metabolic processes and minute changes in pH associated with the respiratory system. Acids are substances that release hydrogen ions. Bases are substances the combine with hydrogen ions in order to neutralize them. Many bases release hydroxide ions that combine with hydrogen ions to form water. Remember that the pH scale ranges from 0 to 14. It is a logarithmic scale that measures tenfold increases in hydrogen ion concentration. A pH of 7 is neutral while a pH below 7 is considered acidic and a pH above 7 is considered basic or alkaline.

There are strong acids that completely dissociate in solution. For example, hydrochloric acid is considered a strong acid:

$HCl \rightarrow H^+ + Cl^-$

Weak acids do not completely dissociate in solution. For example, carbonic acid is considered a weak acid:

$H_2CO_3 \longleftrightarrow H^+ + HCO_3^-$

Notice the double arrow that indicates that the reaction reaches equilibrium in solution. This means that only a portion of the carbonic acid molecules will dissociate. Acid–base balance is maintained by the respiratory and urinary systems as well as buffer systems in the blood. The kidneys secrete hydrogen ions and reabsorb bicarbonate ions. These actions help to regulate pH. The respiratory system works to adjust pH by carbon dioxide storage. Although the kidneys have a large effect on pH, they tend to work slowly over a period of hours or days. Buffer systems work instantly to adjust pH. Most metabolic reactions in the body tend to release more hydrogen ions than combine with them. Hydrogen ions are released in the aerobic and anaerobic respiration of glucose, the incomplete oxidation of fatty acids, oxidation of amino acids containing sulfur, and the hydrolysis of phosphoproteins and nucleic acids.

Buffer systems are bidirectional chemical reactions that either release or combine with hydrogen ions in order to control pH. There are several important buffer systems in the body. These include the carbonic acid system, proteins, phosphates, and ammonium compounds.

Let's take a look at the carbonic acid system.

$$H_2CO_3 \longleftrightarrow H^+ + HCO_3^-$$

When there is an excess of hydrogen ions, the reaction moves to the left. In other words, the hydrogen ions combine with bicarbonate ions to form carbonic acid. This helps to raise pH. Likewise when pH increases the reaction moves to the right. Carbonic acid tends to dissociate into hydrogen ions and bicarbonate. The carbonic acid buffer system reacts quickly to the addition of substances such as lactic acid and carbon dioxide during periods of intense activity. Proteins also play a role in acting as a buffer system. These include cellular proteins such as histones and plasma proteins such as hemoglobin. The ability of proteins to act as buffers has to do with the presence of the carboxyl and amine groups. The carboxyl group acts as a weak acid, while the amine group acts as a weak base. When conditions become acidic, hydrogen ions bind to the amine group. Likewise when conditions become basic, hydrogen ions are released from the carboxyl group. Phosphates also act as buffers. Dihydrogen phosphate acts like a weak acid that dissociates according to the following reaction:

$$H_2PO_4^- \longleftrightarrow H^+ + HPO_4^{-2}$$

When conditions become acidic, monohydrogen phosphate (HPO_4^{-2}) combines with hydrogen ions to form dihydrogen phosphate ($H_2PO_4^-$). Phosphates are located in ATP, DNA, and RNA.

Ammonium also acts as a weak acid.

$$NH_4 \longleftrightarrow H^+ + NH_3$$

When conditions are acidic, hydrogen ions combine with ammonia to form ammonium. When conditions are basic, ammonium tends to dissociate into hydrogen ions and ammonia.

Respiratory System Regulation of Acid–Base Balance

We saw that carbon dioxide was stored in bicarbonate by the following reaction:

$$CO_2 + H_2O \longleftrightarrow H_2CO_3 \longleftrightarrow H^+ + HCO_3^-$$

In areas of higher concentrations of carbon dioxide, the reaction moves to the right. Carbon dioxide combines with water to form the intermediate carbonic acid, which dissociates into hydrogen ions and bicarbonate. Likewise when the concentration of carbon dioxide decreases, the reaction moves to the left. Typically the reaction moves to the right in the tissues as they produce higher levels of carbon dioxide. The reaction moves to the left in the alveoli, where carbon dioxide levels are lower, causing the release of carbon dioxide so that it can be removed by the lungs. As the blood becomes more acidic, respiratory centers in the brainstem respond by increasing the rate and depth of breathing. This causes the reaction to move more readily to the left. Hydrogen ions are removed by recombining with bicarbonate to form carbonic acid and subsequent carbon dioxide. As blood becomes more basic, respiratory centers are inhibited, causing an increased concentration of carbon dioxide. This causes the reaction to move to the right, allowing the release of hydrogen ions. Thus respiratory rate and depth of breathing act to maintain pH.

Urinary System Regulation of Acid–Base Balance

As conditions become acidic, hydrogen ions combine with bicarbonate to eventually form carbon dioxide and water. If this situation continues, then total body bicarbonate will decrease. Thus bicarbonate needs to be replaced by the kidneys. The opposite also applies. If pH gets too high (basic), bicarbonate is produced through the dissociation of carbonic acid. Thus excess bicarbonate needs to be removed by the kidneys. Bicarbonate is a small molecule that passes through the filter in the kidney. If bicarbonate were merely filtered, the total body bicarbonate would decrease to dangerous levels. Bicarbonate must be reabsorbed in the kidney tubules. The reabsorption of bicarbonate occurs in several steps. Carbon dioxide in the filtrate diffuses into kidney tubule cells. Inside the tubule cells, carbon dioxide and water combine to form carbonic acid, which dissociates into hydrogen ions and bicarbonate. Hydrogen ions are then secreted into the tubule by way of a sodium/hydrogen antiporter protein. Once in the tubule, hydrogen ions combine with filtered bicarbonate to form carbonic acid, which dissociates into carbon dioxide and water. The carbon dioxide can diffuse into the tubule cell to complete the circle. Bicarbonate ions inside the tubule cells are transported to the interstitium and blood by way of a sodium/bicarbonate symporter. So for each hydrogen ion secreted, one bicarbonate has been added to the blood. Generally as the filtrate moves through the tubules, most of the bicarbonate is reabsorbed. If there is an excess amount of hydrogen ions secreted, hydrogen ions can combine with non-bicarbonate molecules (such as phosphates). The carbon dioxide used to produce bicarbonate inside of the tubule cell then originates from the blood and enters the tubule cell by diffusion. Bicarbonate is still produced in the tubule cell and ends up being reabsorbed. However, this system produces a new bicarbonate for every hydrogen ion secreted. Besides phosphates, the amino acid glutamate can be a source of new bicarbonate for the body. Glutamate is broken down into bicarbonate and ammonium. The bicarbonate is reabsorbed into the blood. Glutamate produces about 50% of the new bicarbonate in the body each day.

Acid–Base Imbalances

Respiratory acidosis generally occurs from an inability of the lungs to get rid of carbon dioxide. Carbon dioxide then builds up in the blood and is converted to bicarbonate and hydrogen ions. The increased concentration of hydrogen ions causes the acidosis. Respiratory acidosis is caused by injuries to the respiratory centers in the brainstem, obstructions in air passages, and decreases in gas exchange such as with emphysema and pneumonia. The symptoms of respiratory acidosis include drowsiness, disorientation, stupor, coma, and even death in severe cases. Respiratory acidosis is treated with an intravenous infusion of sodium lactate. The lactate ions are converted to bicarbonate ions in the liver and the bicarbonate ions help to buffer the hydrogen ions.

Respiratory alkalosis can develop from fever, hyperventilation, and salicylate (aspirin) poisoning. An increased amount of carbon dioxide is removed from the lungs, decreasing the hydrogen ion concentration in the blood. Respiratory alkalosis results from aspirin poisoning because aspirin stimulates the respiratory centers in the medulla, causing an increased respiratory rate. This causes a subsequent increase in removal of carbon dioxide from the lungs. Symptoms of respiratory alkalosis include lightheadedness, dizziness, tingling sensations in the hands and feet, and tetany of muscles in severe cases. Not all acidosis is from the respiratory system. Metabolic acidosis can occur from an accumulation of acids in or loss of bases from the body. Examples of conditions that can cause metabolic acidosis are kidney disease (kidneys fail to secrete acids), prolonged vomiting (lose alkaline substances from GI tract), and prolonged diarrhea and diabetes mellitus (production of ketone bodies that lower pH). The symptoms are the same as respiratory acidosis and so is the treatment. The underlying cause of the problem must be treated as well as the symptoms. Metabolic alkalosis occurs from a loss of hydrogen ions or gain of bases in the body. Examples of conditions that can cause alkalosis are gastric lavage, prolonged vomiting, diuretics, and taking too much antacid. The symptoms are the same as respiratory alkalosis and so is the treatment.

Acid–base balance can be determined by a blood test called an arterial blood gas (ABG). The normal values are:

pH between 7.35 and 7.45

PCO_2 between 35 and 45 mm Hg

[H^+] between 35 and 45 nanomoles per liter

[HCO_3^-] between 22 and 26 nanomoles per liter

In order to determine an acid–base abnormality, the pH is examined first. If pH is low, we know acidosis exists. Likewise if pH is high, a state of alkalosis exists. Next we look at PCO_2. For example, a low pH indicates acidosis. If we combine this information with a high PCO_2, then we know that the respiratory system is responsible for the acidosis. If the PCO_2 is low, then we can look at bicarbonate ion concentration. If bicarbonate is high, then we know a state of metabolic acidosis exists. In acidosis, bicarbonate ions buffer the excess hydrogen ions. We can use the same rationale for determining alkalosis. For example, if pH is high and PCO_2 is low, we know a state of respiratory alkalosis exists. However, if PCO_2 is normal and bicarbonate is high, we know a state of metabolic alkalosis exists.

CRITICAL-THINKING QUESTIONS

1. Plasma contains more sodium than it does chloride. How can this be if individual ions of sodium and chloride exactly balance each other out, and plasma is electrically neutral?
2. Explain how the CO2 generated by cells and exhaled in the lungs is carried as bicarbonate in the blood.
3. How can one have an imbalance in a substance, but not actually have elevated or deficient levels of that substance in the body?
4. Describe the conservation of bicarbonate ions in the renal system.
5. Describe the control of blood carbonic acid levels through the respiratory system.
6. Aldosterone can stimulate the secretion of hydrogen ions by the kidney. Predict the consequences of increased aldosterone secretion on body fluid pH.
7. A patient suffering from diabetes mellitus often develops a condition called ketoacidosis (an acidosis caused by an accumulation of acidic ketone bodies in the plasma). A patient in this situation will tend to hyperventilate in order to compensate for the acidosis. Explain why hyperventilation might help this situation.
8. A person suffering from second- and third-degree burns over 60% of their body develops hyperkalemia. Why does this happen?

9. The respiratory rate of a hospitalized patient increased from 12 per minute to 38 per minute. This increase in respiratory rate continued for 5 minutes. Would the following blood parameters increase, decrease, or remain the same?
   A. pH
   B. hydrogen ion concentration
   C. carbon dioxide concentration
10. A patient has suffered lesions in the hypothalamus and lost the ability to produce ADH. Despite this loss, plasma sodium levels are maintained at normal or near-normal levels.
   A. Speculate why the urine output of this individual increases dramatically.
   B. If the urine output increases dramatically, what must happen to fluid intake in order to maintain homeostasis?

Section Six
Reproduction

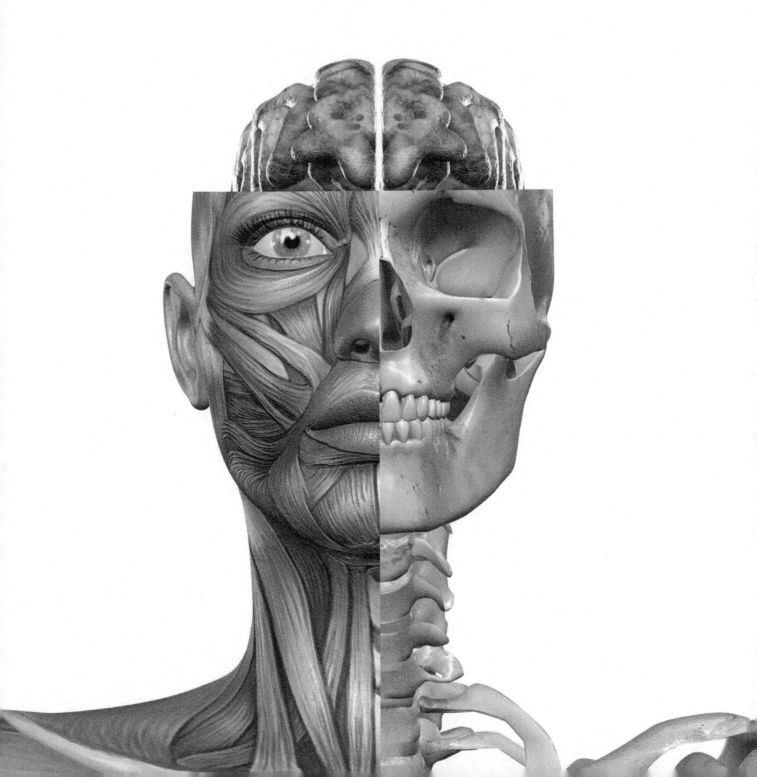

Chapter Twenty

Reproductive Systems

Both male and female reproductive systems may seem somewhat isolated from other body systems in that their purpose is to create new life and not just to maintain existing life. There are, however, significant relationships between the reproductive system and other body systems. All systems relate in one way or another to help our bodies maintain homeostasis. In simple terms, reproduction is the process by which organisms create descendants. This is a characteristic that all living things have in common and sets them apart from nonliving things. Even though the reproductive system is essential to keeping a species alive, it is not essential to keeping an individual alive. In human reproduction, two kinds of sex cells or gametes are involved. Sperm, the male gamete, and a secondary oocyte (along with first polar body and corona radiata), the female gamete, must meet in the female reproductive system to create a new individual. For reproduction to occur, both the female and male reproductive systems are essential. It is a common misnomer to refer to a woman's gametic cell as an egg or ovum, but this is impossible. A secondary oocyte must be fertilized by the male gamete before it becomes an "ovum" or "egg." While both the female and male reproductive systems are involved with producing, nourishing, and transporting either the oocyte or sperm, they are different in shape and structure. The male has reproductive organs, or genitals, that are both inside and outside the pelvis, while the female has reproductive organs entirely within the pelvis.

Male Reproductive System

The male reproductive system can be thought of as having two divisions. The primary organs are the testes and the rest of the organs are considered secondary. The function of the male system is to produce and develop sperm cells, transport the sperm to the female, and produce and secrete sex hormones (Fig. 20.1). The testes are considered the primary sex organs of the male. The testes develop in utero in a retroperitoneal location. They descend and pass through the inguinal canal to finally reside in the scrotum. A tissue structure called the gubernaculums connects the developing testes with the scrotum.

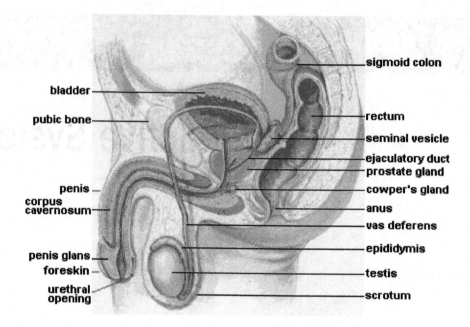

Figure 20.1 Male reproductive system.

As the gubernaculum shortens, the testes descend to the scrotum. This occurs at between 7 and 9 months of fetal development. The secretion of testosterone facilitates this process. The scrotum is a sac located outside of the pelvic cavity. The scrotum consists of skin and is divided into two chambers. It also contains a smooth muscle known as the dartos muscles that can contract and draw the testes closer to the body. The abdominal muscles also connect to the testes by way of the cremaster muscles. These muscles also work to draw the testes closer to the body when they contract. By adjusting the position of the testes the internal temperature of the testes can be adjusted. The testes are considered both endocrine and exocrine glands. They produce hormones that travel through the blood and secrete sperm cells that travel through ducts. The adult male has two testes. The testes are covered by a connective tissue membrane called the tunic albuginea. The inside of the testes is arranged in a series of lobules with tubular structures called seminiferous tubules within. The seminiferous tubules are surrounded by cell known as interstitial cells of Leydig. The Leydig cells secrete testosterone. The seminiferous tubules are coiled structures and empty into straight tubules called tubuli recti, which in turn empty into the rete testes, which constitutes a tubular network. The rete testes empty into the efferent ductules, which in turn move through the tunica albuginea to the epididymis. The testes also contain the sperm cells called spermatogonia as well as sustentacular cells (Sertoli cells). The Sertoli cells are columnar in shape and form a barrier between the testes and the blood. This barrier helps to isolate sperm cells so that the immune system does not attack them. Sertoli cells secrete the hormone called inhibin as well as substances to help sperm mature. Sperm cells have different antigens on their surface than body cells have that could trigger an immune response. The epididymis is a tubular structure that resides on the superior-posterior surface of the testes. These paired structures each have three portions consisting of a head, body, and tail. The head connects with the efferent ductules of the testes. The epididymis works to help sperm cells mature as they spend up to 3 weeks in the tubule system within the epididymis. Sperm move through the epididymis to the vas deferens. The vas deferens or ductus deferens is a tubular structure that is consistent with the tail of the epididymis. The vas deferens has three muscular layers: inner and outer longitudinal layers and a middle circular layer. The muscular layers help to propel sperm cells through the tube. Each vas deferens moves superiorly through the inguinal canal and travels through the abdominal cavity and over the top of the bladder to the seminal vesicle. The vas deferens travels in the spermatic cord, which is a connective tissue sheath that also contains blood vessels and lymphatics. As the vas deferens nears the seminal vesicle, the tube widens into an ampulla. The seminal vesicles are located posterior to the bladder and anterior to the rectum. They contain epithelium that secretes an alkaline substance, fructose, and prostaglandins. The prostate is a walnut-shaped gland just inferior to the bladder. The prostate gland secretes an acidic milky fluid that helps to nourish and mobilize sperm. The fluid also contains enzymes (hyaluronidase) and prostate specific antigen (PSA).

The urethra (prostatic urethra) passes through the prostate gland. The prostate also contains another set of paired ducts that connect the seminal vesicles to the urethra called the ejaculatory ducts. The paired bulbourethral glands (Cowper's glands) are pea-shaped glands that secrete an alkaline substance and mucus to help protect and transport the sperm. The urethra begins at the base of the urinary bladder and passes through the prostate gland and through the penis, ending at the urinary meatus of the penis. The urethra is lined with a mucous membrane. There are three parts to the male urethra. These are the portion traveling through the urethra (prostatic urethra), the portion extending from the base of the prostate gland to the penis (membranous urethra), and the portion running through the center of the penis (penile urethra). The penis consists of three columns of tissue called erectile columns, surrounded by fibrous coverings that are surrounded by skin. The two superior columns are called the

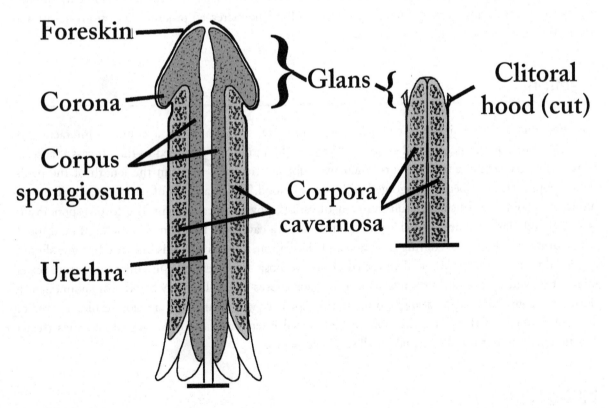

Figure 20.2 Penis.

corpora cavernosa and the lower column is called the corpus spongiosum. Each corpus cavernosum contains a deep artery and is surrounded by a fibrous covering called a tunica albuginea. The corpus spongiosum contains the urethra. The distal portion of the penis contains a slightly larger structure called the glans penis. The glans penis is covered by loose skin called the prepuce, which is sometimes removed by circumcision.

Genitalia

A boy's penis grows little from the fourth year of life until puberty. Average prepubertal penile length is 4 cm. The prepubertal genitalia are described as stage 1. Within months after growth of the testes begins, rising levels of testosterone promote growth of the penis and scrotum. This earliest discernible beginning of pubertal growth of the genitalia is referred to as stage 2. The penis continues to grow until about 18 years of age, reaching an average adult size of about 10–16 cm.

Although erections and orgasm can occur in prepubertal boys, they become much more common during puberty, accompanied by development of *libido* (sexual desire). Ejaculation becomes possible early in puberty; prior to this boys may experience dry orgasms. Emission of seminal fluid may occur due to masturbation or spontaneously during sleep (commonly termed a *wet dream*, and more clinically called a *nocturnal emission*). The ability to ejaculate is a fairly early event in puberty compared to the other characteristics, and can occur even before reproductive capacity itself. In parallel to the irregularity of the first few periods of a girl, for the first one or two years after a boy's first ejaculation, his seminal fluid may contain few active sperm. If the foreskin does not become retractable during childhood, it normally begins to retract during puberty. This occurs as a result of the increased production of testosterone and other hormones in the body.

Erection

The erection of the penis is its enlarged and firm state. It depends on a complex interaction of psychological, neural, vascular, and endocrine factors. The term is also applied to the process that leads to this state. A penile erection occurs when two tubular structures that run the length of the penis, the corpora cavernosa, become engorged with venous blood. This is a result of parasympathetic nerve–induced vasodilation. This may result from any of various physiological stimuli. The corpus spongiosum is a single tubular structure located just below the corpora cavernosa that contains the urethra, through which urine and semen pass during urination and ejaculation, respectively. This may also become slightly engorged with blood, but less so than the corpora cavernosa. Penile erection usually results from sexual stimulation and/or arousal, but can also occur by such causes as a full urinary bladder or spontaneously during the course of a day or at night, often during erotic or wet dreams. An erection results in swelling and enlargement of the penis. Erection enables sexual intercourse and other sexual activities (sexual functions), though it is not essential for all sexual activities.

Ejaculation

"Emission" is the term used when sperm moves into the urethra. "Ejaculation" is the term used when sperm is forced out of the urethra and the penis. These are both stimulated by sympathetic nerves.

Puberty

In addition to producing sperm, the male reproductive system also produces sex hormones, which help a boy develop into a sexually mature man during puberty. When a baby boy is born, he has all the parts of his reproductive system in place, but it isn't until puberty that his reproductive organs mature and become fully functional. As an newborn, FSH and LH levels are high and after a few weeks, levels drop to extremely low. When puberty begins, usually between the ages of 10 and 14, the pituitary gland—which is located in the brain—secretes hormones that stimulate the testicles to produce testosterone. The production of testosterone brings about many physical changes. Although the timing of these changes is different for each individual male, the stages of puberty generally follow a set sequence. First stage: the scrotum and testes grow larger, the *apocrine glands* develop. Second stage: the penis becomes longer, and the seminal vesicles and prostate gland grow. Hair begins to grow in the pubic region. Reproductive capacity has usually developed by this stage. Third stage: hair begins to appear on the face and underarms. During this time, a male's voice also deepens. Fertility continues to increase.

Testicular size, function, and fertility

In boys, testicular enlargement is the first physical manifestation of puberty (and is termed "gonadarche"). Testes in prepubertal boys change little in size from about 1 year of age to the onset of puberty, averaging about 2–3 cc in volume and about 1.5–2 cm in length. Testicular size continues to increase throughout puberty, reaching maximal adult size about 6 years later. While 18–20 cm is reportedly an average adult size, there is wide variation in the normal population.

The testes have two primary functions: to produce hormones and to produce sperm. The Leydig cells produce testosterone, which in turn produces most of the changes of male puberty. However, most of the increasing bulk of testicular tissue is spermatogenic tissue (primarily Sertoli and interstitial cells). Sperm can be detected in the morning urine of most boys after the first year of pubertal changes (and occasionally earlier).

Pubic hair

Pubic hair often appears on a boy shortly after the genitalia begin to grow. As in girls, the first appearance of pubic hair is termed "pubarche" and the pubic hairs are usually first visible at the dorsal (abdominal) base of the penis. The first few hairs are described as stage 2. Stage 3 is usually reached within another 6 to 12 months, when the hairs are too numerous to count. By stage 4, the pubic hairs densely fill the pubic triangle. Stage 5 refers to spread of pubic hair to the inner thighs and upward toward the umbilicus as part of the developing abdominal hair.

Aging

For most men, testosterone secretion continues throughout life, as does sperm production, though both diminish with advancing age. Probably the most common reproductive problem for older men is prostatic hypertrophy, enlargement of the prostate gland. This causes the urethra to compress and urination becomes difficult. Residual urine in the bladder increases the chance of urinary tract infections. Prostate hypertrophy is usually benign, but cancer of the prostate is one of the more common cancers in elderly men. A TURP is commonly used to correct this problem if the symptoms do not improve in response to home treatment and medication. Erectile dysfunction (ED) is another common problem seen in aging males. In older men, ED usually has a physical cause, such as disease, injury, or side effects of drugs. Any disorder that impairs blood flow in the penis or causes injury to the nerves has the potential to cause ED. Although it is not an inevitable part of aging, incidences increases with age: about 5 percent of 40-year-old men and between 15 and 25 percent of 65-year-old men experience ED. As discouraging as erectile dysfunction may be, it is treatable at any age, and awareness of this fact has been growing. More men have been seeking help and returning to normal sexual activity because of improved, successful treatments for ED.

Male Reproductive System Abnormalities

Boys may sometimes experience reproductive system problems. Below are some examples of disorders that affect the male reproductive system (disorders of the scrotum, testicles, or epididymis). Conditions affecting the scrotal contents may involve the testicles, epididymis, or the scrotum itself.

Testicular trauma. Even a mild injury to the testicles can cause severe pain, bruising, or swelling. Most testicular injuries occur when the testicles are struck, hit, kicked, or crushed, usually during sports or due to other trauma. Testicular torsion, when one of the testicles twists around, cutting off the blood supply, is also a problem that some teen males experience—although it's not common. Surgery is needed to untwist the cord and save the testicle.

Varicocele. This is a varicose vein (an abnormally swollen vein) in the network of veins that run from the testicles. Varicoceles commonly develop while a boy is going through puberty. A varicocele is usually not harmful, although in some people it may damage the testicle or decrease sperm production, so it helps for you to take your child to see his doctor if he is concerned about changes in his testicles.

Testicular cancer. This is one of the most common cancers in men younger than 40. It occurs when cells in the testicle divide abnormally and form a tumor. Testicular cancer can spread to other parts of the body, but if it's detected early, the cure rate is excellent. Teen boys should be encouraged to learn to perform testicular self-examinations.

Epididymitis is inflammation of the epididymis, the coiled tubes that connect the testes with the vas deferens. It is usually caused by an infection, such as the sexually transmitted disease chlamydia, and results in pain and swelling next to one of the testicles.

Hydrocele. A hydrocele occurs when fluid collects in the membranes surrounding the testes. Hydroceles may cause swelling of the testicle but are generally painless. In some cases, surgery may be needed to correct the condition.

Inguinal hernia. When a portion of the intestines pushes through an abnormal opening or weakening of the abdominal wall and into the groin or scrotum, it is known as an inguinal hernia. The hernia may look like a bulge or swelling in the groin area. It can be corrected with surgery.

Disorders of Penis

Disorders of the penis include the following:

Inflammation of the penis. Symptoms of penile inflammation include redness, itching, swelling, and pain. Balanitis occurs when the glans (the head of the penis) becomes inflamed. Posthitis is foreskin inflammation, which is usually due to a yeast or bacterial infection.

Hypospadias. This is a disorder in which the urethra opens on the underside of the penis, not at the tip.

Phimosis. This is a tightness of the foreskin of the penis and is common in newborns and young children. It usually resolves itself without treatment. If it interferes with urination, circumcision (removal of the foreskin) may be recommended.

Paraphimosis. This may develop when a boy's uncircumcised penis is retracted but doesn't return to the unretracted position. As a result, blood flow to the penis may be impaired, and your child may experience pain and swelling. A doctor may try to use lubricant to make a small incision so the foreskin can be pulled forward. If that doesn't work, circumcision may be recommended.

Ambiguous genitalia. This occurs when a child is born with genitals that aren't clearly male or female. In most boys born with this disorder, the penis may be very small or nonexistent, but testicular tissue is present. In a small number of cases, the child may have both testicular and ovarian tissue.

Micro penis. This is a disorder in which the penis, although normally formed, is well below the average size, as determined by standard measurements.

Sexually transmitted diseases. Sexually transmitted diseases (STDs) that can affect boys include human immunodeficiency virus/acquired immunodeficiency syndrome (HIV/AIDS), human papillomavirus (HPV, or genital warts), syphilis, chlamydia, gonorrhea, genital herpes, and hepatitis B. They are spread from one person to another mainly through sexual intercourse.

Erectile dysfunction. ED is the inability to get or keep an erection firm enough for sexual intercourse. This can also be called impotence. The word "impotence" may also be used to describe other problems that can interfere with sexual intercourse and reproduction, such as problems with ejaculation or orgasm and lack of sexual desire. Using the term erectile dysfunction clarifies that those other problems are not involved.

Contraceptive for Men

Vasectomy: In this procedure the vas deferens of each testis is cut and tied off to prevent the passage of sperm. Sperm is still produced and stored in crypt sites, causing inflammation. Because of this inflammatory response, the immune system acts on them, destroys them, and then has antisperm antibodies. This causes a lower possibility if the vasectomy is reversed to becoming fertile again.

Condoms: A device, usually made of latex, or more recently polyurethane, that is used during sexual intercourse. It is put on a man's penis and physically blocks ejaculated semen from entering the body of a sexual partner. Condoms are used to prevent pregnancy, transmission of sexually transmitted diseases (STDs such as gonorrhea, syphilis, and HIV), or both.

Female Reproductive System

The reproductive systems of the male and female have some basic similarities and some specialized differences. They are the same in that most of the reproductive organs of both sexes develop from similar embryonic tissue, meaning they are homologous. Both systems have gonads that produce (sperm and egg or ovum) and sex organs. And both systems experience maturation of their reproductive organs, which become functional during puberty as a result of the gonads secreting sex hormones.

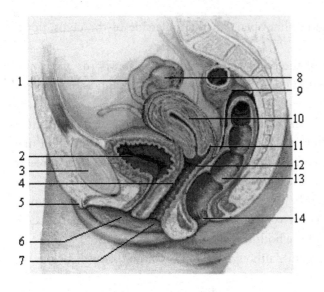

1. Fallopian tube
2. bladder
3. pubic bone (pubic symphysis)
4. g-spot
5. clitoris
6. urethra
7. vagina
8. ovary
9. sigmoid colonti
10. uterus
11. fornix of vagina (including anterior and posterior)
12. cervix
13. rectum
14. anus

Figure 20.3. Female reproductive system.

The differences between the female and male reproductive systems are based on the functions of each individual's role in the reproduction cycle. A male who is healthy and sexually mature continuously produces sperm. The development of women's "eggs" is arrested during fetal development. This means she is born with a predetermined number of oocytes and cannot produce new ones. At about 5 months' gestation, the ovaries contain approximately six to seven million oogonia, which initiate meiosis. The oogonia produce primary oocytes that are arrested in prophase I of meiosis from the time of birth until puberty. After puberty, during each menstrual cycle, one or several oocytes resume meiosis and undergo their first meiotic division during ovulation. This results in the production of a secondary oocyte and one polar body. The meiotic division is arrested in metaphase II. Fertilization triggers completion of the second meiotic division and the result is one ovum and an additional polar body. The ovaries of a newborn girl contain about one million oocytes. This number declines to 400,000 to 500,000 by the time puberty is reached. On average, 500–1,000 oocytes are ovulated during a woman's reproductive lifetime. When a young woman reaches puberty around age 10 to 13, a primary oocyte is discharged from one of the ovaries every 28 days. This continues until the woman reaches menopause, usually around the age of 50 years. Oocytes are present at birth, and age as a woman ages.

The female reproductive system not only carries the genetic information for offspring but is also capable of providing an environment for the early stages of growth and development of the human. The primary sex organs of the female reproductive system are the ovaries. All of the other structures are considered secondary organs. The ovaries are similar in structure to the testes. The ovaries begin in utero as masses of tissue located posterior to the abdominal cavity (retroperitoneal). They descend slightly and

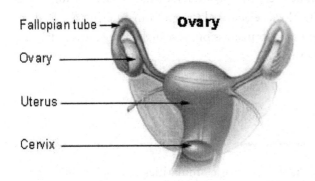

Figure 20.4 Uterus and ovaries.

Uterus and Uterine tubes

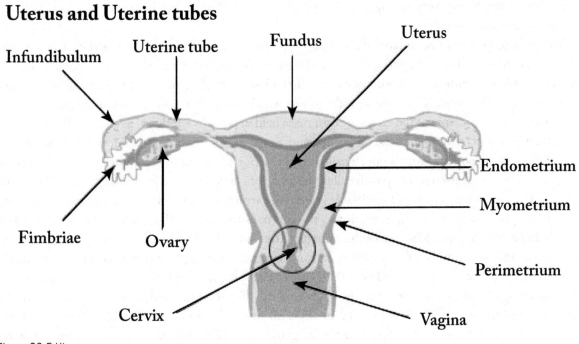

Figure 20.5 Uterus.

reside in the pelvic cavity on each side of the uterus. They are ovoid structures located inferior to the uterine tubes (Fallopian tubes). The ovaries are covered by a thin layer of epithelium called germinal epithelium. The inner region of the ovaries contains structures known as ovarian follicles surrounded by a connective tissue matrix. The ovarian follicle contains the egg cells known as oocytes. The oocytes are released at about halfway through the menstrual cycle in what is known as ovulation. The Fallopian tubes, or uterine tubes, extend from the lateral areas of the uterus and continue to near the ovaries but do not contact it. The tubes contain three layers: an inner mucous membrane, a middle muscular layer, and an outer serous layer. The inner mucous membrane is continuous with the peritoneal membrane surrounding the pelvic cavity. This membrane is also continuous with the walls of the vagina and can be susceptible to infection by microorganisms. The Fallopian tubes have three sections. These are the first third, which extends from the isthmus of the uterus; the second third, which ends in a widened area called the infundibulum; and the final third, which ends in finger-like projections called fimbriae. The Fallopian tubes work to transport the oocyte to the uterus after fertilization and are the sites for fertilization by sperm cells. Most fertilized oocytes move to the uterus but occasionally they will deposit somewhere in the pelvic cavity, causing what is known as an ectopic pregnancy. The uterus is a pear-shaped structure about three inches long and two inches in width. The uterus has two divisions: the body and cervix. The body ends anteriorly as a narrow region called the cervix and posteriorly as a rounded structure called the fundus. The uterus has three layers. The inner layer is called the endometrium. The endometrium varies in thickness and is thinner just after menstruation and thicker at the end of the cycle. The endometrium has an extensive blood supply and contains mucus-secreting cells. The mucus changes its consistency during various times of the menstrual cycle. It is normally thicker during most of the cycle and more watery near the time of ovulation to help move sperm cells through. The middle layer, or myometrium, is a thick smooth muscle layer. The smooth muscle is capable of producing very strong contractions during childbirth. The outer layer or perimetrium consists of a serous membrane.

The body of the uterus lies on top of the bladder in what is called an anteflexed position. The cervix of the uterus connects with the vagina at an upward right angle. This connection allows for pockets around the cervix called the anterior and posterior fornix that allow for pooling of semen to increase the chances of fertilization. The uterus can lie in retroflexion in which the uterus tilts backward. Retroflexion can sometimes cause a prolapse of the uterus. The uterus is held in place by a series of ligaments. These include two broad ligaments; two uterosacral ligaments; and a posterior, anterior, and two round ligaments.

The posterior ligament forms a pouch called the posterior cul de sac or rectouterine pouch. Likewise the anterior ligament also forms a pouch called the anterior cul de sac or vesicouterine pouch. The vagina is located between the rectum and urethra. It is a tubular structure about 3 inches long that opens to the outside and extends superior and posterior to the cervix of the uterus. The vagina is primarily smooth muscle lined with an epithelial mucous membrane. The mucous membrane forms near the opening of the vagina. This structure is called a hymen. In some cases the opening to the vagina can be completely covered by the hymen (imperforate hymen). An imperforate hymen needs to be medically punctured to allow discharge of the menstrual flow. The vulva consists of several externally located structures of the female reproductive system. These include the labia majora and minora, mons pubis, clitoris, vestibule, urinary meatus, and greater and lesser vestibular glands. The labia majora are skin-covered structures consisting of primarily adipose and connective tissue. The outer surface of the labia majora contains hair, while the inner surface does not. They also contain a mucous lining. They are analogous to the scrotum of the male. The labia minora are hairless structures located medially to the labia majora. The space between the labia minor is known as the vestibule. The clitoris is an organ consisting of erectile tissue. It is located just superior and behind the labial junction. The clitoris contains two corpora cavernosa but no corpus spongiosum, and it is similar in structure to the penis. The superior aspect of the clitoris contains a covering of tissue known as the prepuce. Between the clitoris and opening to the vagina (vaginal orifice) is the urinary meatus, which is the external opening of the urethra. On the sides of the vagina are the greater vestibular glands or Bartholin's glands that open into the area between the labia minor and hymen. The lesser vestibular glands or Skene's glands are located near the urinary meatus. The perineum is the area between the vagina and anus. The perineum helps to form the muscular floor of the pelvis and can be torn during vaginal childbirth. The perineum contains the urogenital triangle, which is formed by drawing a line between the ischial tuberosities with the anterior point of the triangle just superior to the prepuce.

The mammary glands or breasts are superficial to the pectoral muscles. Internally they consist of a series of lobes separated by connective tissue. The lobes subdivide into lobules containing secretory cells. The cells are arranged in clusters around a central duct. The smaller ducts combine to form larger ducts called lactiferous ducts for each lobe. The lactiferous ducts open to the outside at the nipple. The breasts also contain suspensory ligaments (of Cooper) that help to support it. Each breast contains a circular pigmented area called an areola. The areola contains sebaceous (oil secreting) glands to help protect the nipple. The breast also contains adipose tissue and lymphatics that drain into the axillary region.

Male Reproductive Physiology

Spermatogenesis encompasses the development of sperm cells in the male reproductive system. Spermatogenesis begins with the undeveloped sex cells called spermatogonia. Spermatogonia reside in the testes and will begin to mature around the age of puberty. They continue to do so throughout an adult male's life. The process begins with the secretion of gonadotropins from the anterior pituitary gland. These include follicle stimulating hormone and luteinizing hormone. Both are secreted in response to the releasing factor gonadotropin releasing hormone secreted by the hypothalamus. Luteinizing hormone (LH) is sometimes referred to as interstitial cell stimulating hormone (ICSH). This hormone targets the interstitial cells (Leydig cells) of the testes and promotes the secretion of testosterone. Follicle stimulating hormone (FSH) targets the sustentacular cells (Sertoli cells) of the testes and promotes their maturation and response to testosterone. Both FSH and testosterone work to facilitate the maturation of spermatogonia. Spermatogonia begin to mature in utero but their maturation is halted until puberty. They will undergo mitosis and develop into primary spermatocytes. Maturation is halted at this stage until puberty and the secretion of the sex hormones, as described above. Once reaching puberty, the primary spermatocytes undergo another type of cell division called meiosis. The two stages of meiosis are meiosis I and meiosis II. During meiosis I, the genetic material is divided in half. The normal adult human has 46 chromosomes (diploid number of chromosomes). Chromosomes form pairs that have the same but not necessarily identical genes. These are called homologous chromosomes. The pairs essentially split into two sections of homologous chromosomes, with each new cell having 23 chromosomes (haploid number of chromosomes). The chromosomes may contain different variants of genes. For example, one cell may contain a different variant for the gene for eye color from that of the other cell.

Steps of Meiosis:

Meiosis I

Prophase I: Chromatin condenses to form chromosomes. The nuclear membrane and nucleoli disappear and the spindle fibers begin to form, much like in mitosis. Homologous chromosomes form pairs. During this process genetic material may be exchanged through what is known as crossing over. In crossing over, chromatids cross over to the other chromosome and vice versa. This allows for a large amount of genetic variability in offspring. Each human has 8 million possible combinations of chromosomes that can combine with millions of combinations from their mate. This works out to more than 70 trillion possible unique human beings.

Metaphase I: Chromosome pairs line up in the midline of the cell and are attached to the spindle. Anaphase I: The chromosome pairs separate and move to each end of the spindle. The chromosome number for each new cell is now reduced by half. Telophase I: The cell cleaves and produces two new cells. The nuclear membrane and nucleolus reappear and the spindle dissipates. The new cells are now ready for meiosis II.

Meiosis II

Meiosis II is very similar to mitosis. Prophase II: Chromosomes reappear, the nuclear membrane and nucleolus disappear, and the spindle begins to form again. Metaphase II: Chromosomes line up in the middle of the cell and attach to spindle fibers. Anaphase II: Chromatids of chromosomes separate and are pulled to opposite ends of the cell. Telophase II: Cell division finishes. The cell cleaves, nuclear membrane and nucleolus reappear, and the chromatids unravel. There are now two cells, each with 23 chromosomes.

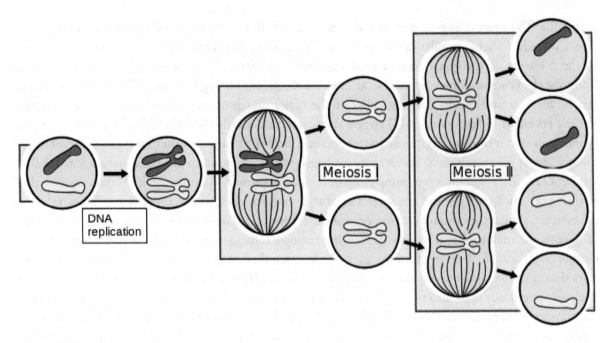

Figure 20.6 Meiosis.

Sperm Cells

Mature sperm cells contain three parts including a head, midpiece, and tail. The head contains the genetic material and has an enzyme-containing structure called an acrosome on its outer surface. The acrosome contains enzymes such as hyaluronidase that help the sperm cell penetrate the egg cell of the female. The midpiece contains many mitochondria that produce a good deal of energy in the form of ATP to power the long tail or flagellum. The tail contains the flagellum, which is constructed of protein microtubules.

Testosterone

Besides facilitating spermatogenesis, testosterone has other important functions in the male reproductive system. Testosterone is one of a group of hormones called androgens (male hormones). Testosterone is a steroid hormone and is converted to another form called dihydrotestosterone in certain cells in the male system, such as in the prostate and seminal vesicles. Testosterone levels are higher during fetal development to help initial development of the male reproductive system and descent of the testes. Testosterone levels then fall during childhood until puberty, when they again rise to essentially finish the job of maturation of the male reproductive system. The actions of testosterone during puberty include the following: enlargement of the vocal cords and deepening of the voice; increased muscular growth; increased body hair on face, axilla, and pubic areas; strengthening of bones; increased metabolism; maturation of the sex organs.

Testosterone is regulated by a feedback mechanism involving hormones from the hypothalamus and anterior pituitary gland. We saw that testosterone is secreted in response to LH being secreted by the anterior pituitary gland. LH is secreted in response to gonadotropin releasing hormone from the

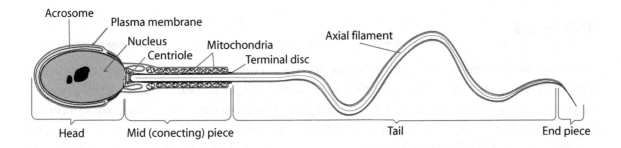

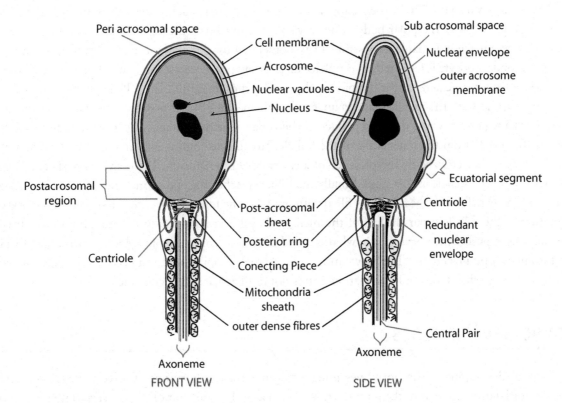

Figure 20.7 Sperm cell morphology.

hypothalamus. Blood concentration of testosterone is monitored by the hypothalamus, which responds through negative feedback to control the secretion of gonadotropin releasing hormone. The testes also secrete a hormone called inhibin, which feeds back to the hypothalamus, exhibiting the same effect as testosterone.

Female Reproductive Physiology

It is interesting to note that a number of the regulatory hormones are the same for both males and females. These include gonadotropin releasing hormone, FSH, and LH.

Oogenesis

Each ovary contains millions of sex cells called oocytes. The oocytes are encased in structures called follicles. At the premature stage, the follicles are known as primordial follicles, and each contains a primary oocyte. The primary oocytes begin meiosis but do not complete it until puberty. The development of oocytes is known as oogenesis. As oogenesis continues, at puberty the primary oocytes finish meiosis I, which results in two cells, each containing the haploid number of chromosomes (23). When the oocytes finish meiosis I, they are called secondary oocytes. Unlike spermatogenesis in the male, the resultant cells consist of one secondary oocyte and a polar body. The polar body is not a viable cell but helps the secondary oocyte to conserve resources to help make it as viable as possible. Development stops at this point unless fertilization occurs. Once the secondary oocyte is fertilized, it completes meiosis and produces a second polar body. The fertilized cell is now called a zygote and has the diploid number of chromosomes (46). The follicle plays an important role in oogenesis as well. The follicle matures under the influence of FSH. It first becomes a primary follicle and contains a region known as the zona pellucida. The zona pellucida contains glycoprotein that gradually separates the oocyte from the inner walls of the follicle. The follicle continues to mature into a secondary follicle, which is characterized by the presence of a cavity called the antrum. The oocyte is pushed against the inner wall of the follicle at this stage. Finally the follicle reaches the end of maturation as it becomes a mature, or Graafian, follicle. The antrum is filled with fluid and the follicle moves to the surface of the ovary. Maturation of the follicle occurs in half of the menstrual cycle. At about midway through the menstrual cycle, the follicle pushes the oocyte out in what is called ovulation. This occurs in response to a surge of LH from the anterior pituitary gland. The oocyte moves toward the Fallopian tube. If it becomes fertilized, it will eventually move to the uterus for implantation. If it is not fertilized, it will degenerate.

Female Sex Hormones

During fetal development, the hormones gonadotropin releasing hormone (GnRH), FSH, and LH cause the initial development of the reproductive system as well as descent of the ovaries to their normal position in the pelvic cavity. Secretion of GnRH then decreases until puberty, which occurs at about age 10. During puberty the levels of these hormones increase, causing the secretion of estrogens and progesterone. Estrogens are a group of molecules with estradiol as the most abundant. Estrogens are secreted by the ovaries as well as the adrenal glands, adipose tissue, and the placenta (during pregnancy). Estrogens promote development of the secondary sex organs of the female. Actions of estrogen include development of the breasts and an increase in adipose tissue under the skin in specific areas of the body (thighs, buttocks, breasts). There are also changes associated with the secretion of androgens during puberty, including increased hair in the genital and axillary regions. Estrogens also provide negative feedback to the hypothalamus and anterior pituitary gland. For example, a rise in estrogen levels works to inhibit the secretion of GnRH, which in turn inhibits the secretion of FSH and LH.

Breast Milk Production

The hormone prolactin works to stimulate milk production after birth. The first milk to appear is called colostrum, which contains some nutrients, including proteins and antibodies. When the infant suckles the breast, the sensory impulses travel to the hypothalamus, which in turn causes the release of oxytocin from the posterior pituitary. Oxytocin causes milk ejection by stimulating contraction of the myoepithelial cells of the breasts.

Menstrual Cycle

The female menstrual cycle begins during puberty (between the ages of 10 and 13 years). It is characterized by changes in the endometrium of the uterus. The first menstrual cycle is called menarche. GnRH is secreted by the hypothalamus, causing the secretion of FSH and LH. FSH causes maturation of the ovarian follicle and secretion of estrogens by the granulose cells. LH also helps the follicle to mature and stimulates the production of estrogens. Estrogens cause an increase in the thickening of the endometrium during the first phase of the menstrual cycle (proliferative phase). During the proliferative phase, the follicle secretes estrogen that works to inhibit the release of LH. Instead of being released by the anterior pituitary, LH is stored. At about the 14th day, LH is released (LH surge), causing the follicle to release the oocyte in what is called ovulation. The follicle then moves toward the Fallopian tube and is either fertilized or not. Following ovulation, the follicle becomes what is known as a corpus luteum. The corpus luteum secretes large amounts of progesterone and estrogens during this second half of the cycle. The estrogens and progesterone inhibit

Figure 20.8 Menstrual cycle.
1 Follicle Stimulating Hormone
2 Estrogens
3 Luteinizing Hormone
4 Progesterone
  A. Maturing Follicle and Corpus Luteum
  B. Hormone Levels

the release of LH and FSH from the anterior pituitary, which in turn keeps other follicles from maturing. Progesterone also facilitates increased vascularization and thickening of the endometrium. If the oocyte is not fertilized, the corpus luteum begins to degenerate near the end of the cycle at around the 24th day. What is left of the degenerated corpus luteum is called a corpus albicans. When the follicle goes from the corpus luteum stage to the corpus albicans stage, the secretions of estrogen and progesterone diminish. This causes the thickened endometrium to slough off. The endometrium and accompanying blood constitute the menstrual flow, which continues for about 3–5 days. Menstruation continues throughout the female lifespan until the late 40s or early 50s, when it begins to become irregular and eventually stops completely. This process marks the period of menopause. During this time the few remaining follicles no longer respond to FSH and LH. Since the follicles do not mature, there is a subsequent drop in estrogens and progesterone. The consequences of low levels of these hormones include thinning of the vaginal, urethral, and uterine linings; osteoporosis; and thinning of the skin.

Sperm cells that reach the oocyte attempt to penetrate it with the help of the enzymes located in the acrosome. Once a sperm cell penetrates the oocyte, it sheds its tail and the oocyte becomes unresponsive to other sperm. The nucleus of the sperm cell enters the oocyte and the oocyte undergoes meiosis II, creating a second polar body. The genetic material from sperm and oocyte combine and the resultant cell is called a zygote. The zygote continues mitotic division to form a group of cells that migrates to the uterus. The Fallopian tube helps the migration along with its ciliated epithelial lining and smooth muscle contractions. The cells eventually implant in the wall of the uterus. The layer surrounding the embryo secretes a hormone called human chorionic gonadotropin, which helps to maintain the corpus luteum throughout the pregnancy. This results in high levels of estrogens and progesterone. After the first trimester, the placenta takes over the job of secreting these hormones.

Signs of ovulation

The female body produces outward signs that can be easily recognized at the time of ovulation. The two main signs are thinning of the cervical mucus and a slight change in body temperature.

Thinning of the Cervical Mucus

After menstruation and right before ovulation, a woman will experience an increase of cervical mucus. At first, it will be thick and yellowish in color and will not be very plentiful. Leading up to ovulation, it will become thinner and clearer. On or around the day of ovulation, the cervical mucus will be very thin, clear, and stretchy. It can be compared to the consistency of raw egg whites. This appearance is known as "spinnbarkeit."

Temperature Change

A woman can also tell the time of ovulation by taking her basal body temperature daily. This is a temperature taken with a very sensitive thermometer first thing in the morning before the woman gets out of bed. The temperature is then tracked to show changes. In the uterine cycle, a normal temperature will be around 97.0–98.0. The day of ovulation, the temperature spikes down, usually into the 96.0–97.0 range, and then the next morning it will spike up to normal of around 98.6 and

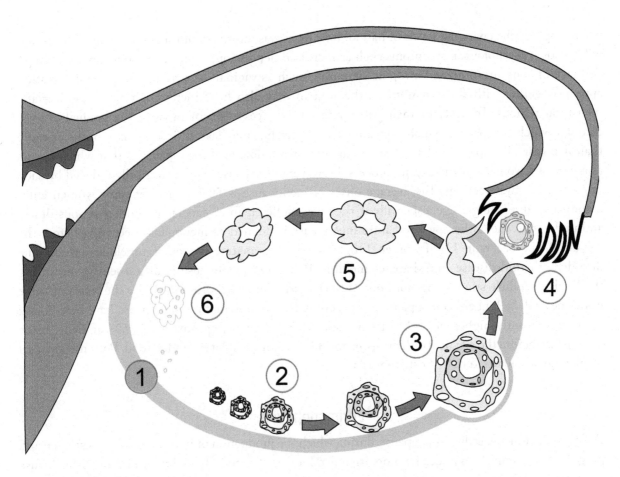

Figure 20.9 Follicle development.
1 - Menstruation
2 - Maturing follicle
3 - Mature follicle
4 - Ovulation
5 - Corpus luteum
6 - Deterioration of corpus luteum

stay in that range until menstruation begins. Both of these methods are used for conception and contraception. They are more efficient in conception due to the fact that sperm can live for two to three days inside of the Fallopian tubes. A woman could be off by a couple of days in her calculations and still become pregnant.

Menopause

Menopause is the physiological cessation of menstrual cycles associated with advancing age. Menopause is sometimes referred to as "change of life" or climacteric. Menopause occurs as the ovaries stop producing estrogen, causing the reproductive system to gradually shut down. As the body adapts to the changing levels of natural hormones, vasomotor symptoms such as hot flashes and palpitations; psychological symptoms such as increased depression, anxiety, irritability, mood swings, and lack of concentration; and atrophic symptoms such as vaginal dryness and urgency of urination appear. Together with these symptoms, the woman may also have increasingly scanty and erratic menstrual periods. Technically, menopause refers to the cessation of menses; the gradual process through which this occurs, which typically takes a year but may last as little as six months or more than five years, is known as climacteric. A natural or physiological menopause is that which occurs as a part of a woman's normal aging process. However, menopause can be surgically induced by such procedures as hysterectomy. The average onset of menopause is 50.5 years, but some women enter menopause at a younger age, especially if they have suffered from cancer or another serious illness and undergone chemotherapy. Premature menopause is defined as menopause occurring before the age of 40, and occurs in 1% of women. Other causes of premature menopause include autoimmune disorders, thyroid disease, and diabetes mellitus. Premature menopause is diagnosed by measuring the levels of follicle stimulating hormone (FSH) and luteinizing hormone (LH). The levels of these hormones will be higher if menopause has occurred. Rates of premature menopause have been found to be significantly higher in both fraternal and identical twins; approximately 5% of twins reach menopause before the age of 40. The reasons for this are not completely understood. Post-menopausal women are at increased risk of osteoporosis.

Perimenopause

Refers to the time preceding menopause, during which the production of hormones such as estrogen and progesterone diminish and become more irregular. During this period fertility diminishes. Menopause is arbitrarily defined as a minimum of twelve months without menstruation. Perimenopause can begin as early as age 35, although it usually begins much later. It can last for a few months or for several years. The duration of perimenopause cannot be predicted.

Premenstrual Syndrome (PMS)

It is common for women to experience some discomfort in the days leading up to their periods. PMS usually is at its worst the seven days before a period starts and can continue through the end of the period. PMS includes both physical and emotional symptoms: acne, bloating, fatigue, backaches, sore breasts, headaches, constipation, diarrhea, food cravings, depression, irritability, difficulty concentrating or handling stress.

Infertility

Infertility is the inability to naturally conceive a child or the inability to carry a pregnancy to term. There are many reasons that a couple may not be able to conceive without medical assistance. Infertility affects approximately 15% of couples. Roughly 40% of cases involve a male contribution or factor, 40% involve a female factor, and the remaining 20% involve both sexes. Healthy couples in their mid-20s having regular sex have a one-in-four chance of getting pregnant in any given month. This is called "fecundity."

Primary vs. Secondary

According to the American Society for Reproductive Medicine, infertility affects about 6.1 million people in the United States, equivalent to 10% of the reproductive-age population. Female infertility accounts for one third of infertility cases, male infertility for another third, combined male and female infertility for another 15%, and the remainder of cases are unexplained. "Secondary infertility" is difficulty conceiving after already having conceived and carried a normal pregnancy. Apart from various medical conditions (e.g., hormonal), this may come as a result of age and stress felt to provide a sibling for their first child. Technically, secondary infertility is not present if there has been a change of partners.

Factors of Infertility

- General factors
- Diabetes mellitus, thyroid disorders, adrenal disease
- Significant liver and/or kidney disease
- Psychological factors
- Hypothalamic-pituitary factors:
- Kallmann syndrome
- Hypothalamic dysfunction
- Hyperprolactinemia
- Hypopituitarism
- Ovarian factors
- Polycystic ovary syndrome
- Anovulation
- Diminished ovarian reserve
- Luteal dysfunction
- Premature menopause
- Gonadal dysgenesis (Turner syndrome)
- Ovarian neoplasm
- Tubal/peritoneal factors
- Endometriosis
- Pelvic adhesions
- Pelvic inflammatory disease (PID, usually due to chlamydia)
- Tubal occlusion
- Uterine factors
- Uterine malformations
- Uterine fibroids (leiomyoma)
- Asherman's Syndrome
- Cervical factors
- Cervical stenosis
- Antisperm antibodies
- Insufficient cervical mucus (for the travel and survival of sperm)
- Vaginal factors

- Vaginismus
- Vaginal obstruction
- Genetic factors
- Various intersexuality|intersexed conditions, such as androgen insensitivity syndrome

Combined Infertility

In some cases, both the man and woman may be infertile or sub-fertile, and the couple's infertility arises from the combination of these factors. In other cases, the cause is suspected to be immunological or genetic; it may be that each partner is independently fertile but the couple cannot conceive together without assistance.

Unexplained Infertility

In about 15% of cases of infertility, investigation will show no abnormalities. In these cases abnormalities are likely to be present but not detected by current methods. Possible problems could be that the egg is not released at the optimum time for fertilization, that it may not enter the Fallopian tube, sperm may not be able to reach the egg, fertilization may fail to occur, transport of the zygote may be disturbed, or implantation fails. It is increasingly recognized that egg quality is of critical importance.

Diagnosis of Infertility

Diagnosis of infertility begins with a medical history and physical exam. The healthcare provider may order tests, including the following:
- an endometrial biopsy, which tests the lining of the uterus
- hormone testing, to measure levels of female hormones
- laparoscopy, which allows the provider to see the pelvic organs
- ovulation testing, which detects the release of an egg from the ovary
- Pap smear, to check for signs of infection
- pelvic exam, to look for abnormalities or infection
- a postcoital test, which is done after sex to check for problems with secretions
- special X-ray tests

Treatment

- Fertility medication that stimulates the ovaries to "ripen" and release eggs (e.g., Clomiphene/clomiphene citrate, which stimulates ovulation)
- Surgery to restore potency of obstructed Fallopian tubes (tuboplasty)

- Donor insemination, which involves the woman being artificially inseminated or artificially inseminated with donor sperm.
- In vitro fertilization (IVF) in which eggs are removed from the woman, fertilized, and then placed in the woman's uterus, bypassing the Fallopian tubes. Variations on IVF:
- Use of donor eggs and/or sperm in IVF. This happens when a couple's eggs and/or sperm are unusable, or to avoid passing on a genetic disease.
- Intracytoplasmic sperm injection (ICSI) in which a single sperm is injected directly into an egg; the fertilized egg is then placed in the woman's uterus as in IVF.
- Zygote intrafallopian transfer (ZIFT) in which eggs are removed from the woman, Fertilized, and then placed in the woman's Fallopian tubes rather than the uterus.
- Gamete intrafallopian transfer(GIFT) in which eggs are removed from the woman and placed in one of the Fallopian tubes, along with the man's sperm. This allows fertilization to take place inside the woman's body.
- Other assisted reproductive technology (ART):
- Assisted hatching
- Fertility preservation
- Freezing (cryopreservation) of sperm, eggs, and reproductive tissue
- Frozen embryo transfer (FET)
- Alternative and complementary treatments
- Acupuncture. Recent controlled trials published in *Fertility and Sterility* have shown acupuncture to increase the success rate of IVF by as much as 60%. Acupuncture was also reported to be effective in the treatment of female anovular infertility, World Health Organization.
- Diet and supplements
- Healthy lifestyle

Diseases and Disorders of the Female Reproductive System

Women commonly deal with many different diseases and disorders that pertain to the reproductive system. Here are some of the most common:

1. Vulvovaginitis (pronounced: vul-vo-vah-juh-ni-tus) is an inflammation of the vulva and vagina. It may be caused by irritating substances such as laundry soap, bubble baths, or poor hygiene, such as wiping from back to front. Symptoms include redness and itching in these areas and sometimes vaginal discharge. It can also be caused by an overgrowth of candida, a fungus normally present in the vagina.
2. Nonmenstrual vaginal bleeding is most commonly due to the presence of a foreign body in the vagina. It may also be due to urethral prolapse, a condition in which the mucous membranes of the urethra protrude into the vagina and forms a tiny, donut-shaped mass of tissue that bleeds easily. It can also be due to a straddle injury or vaginal trauma from sexual abuse.
3. Ectopic pregnancy occurs when a fertilized egg or zygote doesn't travel into the uterus, but instead grows rapidly in the Fallopian tube. Women with this condition can develop severe abdominal pain and should see a doctor because surgery may be necessary.
4. Ovarian tumors, although rare, can occur. Women with ovarian tumors may have abdominal pain and masses that can be felt in the abdomen. Surgery may be needed to remove the tumor.

5. Ovarian cysts are noncancerous sacs filled with fluid or semi-solid material. Although they are common and generally harmless, they can become a problem if they grow very large. Large cysts may push on surrounding organs, causing abdominal pain. In most cases, cysts will pass or disappear on their own and treatment is not necessary. If the cysts are painful and occur frequently, a doctor may prescribe birth control pills to alter their growth and occurrences. Surgery is also an option if they need to be removed.
6. Polycystic ovary syndrome is a hormone disorder in which too many hormones are produced by the ovaries. This condition causes the ovaries to become enlarged and develop many fluid-filled sacs or cysts. It often first appears during the teen years. Depending on the type and the severity of the condition, it may be treated with drugs to regulate hormone balance and menstruation.
7. Trichomonas vaginalis is an inflammatory condition of the vagina, usually a bacterial infection, also called vaginosis.
8. Dysmenorrhea is painful periods.
9. Menorrhagia is when a woman has very heavy periods with excess bleeding.
10. Oligomenorrhea is when a woman misses or has infrequent periods, even though she has been menstruating for a while and is not pregnant.
11. Amenorrhea is when a girl has not started her period by the time she is 16 years old or 3 years after puberty has started, has not developed signs of puberty by 14, or has had normal periods but has stopped menstruating for some reasons other than pregnancy.
12. Toxic shock syndrome is caused by toxins released into the body during a type of bacterial infection that is more likely to develop if a tampon is left in too long. It can produce high fever, diarrhea, vomiting, and shock.
13. Candidiasis: symptoms of yeast infections include itching, burning and discharge. Yeast organisms are always present in all people, but are usually prevented from overgrowth (uncontrolled multiplication resulting in symptoms) by naturally occurring microorganisms.

Endometriosis

Endometriosis is the most common gynecological diseases, affecting more than 5.5 million women in North America alone! The two most common symptoms are pain and infertility. In this disease a specialized type of tissue, the endometrium, which normally lines the uterus, becomes implanted outside the uterus, most commonly on the Fallopian tubes, ovaries, or the tissue lining the pelvis. During the menstrual cycle, hormones signal the lining of the uterus to thicken to prepare for possible pregnancy. If a pregnancy doesn't occur, the hormone levels decrease, causing the thickened lining to shed. When endometrial tissue is located in other parts, it continues to act in its normal way; it thickens, breaks down, and bleeds each month as the hormone levels rise and fall. However, because there's nowhere for the blood from this mislocated tissue to exit the body, it becomes trapped and surrounding tissue becomes irritated. Trapped blood may lead to growth of cysts. Cysts in turn may form scar tissue and adhesions. This causes pain in the area of the misplaced tissue, usually the pelvis. Endometriosis can cause fertility problems. In fact, scars and adhesions on the ovaries or Fallopian tubes can prevent pregnancy. Endometriosis can be mild, moderate, or severe, and tends to get worse over time without treatment. The most common symptoms are:

1. <u>Painful periods</u>: Pelvic pain and severe cramping, intense back and abdominal pain.
2. <u>Pain at other times</u>: Women may experience pelvic pain during ovulation, sharp deep pain in pelvis during intercourse, or pain during bowel movements or urination.
3. <u>Excessive bleeding</u>: Heavy periods or bleeding between periods.
4. <u>Infertility</u>: Approximately 30–40% of women.

The cause of endometriosis remains mysterious. Scientists are studying the roles that hormones and the immune system play in this condition. One theory holds that menstrual blood, which contains endometrial cells, flows back through the Fallopian tubes, takes root, and grows. Another hypothesis proposes that the bloodstream carries endometrial cells to other sites in the body. Still another theory speculates that a predisposition toward endometriosis may be carried in the genes of certain families.

Other researchers believe that certain cells present within the abdomen in some women retain their ability to specialize into endometrial cells. These same cells were responsible for the growth of the woman's reproductive organs when she was an embryo/fetus. It is believed that genetic or environmental influences in later life allow these cells to give rise to endometrial tissue outside the uterus. Experts estimate that up to one in ten American women of childbearing age have endometriosis. There is some thinking that previous damage to cells that line the pelvis can lead to endometriosis. There are several ways to diagnose endometriosis:

1. Pelvic exam
2. Ultrasound
3. Laparoscopy: Usually used, most correct diagnosis
4. Blood test

Endometriosis can be treated with:

1. Pain medication
2. Hormone therapy
    a) Oral contraceptives
    b) Gonadotropin-releasing hormone (Gn-Rh) agonists and antagonists
    c) Danazol (Danocrine)
    d) Medroxyprogesterone (Depo-Provera)
3. Conservative surgery which removes endometrial growths.
4. Hysterectomy

CRITICAL-THINKING QUESTIONS

1. Briefly explain why mature gametes carry only one set of chromosomes.
2. What special features are evident in sperm cells but not in somatic cells, and how do these specializations function?
3. What do each of the three male accessory glands contribute to the semen?

4. Describe how penile erection occurs.
5. While anabolic steroids (synthetic testosterone) bulk up muscles, they can also affect testosterone production in the testis. Using what you know about negative feedback, describe what would happen to testosterone production in the testes if a male takes large amounts of synthetic testosterone.
6. Follow the path of ejaculated sperm from the vagina to the oocyte. Include all structures of the female reproductive tract that the sperm must swim through to reach the egg.
7. Identify some differences between meiosis in men and women.
8. Explain the hormonal regulation of the phases of the menstrual cycle.
9. Endometriosis is a disorder in which endometrial cells implant and proliferate outside of the uterus—in the uterine tubes, on the ovaries, or even in the pelvic cavity. Offer a theory as to why endometriosis increases a woman's risk of infertility.
10. Identify the changes in sensitivity that occur in the hypothalamus, pituitary, and gonads as a boy or girl approaches puberty. Explain how these changes lead to the increases of sex steroid hormone secretions that drive many pubertal changes.
11. Two teenage girls wanted to make a douche (a solution used to rinse and remove sperm from the vagina following intercourse). One girl suggested using Classic Coke, and the other girl wanted to use baking soda solution. Which solution would be most likely to succeed in preventing a pregnancy? Explain your answer.
12. Why are LH and FSH levels elevated in menopausal females?
13. Why might the events that occur in the mature follicle as a result of the LH surge be compared to an inflammatory reaction?

Credits

Chapter One

Figure 1.1 Body cavities. Copyright in the Public Domain.

Figure 1.2 Anatomical position. Copyright in the Public Domain.

Figure 1.3 Abdominal quadrants. Copyright in the Public Domain.

Figure 1.4 Planes dividing the abdominal region into 9 areas. Copyright in the Public Domain.

Figure 1.5 MRI image of the knee. Copyright © Michael R Carmont (CC by 2.0) at http://commons.wikimedia.org/wiki/File:Osteochondroma_MRI.JPEG

Figure 1.6 Anatomical planes. Copyright in the Public Domain.

Figure 1.7 Coronal plane. Copyright in the Public Domain.

Chapter Two

Figure 2.1 Periodic table of the elements. Copyright © (CC by 3.0) at http://commons.wikimedia.org/wiki/Periodic_table

Figure 2.2 The atom. Copyright in the Public Domain.

Figure 2.3 A diagram of an atom. Copyright in the Public Domain.

Fig. 2.4 Sodium (Na) loses an electron. Copyright ©Faraaz Damji (CC by 2.0) at http://commons.wikimedia.org/wiki/Image:NaCl_ionic.png

Figure 2.5 Electrons are shared in a covalent bond. Copyright © Jacek FH (CC by 3.0) at http://commons.wikimedia.org/wiki/Image:Covalent_bond_hydrogen.svg

Figure 2.6 Water has a partial positive charge on the hydrogen side. Copyright in the Public Domain.

Figure 2.7 Copyright in the Public Domain.

Figure 2.8 Copyright in the Public Domain.

Figure 2.9. The pH scale ranges from 0 to 14. Copyright © (CC by 2.5) PatríciaR at http://commons.wikimedia.org/wiki/Image:PHscalenolang.png

Figure 2.10 Copyright in the Public Domain.

Chapter Three

Figure 3.1 The cell contains a variety of organelles. Copyright in the Public Domain.

Figure 3.2 Phospholipids contain a phosphate head and a lipid tail. Copyright © Bryan Derksen (CC by 3.0) at http://commons.wikimedia.org/wiki/Image:Phospholipid_structure.png

Figure 3.3 Phospholipids arrange themselves into a bilayer.

Copyright © Jerome Walker (CC by 3.0) at http://commons.wikimedia.org/wiki/Image:Fluid_Mosaic.svg

Figure 3.4 The endoplasmic reticulum. Copyright in the Public Domain.

Figure 3.5 The mitochondrion. Copyright in the Public Domain.

Figure 3.6. A centriole. Copyright in the Public Domain.

Figure 3.7 Cilia. Copyright © Shigenori Nonaka et al (CC by 3.0) at http://commons.wikimedia.org/wiki/File:Nodal_cilia.jpg

Figure 3.8 Myosin. Copyright © Boumphreyfr (CC by 3.0) at http://commons.wikimedia.org/wiki/File:Microtubules.png

Figure 3.9 The nucleus. Copyright in the Public Domain.

Figure 3.10 Carrier proteins. Copyright in the Public Domain.

Figure 3.11 Hypertonic, isotonic, hypotonic. Copyright in the Public Domain.

Figure 3.12 The sodium-potassium pump. Copyright in the Public Domain.

Figure 3.13 Copyright in the Public Domain.

Figure 3.14. Prophase. Copyright in the Public Domain.

Figure 3.15. Metaphase. Copyright in the Public Domain.

Figure 3.16. Anaphase. Copyright in the Public Domain.

Figure 3.17 Telophase. Copyright in the Public Domain.

Figure 3.18 Copyright © Madprime (CC by 3.0) at http://commons.wikimedia.org/wiki/Image:Genetic_code.svg

Figure 3.19 Copyright © Jculbert (CC by 3.0) at http://commons.wikimedia.org/wiki/Image:DNA-labels.png

Figure 3.20 DNA is coiled into chromosomes. Copyright in the Public Domain.

Figure 3.21 Transcription. Copyright in the Public Domain.

Figure 3.22. Copyright in the Public Domain.

Chapter Four

Figure 4.1 Krebs (Citric Acid cycle). Copyright © YassineMrabet (CC by 3.0) at http://commons.wikimedia.org/wiki/File:Citricacidcycle_ball2.png

Figure 4.2 Copyright in the Public Domain.

Chapter Five

Figure 5.1 Example of epithelium. Copyright in the Public Domain.

Figure 5.2 Different types of epithelia. Copyright in the Public Domain.

Figure 5.3 Simple cuboidal epithelium lines the tubules in the kidneys. Source: http://commons.wikimedia.org/wiki/File:Glomerulus_pas.JPG

http://upload.wikimedia.org/wikipedia/commons/4/4c/Glomerulus_pas.JPG

Figure 5.4 Stratified squamous epithelium. Copyright © (CC by 3.0) at http://commons.wikimedia.org/wiki/File:Spongiotic_dermatitis_%282%29_Dyshidrotic_.JPG

Figure 5.5 Pseudostratified columnar epithelium. Copyright in the Public Domain.

Figure 5.6 Adipose connective tissue consists of adipocytes. Copyright © Department of Histology, Jagiellonian University Medical College (CC by 3.0) at http://commons.wikimedia.org/wiki/Image:Yellow_adipose_tissue_in_paraffin_section_-_lipids_washed_out.jpg

Figure 5.7 Copyright © Department of Histology, Jagiellonian University Medical College (CC by 3.0) at http://commons.wikimedia.org/wiki/Image:Blood_smear.jpg

Figure 5.8 Copyright in the Public Domain.

Figure 5.9 Hyaline cartilage has a smooth matrix. Copyright © Department of Histology, Jagiellonian University Medical College (CC by 3.0) at http://commons.wikimedia.org/wiki/Image:Hyaline_cartilage.jpg

Figure 5.10 Three types of muscle tissue. Copyright in the Public Domain.

Figure 5.11 Nervous tissue. Copyright © Fanny CASTETS (CC by 3.0) at http://commons.wikimedia.org/wiki/Image:Neuronehisto.jpg

Chapter Six

Figure 6.1 The layers of the epidermis. Copyright in the Public Domain.

Figure 6.2 Source: http://commons.wikimedia.org/wiki/Image:Calcitriol-Synthesis.png

Figure 6.3 The integument. Copyright in the Public Domain.

Figure 6.4 The integument. Copyright in the Public Domain.

Chapter Seven

Figure 7.1 Parts of a long bone. Copyright in the Public Domain.

Figure 7.2 Haversian system. Copyright © BDB (CC by 2.5) at http://commons.wikimedia.org/wiki/Image:Transverse_Section_Of_Bone.png

Figure 7.3 Osteocytes. Copyright in the Public Domain.

Figure 7.4 Endochondral ossification. Copyright in the Public Domain.

Figure 7.5 Greenstick fractures. Copyright © Lucien Monfils (CC by 3.0) at http://commons.wikimedia.org/wiki/File:Greenstick.jpg

Figure 7.6 Copyright © Tdvorak (CC by 3.0) at http://commons.wikimedia.org/wiki/Image:Heterotopic_Ossification_Elbow2.JPG

Figure 7.7 Fracture of the clavicle. Copyright © Rbaer (CC by 3.0) at http://commons.wikimedia.org/wiki/Image:Claviculafraktur.JPG

Figure 7.8. The skeleton. Copyright in the Public Domain.

Figure 7.9 Lateral view of the skull. Copyright in the Public Domain.

Figure 7.10 Anterior view of the skull. Copyright in the Public Domain.

Figure 7.11 Sutures of the skull. Copyright © RosarioVanTulpe (CC by 3.0) at http://commons.wikimedia.org/wiki/Image:SkullSchaedelSeitlich1.png

Figure 7.12 The coronal suture unites the frontal and parietal bones. Copyright in the Public Domain.
Figure 7.13 Frontal bone. Copyright in the Public Domain.
Figure 7.14 Parietal bone. Copyright in the Public Domain.
Figure 7.15 Occipital bone. Copyright in the Public Domain.
Figure 7.16 Temporal bone. Copyright in the Public Domain.
Figure 7.17 The sphenoid bone. Copyright in the Public Domain.
Figure 7.18 Ethmoid bone. Copyright in the Public Domain.
Figure 7.19 Maxilla. Copyright in the Public Domain.
Figure 7.20 Vomer bone. Copyright in the Public Domain.
Figure 7.21 Superior view of fontanels. Copyright in the Public Domain.
Figure 7.22 The spine. Copyright in the Public Domain.
Figure 7.23 The spine is divided into cervical, thoracic, lumbar, and pelvic sections. Copyright in the Public Domain.
Figure 7.24 Scoliosis. Copyright © Gkiokas A et al (CC by 2.0) at http://commons.wikimedia.org/wiki/Image:Scoliosis_recklinghausen.jpg
Figure 7.25 The atlas. Copyright in the Public Domain.
Figure 7.26 The axis (C2). Copyright in the Public Domain.
Figure 7.27 Typical cervical vertebra. Copyright in the Public Domain.
Figure 7.28 Typical thoracic vertebra. Copyright in the Public Domain.
Figure 7.29 Typical lumbar vertebra. Copyright in the Public Domain.
Figure 7.30 Pelvis and sacrum (anterior view). Copyright in the Public Domain.
Figure 7.31. Ribcage. Copyright in the Public Domain.
Figure 7.32 Hyoid bone. Copyright in the Public Domain.
Figure 7.33 The pectoral girdle. Copyright in the Public Domain.
Figure 7.34 Anterior view of the ulna. Copyright © (CC by 3.0) at http://commons.wikimedia.org/wiki/Image:Ulna_ant.jpg
Figure 7.35 Medial view of the ulna. Copyright © (CC by 3.0) at http://commons.wikimedia.org/wiki/Image:Ulna_med.jpg
Figure 7.36 Anterior radius. Copyright © (CC by 3.0) at http://commons.wikimedia.org/wiki/Image:Radius_ant.jpg
Figure 7.37 Posterior radius. Copyright © (CC by 3.0) at http://commons.wikimedia.org/wiki/Image:Radius_post.jpg
Figure 7.38 Carpals of the right hand. Copyright © Zoph (CC by 3.0) at http://commons.wikimedia.org/wiki/Image:Carpus.png
Figure 7.40 Bones of the hand. Copyright in the Public Domain.
Figure 7.41. Coxal bones of the pelvis. Copyright in the Public Domain.
Figure 7.42 Anterior femur. Copyright in the Public Domain.
Figure 7.43 Posterior view of the femur. Copyright in the Public Domain.
Figure 7.44 Anterior view of the tibia. Copyright in the Public Domain.
Figure 7.45 Anterior view of fibula. Copyright in the Public Domain.
Figure 7.46 Foot and ankle. Copyright © Minervaaa (CC by 3.0) at http://commons.wikimedia.org/wiki/File:Skeleton_foot.JPG

Chapter Eight

Figure 8.1 Types of synovial joints. Copyright © Produnis (CC by 3.0) at http://commons.wikimedia.org/wiki/Image:Gelenke_Zeichnung01.jpg

Figure 8.2 Shoulder. Copyright in the Public Domain. Figure 8.3. Dislocated shoulder. Copyright © MB (CC by 2.5) at http://commons.wikimedia.org/wiki/Image:Luxation_epaule.PNG

Figure 8.4 Elbow (medial projection). Copyright in the Public Domain.

Figure 8.5 Elbow (lateral projection). Copyright in the Public Domain.

Figure 8.6 Hip (anterior view). Copyright in the Public Domain.

Figure 8.7 Hip (posterior view). Copyright in the Public Domain.

Figure 8.8 Knee ligaments. Copyright in the Public Domain.

Figure 8.9 Osteoarthritis and joint replacement. Copyright in the Public Domain.

Chapter Nine

Figure 9.1 Muscle shapes. Copyright in the Public Domain.

Figure 9.2 Overview of anterior muscles. Copyright in the Public Domain.

Figure 9.3 Overview of posterior muscles. Copyright in the Public Domain.

Figure 9.4 Muscles of facial expression. Copyright © Patrick J. Lynch (CC by 2.5) at http://commons.wikimedia.org/wiki/Image:Face_anatomy_superficial.jpg

Figure 9.5 Facial muscles. Copyright © Patrick J. Lynch (CC by 2.5) at http://commons.wikimedia.org/wiki/Image:Head_ap_anatomy.jpg

Figure 9.6 Lateral view of neck muscles. Copyright in the Public Domain.

Figure 9.7 Medial and lateral pterygoids. Copyright © Berichard (CC by 3.0) at http://commons.wikimedia.org/wiki/Image:Muscle_pterygoidien_lateral.png

Figure 9.8 Erector spinae muscles. Copyright in the Public Domain.

Figure 9.9 Muscles of the tongue. Copyright in the Public Domain.

Figure 9.10 Posterior muscles of the thorax. Copyright in the Public Domain.

Figure 9.11 Muscles of anterior thorax and arm. Copyright in the Public Domain.

Figure 9.12 Deep muscles of the thorax and shoulder. Copyright in the Public Domain.

Figure 9.13 Posterior muscles of the arm. Copyright in the Public Domain.

Figure 9.14 Anterior forearm muscles. Copyright in the Public Domain.

Figure 9.15 Deep anterior forearm muscles. Copyright in the Public Domain.

Figure 9.16 Posterior forearm muscles deep. Copyright in the Public Domain.

Figure 9.17 Superficial abdominal muscles. Copyright in the Public Domain.

Figure 9.18 Rectus abdominus. Copyright © sv:Anvãndare:Chrizz (CC by 3.0) at http://commons.wikimedia.org/wiki/Image:Rectus_abdominis.png

Figure 9.19 The transverse abdominus (the deepest abdominal muscle). Copyright in the Public Domain.

Figure 9.20 Muscles of the pelvic outlet (female). Copyright in the Public Domain.

Figure 9.21 Pelvic outlet muscles (male). Copyright in the Public Domain.

Figure 9.22 Gluteal muscles. The gluteus medius is deep to the gluteus maximus. Copyright © Beth Ohara (CC by 3.0) at http://commons.wikimedia.org/wiki/Image:Posterior_Hip_Muscles_3.PNG

Figure 9.23 Deep muscles of the posterior pelvis. Copyright © Beth Ohara (CC by 3.0) at http://commons.wikimedia.org/wiki/Image:Posterior_Hip_Muscles_1.PNG

Figure 9.24 Deep muscles of the posterior pelvis. Copyright in the Public Domain.

Figure 9.25 Muscles of the anterior thigh. Copyright in the Public Domain.

Figure 9.26 Muscles of the posterior thigh. Copyright in the Public Domain.

Figure 9.27 Anterior lower leg muscles. Copyright in the Public Domain.

Figure 9.28 The gastrocnemius and soleus attach to the Achilles tendon. Copyright in the Public Domain.

Figure 9.29 Dorsum of foot. Copyright in the Public Domain.

Figure 9.30 Deep muscles of dorsum of foot. Copyright in the Public Domain.

Figure 9.31 Skeletal muscle is characterized by long cells and has a striated appearance. Copyright © Department of Histology, Jagiellonian University Medical College (CC by 3.0) at http://commons.wikimedia.org/wiki/Image:Skeletal_muscle_-_longitudinal_section.jpg

Figure 9.32 Structure of skeletal muscle. Copyright in the Public Domain.

Figure 9.33 Actin and myosin. Copyright © Moralapostel (CC by 3.0) at http://commons.wikimedia.org/wiki/Image:Querbr%C3%BCckenzyklus_4.png

Figure 9.34 Diagram of a sarcomere. Copyright © SlothMcCarty (CC by 3.0) at http://commons.wikimedia.org/wiki/File:Sarcomere_diagram.svg

Figure 9.35 Skeletal muscle structure. Copyright © Raul654 (CC by 3.0) at http://commons.wikimedia.org/wiki/Image:Skeletal_muscle_diagram.jpg

Figure 9.36 Motor neuron and muscle. Copyright © Dake (CC by 3.0) at http://commons.wikimedia.org/wiki/Image:Synapse_diag3.png

Figure 9.37 Neuromuscular junction. Copyright © Nrets (CC by 3.0) at http://commons.wikimedia.org/wiki/Image:Synapse_diag4.png

Figure 9.38 Histology of cardiac muscle. Copyright © Netha Hussain (CC by 3.0) at http://commons.wikimedia.org/wiki/File:Cardiac_muscle_histology.jpg

Figure 9.39 Histology of smooth muscle cells. Copyright © Polarlys (CC by 3.0) at http://commons.wikimedia.org/wiki/File:Glatte_Muskelzellen.jpg

Chapter Ten

Figure 10.1 The nervous system. Copyright in the Public Domain.

Figure 10.2 Spinal cord anatomy. Copyright © debivort (CC by 3.0) at http://commons.wikimedia.org/wiki/File:Cervical_vertebra_english.png

Figure 10.3 The cauda equina. Copyright © John A Beal (CC by 2.5) at http://commons.wikimedia.org/wiki/File:Human_caudal_spinal_cord_anterior_view.jpg

Figure 10.4 Meninges covering the spine. Copyright in the Public Domain.

Figure 10.5 Transverse section of spinal cord. Copyright © Polarlys (CC by 3.0) at http://commons.wikimedia.org/wiki/Image:Medulla_spinalis_-_Section_-_English.svg

Figure 10.6 Spinal tracts. Copyright © Polarlys (CC by 3.0) at http://commons.wikimedia.org/wiki/Image:Medulla_spinalis_-_Querschnitt_-_Bahnen_-_German.svg

Figure 10.7 Nerve structure. Copyright in the Public Domain.

Figure 10.8 Spinal nerve. Copyright © Mysid (CC by 3.0) at http://commons.wikimedia.org/wiki/Image:Spinal_nerve.svg

Figure 10.9 Dermatomes are defined areas of the body that carry sensation by spinal nerves. Copyright in the Public Domain.

Figure 10.10 Brachial plexus. Copyright in the Public Domain.

Figure 10.11 Cervical plexus. Copyright in the Public Domain.

Figure 10.12 Lumbar plexus. Copyright in the Public Domain.

Figure 10.13 Developing CNS. Copyright © Nrets (CC by 3.0) at http://commons.wikimedia.org/wiki/Image:Encephalon.png

Figure 10.14 The neural plate develops folds that unite in the center to produce the neural tube. Copyright in the Public Domain.

Figure 10.15 MRI of the brain. Copyright in the Public Domain.

Figure 10.16 Brainstem, posterior view. Copyright © John A Beal (CC by 2.5) at http://commons.wikimedia.org/wiki/File:Human_brainstem-thalamus_posterior_view_description.JPG

Figure 10.17 Brainstem, lateral view. Copyright © John A Beal (CC by 2.5) at http://commons.wikimedia.org/wiki/File:Human_brain_left_midsagitttal_view_closeup_description_3.JPG

Figure 10.18 Brainstem, coronal view. Copyright in the Public Domain.

Figure 10.19 Brainstem, posterior view. Copyright © John A Beal (CC by 2.5) at http://commons.wikimedia.org/wiki/File:Human_brainstem-thalamus_posterior-inferior_view.JPG

Figure 10.20 Brainstem, anterior view showing locations of cranial nerves. Copyright © John A Beal (CC by 2.5) at http://commons.wikimedia.org/wiki/File:Human_brainstem_anterior_view_description.JPG

Figure 10.21 The cerebellum. Copyright in the Public Domain.

Figure 10.22 Cerebellum, inferior view. Copyright © John A Beal (CC by 2.5) at http://commons.wikimedia.org/wiki/Image:Human_cerebellum_posterior_view_description.JPG

Figure 10.23 Highly branching Purkinje cells found in the cerebellum. Copyright in the Public Domain.

Figure 10.24 Cerebellum, sagittal view. Copyright © John A Beal (CC by 2.5) at http://commons.wikimedia.org/wiki/Image:Human_brain_midsagittal_view_description.JPG

Figure 10.25 Diencephalon, thalamus, and hypothalamus. Copyright © John A Beal (CC by 2.5) at http://commons.wikimedia.org/wiki/File:Human_brain_midsagittal_view_description.JPG

Figure 10.26 The pineal gland (highlighted area) is located in the posterior region of the diencephalon. Copyright © Martin Hasselblatt (CC by 2.5) at http://commons.wikimedia.org/wiki/Image:PTPR_MRI.jpg

Figure 10.27 MRI showing location of thalamus. Copyright © (CC by 3.0) at http://commons.wikimedia.org/wiki/Image:Brain_chrischan_thalamus.jpg

Figure 10.28 MRI showing location of hypothalamus. Copyright in the Public Domain.

Figure 10.29 Cerebrum superior view. Copyright © (CC by 3.0) at http://commons.wikimedia.org/wiki/Image:Cerebral_lobes.png

Figure 10.30 Lobes of cerebrum, lateral view. Copyright © RobinH (CC by 3.0) at http://commons.wikimedia.org/wiki/Image:Main_brain_lobes.gif

Figure 10.31 Limbic system. Copyright in the Public Domain.

Figure 10.32 The brain has 3 layers of meninges. Copyright in the Public Domain.

Figure 10.33 Lateral ventricles of brain. Copyright in the Public Domain.

Figure 10.34 Third and fourth ventricles of the brain. Copyright © John A Beal (CC by 2.5) at http://commons.wikimedia.org/wiki/File:Human_brain_left_midsagitttal_view_closeup_description_3.JPG

Figure 10.35 CSF circulatory structures. Copyright © John A Beal (CC by 2.5) at http://commons.wikimedia.org/wiki/File:Human_brain_inferior-medial_view_description_3.JPG

Figure 10.36 Olfactory nerve. Copyright © Patrick J. Lynch (CC by 2.5) at http://commons.wikimedia.org/wiki/Image:Head_olfactory_nerve.jpg

Figure 10.37 Olfactory bulb (highlighted in red). Copyright in the Public Domain.

Figure 10.38 Optic nerve. Copyright in the Public Domain. Figure 10.39 Oculomotor nerve. Copyright © Patrick J. Lynch (CC by 2.5) at http://commons.wikimedia.org/wiki/Image:Cranial_Nerve_III_somatic.svg

Figure 10.40 Trigeminal nerve. Copyright © Patrick J. Lynch (CC by 2.5) at http://commons.wikimedia.org/wiki/Image:Head_deep_facial_trigeminal.jpg

Figure 10.41 Facial nerve. Copyright © Patrick J. Lynch (CC by 2.5) at http://commons.wikimedia.org/wiki/Image:Head_facial_nerve_branches.jpg

Figure 10.42 Glossopharyngeal nerve (yellow). Copyright in the Public Domain.

Figure 10.43 Autonomic nervous system. Copyright in the Public Domain.

Figure 10.44 The abdominal aortic plexus contains the celiac and mesenteric ganglia. Copyright in the Public Domain. Figure 10.45 Structure of a typical neuron. Copyright © NickGorton (CC by 3.0) at http://commons.wikimedia.org/wiki/Image:Neuron1.jpg

Figure 10.46 Various types of neurons. Copyright © Juoj8 (CC by 3.0) at http://commons.wikimedia.org/wiki/File:Neurons_uni_bi_multi_pseudouni.svg

Figure 10.47 Astrocytes (green) in the cortex. Copyright © Mhisted (CC by 3.0) at http://commons.wikimedia.org/wiki/Image:Astrocytes-mouse-cortex.png

Figure 10.48 Events of an action potential. Copyright © Chris 73 (CC by 3.0) at http://commons.wikimedia.org/wiki/Image:ActionPotential.png

Figure 10.49 Propagation of an action potential. Copyright © John Schmidt (CC by 3.0) at http://commons.wikimedia.org/wiki/Image:Action_potential_propagation_animation.gif

Figure 10.50 Neurotransmitter reuptake. Copyright in the Public Domain.

Figure 10.51 Reflex arc. Copyright © Mysid (CC by 3.0) at http://commons.wikimedia.org/wiki/Image:Spinal_nerve.svg

Figure 10.52 Homunculus showing both sensory (blue) and motor (red) areas. Copyright in the Public Domain.

Figure 10.53 Broca's and Wernicke's areas. Copyright © James.mcd.nz (CC by 3.0) at http://commons.wikimedia.org/wiki/Image:Brain_Surface_Gyri.SVG

Figure 10.54 The reticular activating system. Copyright © (CC by 2.5) at http://commons.wikimedia.org/wiki/File:Brain_bulbar_region.PNG

Figure 10.55 An EEG displaying brain waves. Copyright in the Public Domain.

Chapter Eleven

Figure 11.1 Meissner's corpuscle located in the superficial dermis (the superficial layers of the skin are located at the bottom of the slide). Copyright © Wbensmith (CC by 3.0) at http://commons.wikimedia.org/wiki/Image:WVSOM_Meissner%27s_corpuslce.JPG

Figure 11.2 Pacinian corpuscles are located in the deeper areas of the dermis. Copyright in the Public Domain.

Figure 11.3 Muscle spindle. Copyright © Hati (CC by 2.5) at http://commons.wikimedia.org/wiki/File:MuscleSpindle.svg

Figure 11.4 Olfactory nerve fibers. Copyright in the Public Domain.

Figure 11.5 Olfactory bulbs (red). Copyright in the Public Domain.

Figure 11.6 Taste bud. Copyright © NEUROtiker (CC by 3.0) at http://commons.wikimedia.org/wiki/Image:Taste_bud.svg

Figure 11.7 Taste buds for bitter. Copyright in the Public Domain.

Figure 11.7 Taste buds for salty. Copyright in the Public Domain.

Figure 11.7 Taste buds for sour. Copyright in the Public Domain.

Figure 11.7 Taste buds for sweet. Copyright in the Public Domain.

Figure 11.8 Conjunctivitis. Copyright © (CC by 1.0) at http://commons.wikimedia.org/wiki/Image:Vernal.jpg

Figure 11.9 Lacrimal apparatus. Copyright © Erin_Silversmith (CC by 2.5) at http://commons.wikimedia.org/wiki/Image:Tear_system.svg

Figure 11.10. Eye muscles. Copyright in the Public Domain. Figure 11.11 Structures of the eye. http://commons. Copyright © Chabacano (CC by 3.0) at http://commons.wikimedia.org/wiki/Image:Eye-diagram_no_circles_border_1.svg

Figure 11.12. A concave lens causes light rays to diverge. Copyright © (CC by 2.5) at http://commons.wikimedia.org/wiki/Image:ConcaveFocalLength.png

Figure 11.13 A convex lens causes light rays to converge. Copyright © (CC by 2.5) at http://commons.wikimedia.org/wiki/Image:FocalLength.png

Figure 11.14 Myopia is corrected with a concave lens. Copyright © A. Baris Toprak (CC by 1.0) at http://commons.wikimedia.org/wiki/Image:Myopia.png

Figure 11.15 Accommodation is the change in shape of the lens. Copyright © A. Baris Toprak (CC by 1.0) at http://commons.wikimedia.org/wiki/File:Accomodation.png

Figure 11.16 Layers of the retina.

Copyright © Peter Hartmann (CC by 3.0) at http://commons.wikimedia.org/wiki/File:Retina.jpg

Figure 11.17 Retinal cis (top) and trans (bottom) configurations. Copyright © V8rik (CC by 3.0) at http://commons.wikimedia.org/wiki/File:RetinalCisandTrans.png

Figure 11.18 Ishihara color blindness test object. Copyright in the Public Domain.

Figure 11.19 Cataract. Copyright © Rakesh Ahuja (CC by 3.0) at http://commons.wikimedia.org/wiki/File:Cataract_in_human_eye.png

Figure 11.20 Diabetic macular edema. Copyright in the Public Domain.

Figure 11.21 Retinoblastoma. Copyright © Aerts, I, Lumbroso-Le Rouic, L, Marion Gauthier-Villars, M, Brisse, H, Doz, F, Desjardins, L. Retinoblastoma (CC by 2.0) at http://commons.wikimedia.org/wiki/File:Fundus_retinoblastoma.jpg

Figure 11.22 Ear anatomy. Source: http://commons.wikimedia.org/wiki/File:HumanEar_svenska.png

Figure 11.23 Structures of the inner ear. Copyright © Patrick J. Lynch (CC by 2.5) at http://commons.wikimedia.org/wiki/File:Ear_internal_anatomy_numbered.svg

Figure 11.24 Chambers of the cochlea. Copyright © welleschik (CC by 3.0) at http://commons.wikimedia.org/wiki/File:Ductus_cochlearis_schema.jpg

Figure 11.25 Inner ear structures. Copyright © Oarih (CC by 3.0) at http://commons.wikimedia.org/wiki/File:Cochlea-crosssection.png

Figure 11.26 Pure tones are sine waves. Copyright © (CC by 3.0) at http://commons.wikimedia.org/wiki/File:Sinus_amplitude_en.svg

Figure 11.27 Sound is a complex wave consisting of areas of high and low pressure. Copyright © Omegatron (CC by 3.0) at http://commons.wikimedia.org/wiki/File:ALC_-12dB_clipped_closeup.png

Figure 11.28. Vestibule. Copyright in the Public Domain. Figure 11.29 Nystagmus is caused by endolymph movement in the semicircular canals. Copyright in the Public Domain. Figure 11.30 Acute otitis media. Copyright © B. Welleschik (CC by 3.0) at http://commons.wikimedia.org/wiki/File:Otitis_media_entdifferenziert2.jpg

Chapter Twelve

Figure 12.1 Endocrine system. Copyright in the Public Domain.

Figure 12.2 A negative feedback pathway. Copyright © DRosenbach (CC by 3.0) at http://commons.wikimedia.org/wiki/File:ACTH_Negative_Feedback.jpg

Figure 12.3 Location of the thyroid gland. Copyright in the Public Domain.

Figure 12.4 Adrenal glands. Copyright in the Public Domain.

Figure 12.5 Pancreas. Copyright in the Public Domain.

Figure 12.6 Pineal gland. Copyright in the Public Domain.

Figure 12.7 Thymus gland. Copyright in the Public Domain.

Chapter Thirteen

Figure 13.1 Hematopoiesis. Copyright © A. Rad (CC by 3.0) at http://commons.wikimedia.org/wiki/File:Hematopoiesis_simple.png

Figure 13.2 Red blood cells have a unique biconcave disc shape. Copyright in the Public Domain.

Figure 13.3 Hemoglobin molecule. Copyright © Kku (CC by 3.0) at http://commons.wikimedia.org/wiki/File:Hemoglobin.jpg

Figure 13.4 Neutrophil. Copyright in the Public Domain.

Figure 13.5 Neutrophils move from the blood into tissue and phagocytize bacteria. Copyright © Uwe Thormann (CC by 3.0) at http://commons.wikimedia.org/wiki/File:NeutrophilerAktion.png

Figure 13.6 Eosinophil. Copyright © (CC by 3.0) at http://commons.wikimedia.org/wiki/File:Eosinophil_1.png

Figure 13.7 Basophil. Copyright © Department of Histology, Jagiellonian University Medical College (CC by 3.0) at http://commons.wikimedia.org/wiki/File:Basophil.jpg

Figure 13.8 Monocyte. Copyright © Department of Histology, Jagiellonian University Medical College (CC by 3.0) at http://commons.wikimedia.org/wiki/File:Monocyte.jpg

Figure 13.9 Lymphocyte. Copyright © Department of Histology, Jagiellonian University Medical College (CC by 3.0) at http://commons.wikimedia.org/wiki/File:Lymphocyte.jpg

Figure 13.10 Platelets. Copyright © (CC by 3.0) at http://commons.wikimedia.org/wiki/File:Lymphocyte2.jpg

Figure 13.11 Blood types. Copyright in the Public Domain.

Chapter Fourteen

Figure 14.1 The mediastinum. Copyright in the Public Domain.

Figure 14.2 Base and apex of heart. Copyright in the Public Domain.

Figure 14.3. Blood flow through the heart. Copyright © (CC by 3.0) at http://commons.wikimedia.org/wiki/Image:Diagram_of_the_human_heart_(cropped).svg

Figure 14.4 Blood passively flows into the heart during the rest phase, and then is pushed into the ventricles during atrial systole. Copyright © Wapcaplet (CC by 3.0) at http://commons.wikimedia.org/wiki/File:Heart_diastole.png

Figure 14.5 Blood moves from the ventricles to the pulmonary trunk and aorta during ventricular systole.

Figure 14.6. Events of the cardiac cycle. Copyright © Wapcaplet, Reytan, Mtcv (CC by 3.0) at http://commons.wikimedia.org/wiki/File:Heart_systole.svg

Figure 14.7 Artery. Copyright in the Public Domain.

Figure 14.8 Capillary bed. Copyright in the Public Domain.

Figure 14.9 Vein. Copyright © Pdefer (CC by 3.0) at http://commons.wikimedia.org/wiki/File:Veincrosssection.png

Figure 14.10 Circulatory system. Copyright in the Public Domain.

Figure 14.11 Aorta. Copyright in the Public Domain.

Figure 14.12 Circle of Willis. Copyright in the Public Domain.

Figure 14.13 Action potential in cardiac muscle. Copyright © Ksheka (CC by 3.0) at http://commons.wikimedia.org/wiki/File:Action_potential.png

Figure 14.14 Cardiac conduction system. Copyright © tekksavvy (CC by 3.0) at http://commons.wikimedia.org/wiki/File:Cardiac_conduction_system_of_the_heart.jpg

Figure 14.15 ECG. Copyright in the Public Domain.

Figure 14.16 Sinus bradycardia. (Note the long interval between beats.) Copyright in the Public Domain.

Figure 14.17 Ventricular tachycardia. Copyright in the Public Domain.

Figure 14.18 Ventricular fibrillation. Note: there is no organized contraction with this arrhythmia. Copyright in the Public Domain.

Figure 14.19 Asystole represents no contraction. Copyright in the Public Domain.

Figure 14.20 The renin-angiotensin-aldosterone system. Copyright © A. Rad (CC by 3.0) at http://commons.wikimedia.org/wiki/File:Renin-angiotensin-aldosterone_system.png

Figure 14.21 Atherosclerosis. Copyright © Grahams Child (CC by 3.0) at http://commons.wikimedia.org/wiki/File:Endo_dysfunction_Athero.PNG

Figure 14.22 Movement of fluid in capillary. Copyright in the Public Domain.

Chapter Fifteen

Figure 15.1 Lymphatic system. Copyright in the Public Domain.

Figure 15.2 Lymphatic capillaries. Copyright in the Public Domain.

Figure 15.3 Lymph node. Copyright in the Public Domain.

Figure 15.4 Lymph node locations. Copyright in the Public Domain.

Figure 15.5 Spleen. Copyright in the Public Domain.

Figure 15.6 Thymus. Copyright in the Public Domain.

Figure 15.7 Complement system. Copyright in the Public Domain.

Figure 15.8 Antibody structure. Copyright in the Public Domain.

Figure 15.9 T-cell activation. Copyright in the Public Domain.

Figure 15.10 B-cell activation. Copyright in the Public Domain.

Figure 15.11 Immune response. Copyright © DO11.10 (CC by 3.0) at http://commons.wikimedia.org/wiki/File:Immune_response.jpg

Figure 15.12 Allergic response. Copyright in the Public Domain.

Chapter Sixteen

Figure 16.1. Respiratory system. Copyright in the Public Domain.

Figure 16.2. Alveolus. Copyright in the Public Domain.

Figure 16.3. Larynx. Copyright © Olek Remesz (CC by 2.5) at http://commons.wikimedia.org/wiki/File:Larynx_external_en.svg

Figure 16.4. Inhalation. Copyright in the Public Domain.

Figure 16.5. Exhalation. Copyright in the Public Domain.

Figure 16.6. Oxygen-hemoglobin saturation curve. Copyright © Diberri (CC by 3.0) at http://commons.wikimedia.org/wiki/File:Hb_saturation_curve.png

Chapter Seventeen

Figure 17.1. The gastrointestinal system. Copyright in the Public Domain.

Figure 17.2. Tooth. Copyright © Asbestos (CC by 2.0) at http://commons.wikimedia.org/wiki/File:ToothSection.jpg

Figure 17.3 Pharynx. Copyright in the Public Domain.

Figure 17.4. Esophagus. Copyright © Olek Remesz (CC by 2.5) at http://commons.wikimedia.org/wiki/File:Tractus_intestinalis_esophagus.svg

Figure 17.5. Stomach. Copyright in the Public Domain.

Figure 17.6. Stomach. Copyright © Olek Remesz(CC by 2.5) at http://commons.wikimedia.org/wiki/File:Ventriculus.svg

Figure 17.7. Intestines and colon. Copyright in the Public Domain.

Figure 17.8. Intestinal villi. Copyright in the Public Domain.

Figure 17.9. Pancreas. Copyright in the Public Domain.

Figure 17.10. Enzymes in the digestive system. Copyright in the Public Domain.

Figure 17.11. Intestinal villus. Copyright in the Public Domain.

Chapter Eighteen

Figure 18.1. Location of the kidneys. Copyright in the Public Domain.

Figure 18.2. Structures of the kidney: Copyright © Piotr Michał Jaworski (CC by 3.0) at http://commons.wikimedia.org/wiki/File:Kidney_PioM.png

Figure 18.3. Nephron. Copyright © Yosi I (CC by 3.0) at http://commons.wikimedia.org/wiki/File:Nephron_blank.svg

Figure 18.4. Urinary bladder. Copyright in the Public Domain.

Figure 18.5. Tubular reabsorption and secretion. Copyright © Yosi I (CC by 3.0) at http://commons.wikimedia.org/wiki/File:Reabsorption.svg

Figure 18.6. Transport proteins in the tubular cells. Copyright © Lennert B (CC by 3.0) at http://commons.wikimedia.org/wiki/File:Sammelrohr.svg

Figure 18.7. Filtrate concentration in the renal tubule. Copyright © Yosi I (CC by 3.0) at http://commons.wikimedia.org/wiki/File:Hypertonic_urine1.svg

Chapter Nineteen

Figure 19.1 Hyponatremia. Copyright in the Public Domain.

Chapter Twenty

Figure 20.1 Male reproductive system. Copyright © (CC by 3.0) at http://commons.wikimedia.org/wiki/File:Male_anatomy.png

Figure 20.2 Penis. Copyright © Esseh (CC by 3.0) at http://commons.wikimedia.org/wiki/File:Penile-Clitoral_Structure.JPG

Figure 20.3. Female reproductive system. Copyright © Elf Sternberg (CC by 3.0) at http://commons.wikimedia.org/wiki/File:Female_reproductive_system_lateral_nolabel.png

Figure 20.4 Uterus and ovaries. Copyright in the Public Domain.

Figure 20.5 Uterus. Copyright in the Public Domain.

Figure 20.6 Meiosis. Copyright in the Public Domain.

Figure 20.7 Sperm cell morphology. Copyright in the Public Domain.

Figure 20.8 Menstrual cycle. Copyright © Shazz (CC by 3.0) at http://commons.wikimedia.org/wiki/File:Hormons_level_-_follicle_%26_corpus_luteum.svg

Figure 20.9 Follicle development. Copyright © (CC by 3.0) at http://commons.wikimedia.org/wiki/File:Order_of_changes_in_ovary.svg

CPSIA information can be obtained
at www.ICGtesting.com
Printed in the USA
LVOW05s0007041115
460971LV00002B/2/P